D1313733

Elements of Physical Chemistry

Elements of Physical Chemistry

J. WILLIAM MONCRIEF
WILLIAM H. JONES
EMORY UNIVERSITY

ADDISON-WESLEY PUBLISHING COMPANY
READING, MASSACHUSETTS · MENLO PARK, CALIFORNIA
LONDON · AMSTERDAM · DON MILLS, ONTARIO · SYDNEY

This book is in the
ADDISON-WESLEY SERIES IN CHEMISTRY

Consulting Editor
FRANCIS T. BONNER

The photograph of Saturn which precedes Chapter 14 was provided by the Fernbank Science Center of Dekalb County, Georgia. All other photographs are by J. William Moncrief.

 Library of Congress Catalog Card No.75–20987.

ISBN 0-201-04897-3
ABCDEFGHIJ-HA-79876

TO BARBARA AND MIRIAM

Preface

We have at Emory a high proportion of students who have an interest in chemistry but who plan neither to go on into graduate work in chemistry nor to pursue a chemical career in industry. These are primarily young men and women who intend to enter medical school or dental school, to teach secondary school science, or to undertake graduate study in biology or in one of the health sciences. Our department has felt, and a growing number of students agree, that these undergraduates can benefit from some acquaintance with physical chemistry. Their schedules, however, must give priority to those courses specified in their major programs, plus those required for professional school admission. Consequently, most cannot afford the time to undertake the full-year course in physical chemistry taught for our B.S. majors. To meet this need we have for a number of years offered an abbreviated single-quarter course in physical chemistry. W. H. J. has taught this course for fifteen years and J. W. M. for four. Interest has increased in recent years due to the growing number of medical schools which require or recommend some physical chemistry. Roughly one out of five Emory seniors now elects to take this course.

Our major problem with this course has been the lack of a suitable text. Most good physical chemistry texts available are much too voluminous for a single-quarter course. There are shorter texts, but they owe their brevity to the omission of whole areas of physical chemistry—quantum chemistry, molecular structure, and spectroscopy, for example. We have tried adopting longer texts, then choosing chapters, topics, and parts of topics to be covered. In other years we have used a shorter text, supplementing it with lecture notes. Neither approach has been as successful as we would like.

We have therefore written our own text, designing it to offer the breadth and depth we feel is needed for this type of course. The book is brief enough to be assimilated in a single term (quarter or semester) by an average student with good motivation and adequate preparation in the physical sciences and mathematics, yet it still treats topics in sufficient detail to be clear. Certain topics found in many physical chemistry texts are either eliminated altogether or treated in sharply reduced detail. This is especially true of areas which are covered in sufficient detail in modern courses in analytical chemistry. Coverage includes the major important and rewarding areas of modern physical chemistry, offering the background necessary for understanding those applications of physical chemistry the student may subsequently encounter. Although it is not designed to prepare the student for graduate work in chemistry, this text may meet the needs of a full-year physical chemistry course in some colleges. Too, it should be a useful supplement for students in courses based on more complex texts.

Although frequent reference is made in this text to applications of physical chemistry in such areas as biochemistry and biology, our primary purpose is to teach the basic principles of physical chemistry. We do not pretend to cover the details of the many important applications of physical chemistry in biology, medicine, biochemistry, and related fields. Equipped with knowledge of the basic principles, students should be prepared to understand those applications they encounter in their subsequent work.

Our presentation is organized into three major sections to mirror the three major areas of physical chemistry: thermodynamics, chemical kinetics, and quantum chemistry. Within these sections other relevant topics are also discussed. In general, topics are presented in logical rather than chronological order.

The discussions of quantum chemistry, molecular structure, spectroscopy, and related subjects have been included in this text because a reasonable aquaintance with these subjects has become essential for the biological and health sciences. Exposure to the principles of molecular structure and molecular interaction provides a helpful basis for learning and understanding many recent advances in these fields. The material in the quantum chemistry section of this book has been carefully selected to provide the necessary fundamentals of this important subject.

Numerous illustrative examples appear throughout the text, and sets of problems for the student to solve are included at the end of each chapter. Both examples and problems are designed for interest and relevance; many concern biological or ecological matters.

If a briefer, less mathematical approach is desired, shaded portions of the text may be omitted by the student. These shaded sections appear principally in the quantum chemistry chapters.

Neither the material nor our method of presenting it is new, for the most part. Our ideas have been shaped over the years by the texts we have read and used (with authors such as Walter J. Moore, Gilbert W. Castellan, Gordon M. Barrow, A. R. Knight, Gilbert N. Lewis, Merle Randall, Jurg Waser, Arthur A. Frost, Ralph G. Pearson, Melvin W. Hanna, Farrington Daniels, and Samuel Glasstone), by the needs and responses of the students we have taught, and by our interactions with our own teachers and with our faculty colleagues (among whom we would particularly like to acknowledge the influences of Leonard K. Nash, Allen Kropf, and Sir Hugh Taylor). We have chosen those approaches that have worked best and with which we feel most comfortable. To them we have added our own ideas and points of view to produce a text which is concise yet, within its defined limits, thorough. We believe this book offers a sound and stimulating introduction to the subject of physical chemistry.

Atlanta, Georgia
September 1976

J. W. M.
W. H. J.

Contents

Elements of Physical Chemistry

PART ONE
Thermodynamics

1 / Introduction and the First Law

1.1. Introduction

We know from chemical experience that some substances react with each other readily and completely, others only partially, and still others not at all. Moreover, the products of a given chemical reaction may differ when the reaction is carried out at a different temperature or a different pressure.

Physical chemists would like to generalize and to quantify such observations and finally to express them as mathematical relationships. They would like to be able to know in advance the outcome of a reaction or to predict the effect of varying the experimental conditions.

Efforts to understand and to predict chemical behavior can be traced to antiquity. To explain the mutual reactivity of reacting substances and the vigor of chemical processes, the qualities of love, hate, eagerness, and reluctance were at one time attributed to the reacting materials. Later the idea developed that the amount of heat given off when substances react was a direct measure of their mutual chemical reactivity. This idea (only partly true) was explored and expanded to become chemical thermodynamics, the science which examines how chemical processes interrelate with energy in all its forms.

By proceeding from the basic laws of thermodynamics with the help of mathematics, we shall be able to predict the results of chemical reactions. We shall ascertain not only whether a given reaction is possible but, if so, what relative amounts of products and unconsumed reactants may be expected. We shall also predict quantitatively the effect of changes in the experimental conditions.

Though such goals fully justify a study of thermodynamics, we shall extend these principles to other chemical problems as well.

1.2. Some Basic Terms and Definitions

Thermodynamics is based upon a set of laws which apply to processes in which a system undergoes a change of state involving heat or work or both.

The *system* is a portion of the physical world which interests us and which we set apart for study. Everything else is called the *surroundings*. An enclosing boundary, either real or mentally constructed, separates the system from the surroundings.

In our thermodynamic study, we shall only be concerned with those properties of a system which can be measured experimentally, such as volume, pressure, mass, and temperature. All such properties are either *extensive* (they vary with the amount of material considered) or *intensive* (they are independent of the amount of material considered). Mass and volume are examples of extensive properties, while pressure, temperature, and density are intensive properties.

The *state* of a system at a given instant is determined by its properties; the state changes when the properties change. As examples of this let us consider particular substances, the ideal gas and real gases.

1.3. The Ideal Gas

Under conditions of low to moderate pressure and moderate to high temperature, all gases behave similarly. Three classic experimental laws predict the volume of a gas under these conditions. The volume V is proportional to the number of moles n present (Avogadro's law), is proportional to the absolute temperature T (Charles's law), and is inversely proportional to the pressure P (Boyle's law). Combining these gives $V = R(nT/P)$, where R is a proportionality constant. This relation,[1] more familiarly written

$$PV = nRT, \tag{1}$$

is known as an *equation of state*. A gas which precisely obeys this equation under all conditions is called an *ideal gas*.

The proportionality constant R is the *universal gas constant*. Depending on the units employed, it may have the following equivalent values:

8.2054×10^{-2} liter-atmospheres mole^{-1} degree^{-1}

8.3144×10^{7} ergs mole^{-1} degree^{-1}

1.9872 calories mole^{-1} degree^{-1}.

The temperature T in the equation of state is the *absolute* or *kelvin* temperature. It is defined to be

$$T = 273.15 + t, \tag{2}$$

where t is the *centigrade* or *celsius* temperature (°C) and T is the absolute or kelvin temperature (°K).

Equation (1) is called an equation of state because it gives the relationship among those properties or variables of state which are of importance in thermodynamic and other physical considerations. By specifying the values of P, V, n, and T, we fix the state of the system. If we change the values of these properties we change the state of the system.

---

[1] This fundamental relationship might indeed be called the "ABC" (Avogadro, Boyle, Charles) of the gaseous state.

Note that all the variables of state are not independent but are interrelated through Eq. (1). Thus if we know any three of them, the fourth is automatically fixed and can be calculated.

1.4. Equations of State for Real Gases

As mentioned before, all gases behave similarly under conditions of low to moderate pressure and moderate to high temperature. Under these conditions, the behavior of any gas follows very closely the equation of state of the hypothetical ideal gas. But can we develop equations of state for real gases under other conditions? Are there equations which accurately relate the state variables of real gases under all conditions?

Over the years, more than 200 equations of state have been proposed for gases. Many of these have no justifying theory but were concocted to make the equation fit the nonideal behavior of the gas as actually observed. Such equations may work well for a particular gas over a limited range of temperature and pressure but may fail badly when applied to other gases or even when applied to the same gas under markedly different conditions.

Some of the equations of state for real gases which have been used with a degree of success are presented in Table 1.1. Notice that, unlike the ideal gas equation, all these equations contain constants which are characteristic of the particular gas. The values of these constants are usually obtained empirically by measuring the values of the state variables under various conditions and then finding the value of each constant which gives the best agreement between the observed and the calculated data.

Compared with the other equations shown, the one proposed by van der Waals is used most because it is simplest and because its constants can be

Table 1.1. Selected Equations of State for Real Gases

van der Waals	$(P + an^2/V^2)(V - bn) = nRT$
Dieterici	$Pe^{an/VRT}(V - bn) = nRT$
Bertholet	$(P + an^2/TV^2)(V - bn) = nRT$
Clausius	$(P + an^2/T(V + cn)^2)(V - bn) = nRT$
Virial Equations	$PV = nRT(1 + Bn/V + Cn^2/V^2 + Dn^3/V^3 + \cdots)$
	$PV = nRT(1 + B'P + C'P^2 + D'P^3 + \cdots)$
Beattie-Bridgeman	$P = nRT/V + n^2\beta/V^2 + n^3\gamma/V^3 + n^4\delta/V^4$

rationalized (see Section 1.13) rather than being totally empirical. Its weakness is that it makes no provision for the temperature dependency of its correction terms. The van der Waals equation of state for real gases is

$$(P + an^2/V^2)(V - bn) = nRT. \tag{3}$$

Values of the constants a and b for a few gases are given in Table 1.2.

Table 1.2. van der Waals Constants, Normal Boiling Points, and Critical Parameters

Gas	a, liter^2 atm mole^{-2}	b, liter mole^{-1}	*Normal boiling point*, °K	T_c, °K	P_c, atm
He	0.03412	0.02370	4.6	5.3	2.26
H_2	0.2444	0.02661	20.7	33.3	12.8
O_2	1.360	0.03183	90.2	154.3	49.7
N_2	1.390	0.03913	77.4	126.1	33.5
CO	1.485	0.03985	81.7	134.4	34.6
CH_4	2.253	0.04278	109.2	190.3	45.6
CO_2	3.592	0.04267	194.7	304.2	72.8
NH_3	4.170	0.03707	239.8	405.6	112.2
H_2O	5.464	0.03049	373.2	647.2	217.7
Cl_2	6.493	0.05622	238.6	417.1	76.1
CH_3OH	9.523	0.06702	338.2	513.1	78.5
C_2H_5OH	12.02	0.08407	351.7	516.2	63.0
$(C_2H_5)_2O$	17.38	0.1344	307.8	—	—
C_6H_6	18.00	0.1154	353.3	561.6	47.9

Figure 1.1 gives the pressure-volume behavior predicted for carbon dioxide by the ideal gas equation and by the van der Waals equation compared with that actually observed experimentally. As can be seen, the van der Waals equation more accurately predicts the actual behavior than does the ideal gas equation.

Let us consider several examples which should help us understand the conditions under which it may or may not be acceptable to assume ideal behavior for a gas.

Example 1 A large desiccator is selected for an animal experiment; its measured volume is 14.70 liters. On a day when room temperature is 27°C and the barometer reads 740 torr, a rat (body volume 202 cc) is placed inside, the lid clamped on, and the outlet closed with a serum stopper. By hypodermic, 10.00 cc of ether is then introduced.

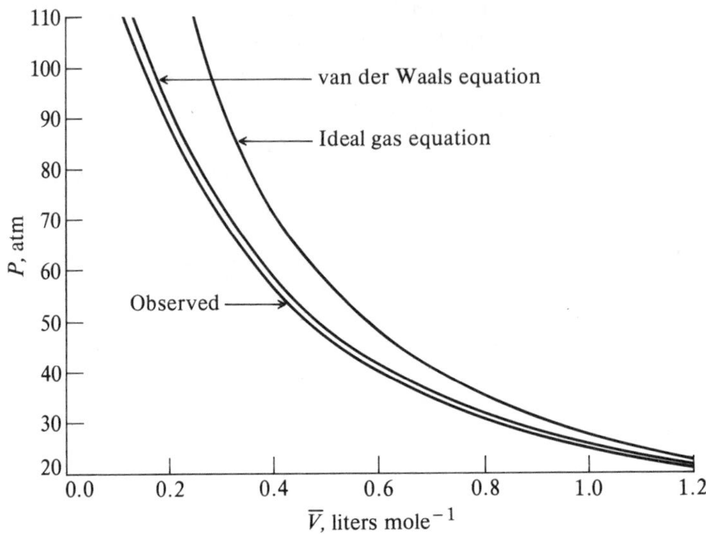

Fig. 1.1. Pressure-volume relations of CO_2 at 70°C.

a) Use the van der Waals equation to find the partial pressure of ether vapor which will be established inside when evaporation is complete. (The density of liquid ether is 0.708 g cm^{-3}; its molecular weight is 74.1 g mole^{-1}.)
b) Taking the van der Waals answer as correct, what error (percentage and sign) would have been introduced by assuming ideal gas behavior?

Answer: a) Since we used 10.00 cm^3 of liquid whose density is 0.708 g cm^{-3}, there will be 10.00 × 0.708 = 7.08 g or 7.08 g ÷ 74.1 g mole^{-1} = 0.0956 mole of ether in the desiccator. Using the van der Waals equation:

$$(P + an^2/V^2)(V - bn) = nRT$$

$$P = nRT/(V - bn) - an^2/V^2,$$

and substituting the values we know (see Table 1.2):

$$P = \frac{(0.0956)(0.0821)(300)}{(14.70 - 0.20) - (0.1344)(0.0956)} - \frac{(17.38)(0.956)^2}{(14.70 - 0.20)^2}$$

$$= 0.1619 \text{ atm or } 123.2 \text{ torr.}$$

(An atmosphere is equal to 760 torr.) Notice that we must subtract the body volume of the rat from the total volume of the container to obtain the volume available to the gas.

b) From the ideal gas equation we would calculate the following pressure:

$$P' = nRT/V$$

$$= \frac{(0.0956 \text{ mole})(0.0821 \text{ liter-atm mole}^{-1} \text{ deg}^{-1})(300 \text{ deg})}{(14.70 \text{ liters} - 0.20 \text{ liter})}$$

$$= 0.1626 \text{ atm.}$$

If we assume the van der Waals answer is correct, the error would be $(P' - P)/P = (0.1626 - 0.1619) \div 0.1619 = 0.0043$ or $+0.43\%$.

Our calculation shows that, for this experiment, the work involved in using the more complicated van der Waals equation is probably not justified. The uncertainty of the data taken and the accuracy goal of the experiment itself combine to suggest that the less-than-half-percent improvement is not worth the extra effort.

Example 2 A partially discharged CO_2 fire extinguisher of 2.500-liter capacity still contains enough gas to exert a pressure of 11.26 atm at 20°C.

a) Calculate the number of moles of CO_2 in the extinguisher using the van der Waals equation.
b) Find the value predicted by the ideal gas equation.
c) Find the error (percentage and sign) by which each of the above answers deviates from the result found experimentally, which was 1.250 moles.

Answer: a) It is quickly apparent that using the van der Waals equation to find P or T is simple, but solving for n or V is more complex since the equation is cubic in these variables. However, as chemists rather than mathematicians, we should take a practical approach to the numerical problems we encounter. Our goal is to get the answer which can be defended as correct, whether the method is mathematically "elegant" or not. We should attack the problem with all the information we have, and take short-cuts without apology.

We can simplify the present calculation somewhat by first finding the number of moles in a single liter, then multiplying by the number of liters. Our equation with $V = 1$ thus becomes

$$(P + an^2)(1 - bn) = nRT,$$

or

$$(ab)n^3 - (a)n^2 + (Pb + RT)n - P = 0.$$

Substituting known values and simplifying, we have

$$n^3 - 23.44n^2 + 159.7n = 73.3.$$

This cubic is most easily solved by the method of successive approximations, sometimes inelegantly called "trial and error." For the first approximation we use the value given by the ideal gas equation:

$$n = \frac{PV}{RT} = \frac{11.26 \times 1.000}{.0821 \times 293} = 0.468 \text{ mole.}$$

It is well to proceed systematically, remembering that the desired value in the last column of the following table is 73.3. We try successive values of n, revising our values in such a way that we eventually close in on the correct one.

When n is	n^2	n^3	$23.44n^2$	$159.7n$	P/ab
0.468	0.219	0.103	−5.13	74.7	69.7
0.500	0.250	0.125	−5.86	79.9	74.1
0.490	0.240	0.118	−5.63	78.3	72.7
0.495	0.245	0.121	−5.74	79.05	73.4
0.494	0.244	0.121	−5.72	78.89	73.3

The desired answer (to sufficient accuracy) is thus 0.494 mole per liter; in the 2.500-liter extinguisher there are 2.500 × .494 = 1.235 moles remaining.

b) The ideal gas prediction is

$$n = \frac{PV}{RT} = \frac{11.26 \times 2.500}{.0821 \times 293} = 1.170 \text{ moles.}$$

c) van der Waals error:

$$\frac{1.235 - 1.250}{1.250} = \frac{-0.015}{1.250} = -1.2\%.$$

Ideal assumption error:

$$\frac{1.170 - 1.250}{1.250} = \frac{-0.080}{1.250} = -6.4\%.$$

Here the van der Waals equation predicts about five times as accurately as the ideal gas equation. One cannot, however, assume that the former equation is superior in every situation; the comparison depends on the gas chosen and on the P, V, and T conditions.

In this text we shall usually be dealing with gases under conditions similar to those of Example 1. Unless otherwise noted, we shall therefore assume that the use of the ideal gas equation as the equation of state for the real gas we are treating is acceptable. As the two foregoing examples indicate, this will not be a bad assumption as long as we confine our attention to gases under moderate conditions.

1.5. Heat

Let us now return to our task of defining the basic terms of thermodynamics. *Heat* can be defined operationally as

$$Q = n_A C_A(T_2 - T_1) = n_A C_A \Delta T. \tag{4a}$$

The amount of heat Q which must be added to a sample of substance A to change its temperature from T_1 to T_2 is proportional to the amount of the material present n_A and to the temperature difference ΔT. The proportionality constant C_A is a characteristic of the particular substance and is called its molar *heat capacity* when the amount of material is given in moles. It is the quantity of heat necessary to raise the temperature of one mole of the material by one degree kelvin or centigrade.

When the size of the sample is expressed in grams, Eq. (4) is written

$$Q = m_A c_A \Delta T, \tag{4b}$$

where m_A is the mass in grams and c_A is the quantity of heat necessary to raise the temperature of 1 g of material by 1°. The constant c_A is called the *specific heat* of substance A.

By convention, Q is positive when heat is added to the substance and negative when heat is lost by the substance. It is positive when T_2 is greater than T_1, negative when T_2 is less than T_1.

As we shall see in Section 1.7, a portion of the heat absorbed by a system may in some cases be used to do work, and thus is not available for raising the system's temperature. When this occurs the total amount of heat contributed to the system cannot be determined by Eq. (4) employing the heat capacity and the temperature change. Determination of the heat in such cases will be considered in later sections.

The quantity of heat is usually measured in *calories*. One calorie is defined as the amount of heat needed to raise the temperature of 1 g of water from 14.5°C to 15.5°C. On this basis the specific heat of water is 1 calorie gram^{-1} degree^{-1}. Since there are 18 g in a mole of water, its heat capacity is 18 cal mole^{-1} degree^{-1}.

Example 3 Suppose a weighed specimen of excised fatty tissue is sealed in a stout vessel, and a surplus of oxygen introduced under pressure. The vessel is then immersed in 1000 ml of water in a calorimeter (see Section 2.1) and combustion initiated by an electric current. The completed reaction is observed to raise the water temperature from 25.0°C to a final 38.8°C. A previous calibration has shown that the heat capacity of those calorimeter parts undergoing temperature rise may be expressed as a "water equivalent" of 210 ml. What is the heat given off or absorbed in this process?

Answer: Here $\Delta T = 13.8°$, $c = 1$ cal g^{-1} deg^{-1}, and $m = 1000 + 210$. Using Eq. (4b), we find the heat absorbed by the surroundings to be

$$(1.000 \text{ cal g}^{-1} \text{ deg}^{-1})(1210 \text{ g})(13.8 \text{ deg}) = 16{,}700 \text{ cal.}$$

This figure represents the heat *lost* by the reacting mixture to the surroundings (water and calorimeter). The combustion process pictured is exothermic, and

$$Q = -16.7 \text{ kcal.}$$

Example 4 A man who weighs 68.0 kg (approximately 150 pounds) produces about 3000 kcal of heat every day as a result of his normal metabolic processes. If we assume his heat capacity to be the same as that of water, how hot would he get if his body retained all the heat produced in a day?

Answer:

$$3000 \text{ kcal} = 3.00 \times 10^6 \text{ cal}$$

$$68 \text{ kg} = 6.80 \times 10^4 \text{ g}$$

$$\Delta T = Q/mc$$

$$= \frac{(3.00 \times 10^6 \text{ cal})}{(6.8 \times 10^4 \text{ g})(1 \text{ cal } g^{-1} \text{ deg}^{-1})}$$

$$= 44°\text{K (or °C).}$$

Since body temperature is normally 37°C (310°K), his temperature at the end of 24 hr would be 37° + 44° = 81°C (or 178°F).

1.6. Work

Mechanical *work* W is operationally defined as the force exerted on an object multiplied by the distance which the object moves as a result of the applied force. Mathematically:

$$W = \int_1^2 \boldsymbol{f} \cdot d\boldsymbol{r}, \tag{5}$$

where the integration is performed over the interval from the initial position or state 1 to the final position or state 2, $\boldsymbol{f}$ is the vector force exerted, and $d\boldsymbol{r}$ is an infinitesimal change in position.

Pressure-volume work is work done when a volume change occurs against an opposing pressure, as when a gas expands against its surroundings. Its value is given by the equation:

$$W = -\int_1^2 P_{\text{ext}}\, dV. \tag{6}$$

P_{ext} is the *external pressure*, the pressure exerted on the system by the surroundings.

When work is done *on* the system, the volume change dV is negative (the system is compressed) and work is positive. When work is done *by* the system, dV is positive (the system expands) and work is negative.[2]

The units of work which we shall employ most frequently are *joules* and *ergs*. One joule is equal to 10^7 ergs or 10^7 g-cm^2 sec^{-2}.

Example 5 Derive Eq. (6).

Answer: Consider a gas contained in a piston-and-cylinder. A is the cross-sectional area of the piston. Recall that pressure is force per unit area, $P_{ext} = f/A$ or $f = P_{ext}A$. If dr is an infinitesimal distance the piston moves, then $Adr = dV$, the change (due to the movement of the piston) in the volume of the part of the cylinder containing the gas, and $dr = dV/A$ (see Fig. 1.2). Using Eq. (5), we have

$$W = \int_1^2 f\,dr = \int_1^2 (P_{ext}\,A)(dV/A) = \int_1^2 P_{ext}\,dV.$$

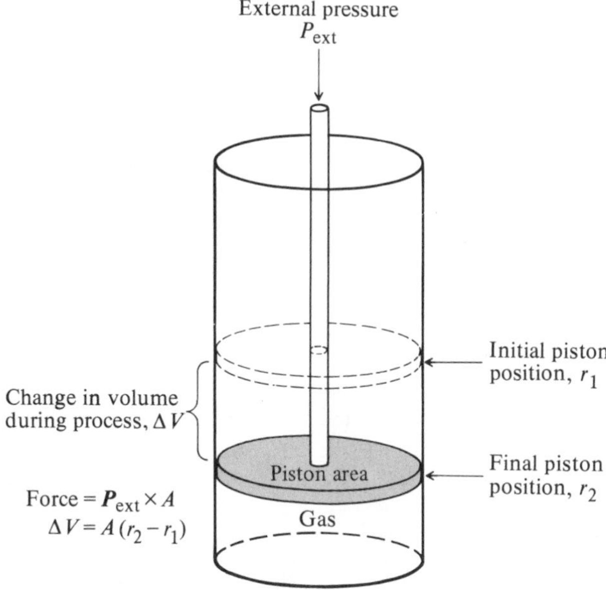

Fig. 1.2. Compression of a gas by a piston.

[2] In some texts the opposite convention is used: work done by the system is positive and work done on the system is negative. In that case, $W = \int P_{ext}\,dV$. Since, however, heat entering the system is considered positive, it would seem more consistent to hold a similar view for work: When the system receives heat or work, the sign is positive; when it emits heat or work, the sign is negative.

Remember, we are defining work done on the gas as positive; work is negative when dV is positive, and thus

$$W = -\int_1^2 P_{ext}\, dV. \tag{6}$$

Example 6 Derive expressions for the pressure-volume work involved in the following processes:
a) any process which occurs with no change in volume
b) expansion of an ideal gas into a vacuum ("free" expansion)
c) any process occurring with a constant external pressure
d) expansion of an ideal gas during which the temperature is held constant and the external pressure is constantly adjusted to match the pressure which the gas itself exerts

Answer:

a) $W = -\int_1^2 P_{ext}\, dV = 0$ since $dV = 0$.

b) $W = -\int_1^2 P_{ext}\, dV = 0$ since $P_{ext} = 0$.

c) $W = -\int_1^2 P_{ext}\, dV = -P_{ext}\int_1^2 dV = -P_{ext}(V_2 - V_1) = -P_{ext}\,\Delta V$.

d) $W = -\int_1^2 P_{ext}\, dV = -\int_1^2 \frac{nRT}{V}\, dV = -nRT\int_1^2 \frac{dV}{V} = -nRT \ln\left(\frac{V_2}{V_1}\right)$,

where ln is the *natural logarithm*.[3] Since the process is isothermal and the gas ideal,

$$\frac{V_2}{V_1} = \frac{nRT/P_2}{nRT/P_1} = \frac{P_1}{P_2}$$

$$W = -nRT \ln\left(\frac{V_2}{V_1}\right) = -nRT \ln\left(\frac{P_1}{P_2}\right).$$

1.7. Equivalence of Heat and Work

At one time heat was regarded as a fluid substance which could neither be created nor destroyed. In the eighteenth century, however, upon noticing that

[3] The natural logarithm of a number x is the power to which e (numerical value 2.718 ...) must be raised to equal x; $\ln x = y$ where $e^y = x$. The natural logarithm and its base e appear often in the mathematics of physical chemistry. For numerical calculations, the natural logarithm is obtained by multiplying the base-10 logarithm by 2.303: $\ln x = 2.303 \log x$.

cannon became quite hot as they were being bored, Count Rumford[4] observed that in this process work was being converted into heat and proposed that heat and work were interconvertible.

The equivalence of heat and work was quantified by James Joule, who experimentally determined the amount of work necessary to produce a given amount of heat. The accepted value of the "mechanical equivalent of heat" is 1 calorie = 4.184 joules.

Example 7 An athlete is 30 lb overweight. He decides to work off this extra fat by lifting weights. The human body can be considered a heat engine that can do work by using the heat given off in physiological chemical reactions. It can derive approximately 9.3 kcal of metabolic energy from each gram of fatty material. Assuming the heat energy derivable from the fat goes only toward the work of lifting, how many times would this athlete have to lift a 100-lb weight to a height of 2 ft to use up the excess fat?

Answer: Converting units, 30.0 lb = 13600 g, 100 lb = 45400 g and 2.0 ft = 61.0 cm. When burned, the 13600 g of fat releases 13600 g $\times$ 9.3 kcal g^{-1} = 1.27×10^5 kcal of heat. This is equivalent to 1.27×10^8 cal $\times$ 4.18 joules cal^{-1} = 5.31×10^8 joules. The work required to lift a 100-lb weight 2 ft is equal to the mass m multiplied by the height it is lifted h and by the gravitational factor g, which is 980.6 cm sec^{-2}.

$$W = mgh = (4.54 \times 10^4 \text{ g})(980 \text{ cm sec}^{-2})(61.0 \text{ cm})$$
$$= 2.7 \times 10^9 \text{ g-cm}^2 \text{ sec}^{-2} \text{ or } 2.7 \times 10^2 \text{ joules per lift.}$$

The athlete must repeat the lift 5.31×10^8 joules $\div$ 2.71×10^2 joules per lift = 1.93×10^6 times!

1.8. Functions of State

A *state variable* or *state function* is one which has a single, definite value for a particular state of the system. When a change in the state of the system occurs, the change in a state variable or function is simply its value in the final state minus its value in the initial state, regardless of the procedure or sequence of steps through which the change in state is brought about.

Consider, for instance, the pressure of one mole of an ideal gas. Initially the volume is V_1 and the temperature T_1. The pressure is a variable of state and

---

[4] Born Benjamin Thompson in Woburn, Massachusetts, this man enjoyed a career filled with such diverse accomplishments as the invention of the modern fireplace and drip coffee-maker, the development of the first poverty program, marriage to the widow of Lavoisier, being knighted by George III of England, being made a Count of the Holy Roman Empire, and tearing down the village church and cutting down all the apple trees in Huntington, Long Island. For a brief but full account of the life of this fascinating eighteenth-century scientist, see *Count Rumford, Physicist Extraordinary* by Sanborn C. Brown (Garden City N.Y.: Anchor Books, 1962).

can be calculated by using the ideal gas equation which gives the relationship among the variables of state: $P_1 = RT_1/V_1$. If the state of the system is changed so that the volume is V_2 and the temperature is T_2, then the pressure in the new state is $P_2 = RT_2/V_2$. The change in pressure is $\Delta P = P_2 - P_1$, completely independent of the method by which the change of state is accomplished.

Since a change in a function or variable of state depends only on the initial and final states of the system, in mathematical terms it is said to be *integrable*. Thus if F is a function of state, then

$$\Delta F = \int_1^2 dF = F_2 - F_1. \tag{7}$$

When this is true, dF is called an *exact differential.*

If F is an extensive function of state (see Section 1.2) the molar value of F, denoted $\bar{F}$, may be written as a function of any two state variables. For instance, choosing volume and temperature as the two variables, $\bar{F}$ can be written as a function of V and T:

$$\bar{F} = f(V, T). \tag{8}$$

Since $\bar{F}$ is a state function and therefore $d\bar{F}$ is an exact differential,

$$d\bar{F} = (\partial\bar{F}/\partial T)_V \, dT + (\partial\bar{F}/\partial V)_T \, dV, \tag{9a}$$

where $(\partial\bar{F}/\partial T)_V$ and $(\partial F/\partial V)_T$ are *partial derivatives.* The notation $(\partial\bar{F}/\partial T)_V$ says that we are looking at the rate at which $\bar{F}$ changes with a change in temperature alone. The subscript V acknowledges that the volume of the system also affects $\bar{F}$ but that the volume will be kept constant during this observation of the effect of temperature. Mathematically speaking, we take the derivative of $\bar{F}$, which is a function of both T and V, with respect to T, treating V as a constant. Similar reasoning applies to $(\partial\bar{F}/\partial V)_T$, with volume as the variable and temperature held constant.

Example 8 Suppose $f(x, y) = x^2 + xy + y^2$. What is the partial derivative of this function with respect to x and with respect to y?

Answer:

$$(\partial f/\partial x)_y = 2x + y$$

$$(\partial f/\partial y)_x = x + 2y.$$

If we choose pressure and temperature as our variables instead of volume and temperature, $\bar{F} = f(P,T)$ and the exact differential is

$$d\bar{F} = (\partial\bar{F}/\partial T)_P \, dT + (\partial\bar{F}/\partial P)_T \, dP. \tag{9b}$$

The defining relation of exact differentials given in Eqs. (9a) and (9b) holds for all functions of state. We shall find it useful in a number of situations.

There are functions which are not state functions. Due to the fact that they can be interchanged, neither work nor heat, taken individually, need correspond exactly with a difference in a state function. This fact can be demonstrated in the following manner.

Suppose you left me alone in a room with a beaker of ethanol at room temperature and returned later to find the ethanol at a higher temperature, that is, in a new state. You could perhaps say that an amount of heat $Q = (T_2 - T_1)nC_{\text{ethanol}}$ had been added while you were out of the room. If heat did correspond exactly to a change in a state function, you would be correct. Since a state function depends only on the condition of the system, the amount of change in the state variable could be calculated, knowing the initial and the final conditions of the system (in this case, the initial and final temperatures).

But I may not have added any heat to increase the temperature. Instead I may have performed on the system an amount of work equivalent to the heat calculated above ($W = Q \times 4.184$ joules calorie^{-1}). I could do this by stirring briskly. Or I could have increased the temperature by adding some heat *and* doing some work. In fact, there is an infinite number of combinations of heat added to and work done on our system which would raise its temperature to the same final value. The only way you could tell me how much heat I had added and how much work I had done would be for you to know precisely what I did to the system during your absence.

Neither the quantity of heat nor the work involved in a change of state may therefore be determined simply by knowing the initial and final states of the system. The method of proceeding between the two states must be specified if the heat and work involved are to be successfully determined. They do not correspond to changes in functions of state.

1.9. The First Law of Thermodynamics; Energy

It is observed in all cases, however, that the heat added to the system plus the work done on the system during a change of state does correspond to a change in a state function. The sum of the heat I add to our ethanol sample and the work I do on it *to produce a specific change in state* will always be the same. This is true even though there is an infinite number of possible combinations of Q and W which will produce the same change in the system.

The state function associated with this total amount of heat and work is called the *thermodynamic energy*. The system is said to possess an amount of energy E_1 before, and an amount E_2 after, a change in state occurs. The change in energy ΔE is equal to the sum of the heat (thermal energy) *added to* the system during the change in state and the work (mechanical energy) *done on* the system during the process:

$$\Delta E = E_2 - E_1 = Q + W. \tag{10}$$

This relationship has been observed to be true for *all* changes in state involving heat and/or work.[5] Regardless of the process we use to change a system from a particular initial state (state 1) to a particular final state (state 2) the sum $Q + W$ is always the same *for a given change in state*. This is a conservation law. Expressed as Eq. (10), it is called the *first law of thermodynamics*.

Example 9 Assuming no heat loss to the surroundings and no work done on the surroundings, what is the temperature difference between the water at the top of a 150-ft waterfall and that at the bottom?

Answer: In going from the top of the falls to the bottom, the water undergoes a change in state. At the top there is an amount of potential energy mgh which it no longer possesses when it reaches the bottom. This $mgh = (1\text{ g})(980\text{ cm sec}^{-2}) \times (4570\text{ cm}) = 4.48 \times 10^6\text{ g-cm}^2\text{ sec}^{-2} = 0.448\text{ joule g}^{-1}$ or 0.107 cal g^{-1}. Since no work is done and no heat exchanged with the surroundings, $\Delta E = Q + W = 0$. The potential energy lost by the system must therefore be converted into another form of energy which remains within the system, so that the net change in energy is zero. The potential energy is changed into heat energy manifested by an increase in temperature:

$$\begin{aligned} mgh &= mc\,\Delta T \\ \Delta T &= 0.107\text{ cal g}^{-1}/1\text{ cal g}^{-1}\text{ deg}^{-1} \\ &= 0.107°\text{C}. \end{aligned}$$

Example 10 In a particular change in state a system absorbed 200 cal of heat. When the system was returned to its initial state by a different process, 100 cal of heat were emitted and the system did the equivalent of 400 cal of work on the surroundings. How much work was involved in the first process?

Answer: The energy change in the initial process was

$$\Delta E_1 = Q_1 + W_1 = 200\text{ cal} + W_1.$$

In the second process

$$\Delta E_2 = Q_2 + W_2 = -100\text{ cal} - 400\text{ cal} = -500\text{ cal}.$$

---

[5] Some authors prefer to write this expression $\Delta E = Q - W$. In that case, however, W is the work done *by* the system in a change of state. For instance, the pressure-volume work would be $\int P\,dV$ if their convention is used instead of $-\int P\,dV$ using our convention. Of course, the same answer for ΔE will be obtained if either convention is used consistently.

We have chosen the $\Delta E = Q + W$ convention for three reasons. (1) It seems more logical to consider both heat and work positive when they are added to the system. (2) Most chemical thermodynamics and physical chemistry texts are now using this convention. (3) The International Union of Pure and Applied Chemistry has recommended its adoption.

Since the second process returned the system to the initial state, the final energy must be equal to the initial energy: $\Delta E_{\text{overall}} = E_{\text{final}} - E_{\text{initial}} = 0$. Consequently the sum of the energy changes in the two steps must also be equal to zero:

$$\Delta E_{\text{overall}} = \Delta E_1 + \Delta E_2 = 0$$

$$(200 \text{ cal} + W_1) + (-500 \text{ cal}) = 0$$

$$W_1 = 300 \text{ cal}.$$

1.10. Energy Changes in Constant Volume and Constant Pressure Processes; Enthalpy

Suppose we change the state of a system in such a way as to keep its volume constant. If work other than pressure-volume work (electrical work, work against gravity, etc.) is excluded during the change in state, there will be no work involved in the process (see Example 6a). We can measure the heat Q_V required in this transformation and, since there is no work, it follows that for a constant volume process

$$\Delta E = Q_V . \tag{11a}$$

For an infinitesimal change in which the volume of the system does not change

$$dE = dQ_V . \tag{11b}$$

Now consider a change in state under constant pressure conditions. We can measure the heat Q_P involved in such a process and, if no work other than pressure-volume work can occur, $W = -P_{\text{ext}}\, \Delta V$ (see Example 6b) and

$$\Delta E = Q_P - P_{\text{ext}}\, \Delta V, \tag{12a}$$

for a constant pressure process. Again, for an infinitesimal change in state

$$dE = dQ_P - P_{\text{ext}}\, dV. \tag{12b}$$

Many chemical processes occur in open containers and are therefore subjected only to the pressure of the surrounding atmosphere. Thus they are carried out under conditions of constant pressure rather than constant volume. We find it helpful therefore to introduce a new state function specifically suited to constant-pressure operating conditions. This new function is the *enthalpy*, H. It is defined as

$$H = E + P_{\text{int}}\, V. \tag{13}$$

Here P_{int} is the *internal pressure*, the pressure exerted by the system on the surroundings. If the system and surroundings are in equilibrium, $P_{\text{ext}} = P_{\text{int}}$.

If the system is not in a state of equilibrium with the surroundings, the internal pressure may, or it may not, be equal to the external pressure.

For any infinitesimal change in state which occurs under constant-pressure conditions with the external pressure equal to the internal pressure:

$$dH = dE + d(PV) = dE + P\,dV. \tag{14}$$

Substituting Eq. (12b) into Eq. (14):

$$dH = (dQ_P - P\,dV) + P\,dV$$

$$dH = dQ_P. \tag{15a}$$

For a finite change in state

$$\Delta H = Q_P. \tag{15b}$$

Note that when the path (constant volume or constant pressure) is specified, Q may be replaced by the change in a state function. Under conditions of constant volume, Q is identified with ΔE; under conditions of constant pressure, with ΔH.

1.11. Heat Capacity

Recall that in our definition of heat in Eq. (4) we introduced a constant C, the heat capacity. This quantity, which is characteristic of the material, is the amount of heat required to change the temperature of one mole of the material by one degree, or $C = Q/n\,\Delta T$. For a constant-volume process (Equation 11)

$$C_V = Q_V/n\,\Delta T = \Delta E/n\,\Delta T$$

Defined in this way C_V is actually the average heat capacity over a finite temperature range ΔT. Since the heat capacity is a function of temperature, it would be better to define it in terms of an infinitesimal change. For one mole of material

$$C_V = (\partial \bar{E}/\partial T)_V. \tag{16}$$

$\bar{E}$ is the molar energy, the energy of one mole of a substance. Note that $(\partial \bar{E}/\partial T)_V$ is a partial derivative with the volume designated as constant.

For a constant pressure process (Eq. 15)

$$C_P = Q_P/n\,\Delta T = \Delta H/n\,\Delta T,$$

or

$$C_P = (\partial \bar{H}/\partial T)_P. \tag{17}$$

Equations (16) and (17) are definitions of the heat capacity under constant volume conditions and under constant pressure conditions, respectively.

1.12. A Microscopic Interpretation of Energy and Temperature: The Kinetic-Molecular Theory of Gases

Classical thermodynamics concerns itself only with relatively large collections of matter rather than with individual molecules. The properties of matter in bulk are called *macroscopic* properties; those associated with the particles and aggregates of atomic and molecular theory are termed *microscopic* properties. If the microscopic theory is correct, the macroscopic properties are observable manifestations of atomic and molecular behavior.

As modern chemists, we like to think in terms of molecules. Even though thermodynamics does not require a microscopic model, we find it satisfying to interpret our thermodynamic results in terms of molecular behavior.

The currently accepted microscopic theory will be discussed later in a section on quantum chemistry. This theory holds that a molecule possesses energy in a variety of ways. The correlation of the energy of the individual molecules with the macroscopic energy of a large collection of molecules requires the use of the rather complex mathematics of statistical mechanics.

Although at this point most systems are much too complex to be fully interpreted in terms of the molecules they contain, we can treat a simple case: the ideal monatomic gas. The results obtained for this system will significantly assist us in interpreting other thermodynamic systems in terms of the molecules comprising them.

The microscopic theory we shall use in our treatment is called the *kinetic-molecular theory*, initially proposed by Bernoulli in the eighteenth century. We make certain postulates for the monatomic ideal gas in terms of atoms. This model will be a good one if the observed physical behavior of such a gas agrees with predictions based on our postulates.

We postulate that the gas is made up of randomly moving point particles (atoms) which occupy no space, which collide elastically with the walls of the confining container, and which do not interact with each other. The only energy which a particle of this type can possess (ignoring the potential energy due to gravity) is its kinetic energy E. This is equal to $(1/2)mv^2$ where m is its mass and v its velocity. The pressure on the walls of the container is due to myriad collisions the particles make with the wall.[6]

Consider one such gas molecule in a rectangular box with sides a, b and c. Set up a cartesian coordinate system with axes x, y and z parallel with a, b and c respectively. Using vector notation the total velocity of the particle may then be written

$$\mathbf{v}_1 = \hat{i}v_x + \hat{j}v_y + \hat{k}v_z,$$

---

[6] The section that follows is shaded to indicate that it goes into more detail than the average student may want to investigate. Such sections are shaded throughout the text.

where $\hat{\imath}, \hat{\jmath}$ and $\hat{k}$ simply denote the direction of the v_x, v_y and v_z components of the total velocity. If this is the velocity of the molecule before an elastic collision with the wall of the box perpendicular to the x direction (the "x wall"), the velocity after such a collision will be

$$\boldsymbol{v}_2 = -\hat{\imath}v_x + \hat{\jmath}v_y + \hat{k}v_z,$$

since the collision does nothing except change the direction of motion relative to the x axis (see Fig. 1.3).

The force on the x wall during the collision is given by

$$f = m\boldsymbol{a} = m\, d\boldsymbol{v}/dt.$$

The change in $\boldsymbol{v}$ during the collision is $\boldsymbol{v}_2 - \boldsymbol{v}_1 = 2\hat{\imath}v_x$ with a magnitude of $2v_x$. The force on the wall *during* the collision is therefore

$$f = m\, dv/dt = m(2v_x).$$

The average force $\bar{f}_x$ exerted on the x wall by one molecule is the force of one collision divided by the time between collisions with that particular wall. The molecule moves in the x direction (parallel to a) with a speed v_x cm sec^{-1}. It will collide with the x wall again after traveling in the x direction to the opposite wall and back, that is, after traveling $2a$ cm (see Fig. 1.4). (Collisions with the other walls have no effect on the collisions with the x wall. It is as if the molecule were only moving in the x direction with speed v_x). The molecule will then collide with the x wall every $(2a/v_x)$ sec. The average force on this one wall exerted by this one molecule is then

$$\begin{aligned}\bar{f}_x &= (\text{force/collision})/(\text{sec/collision}) \\ &= (2v_x m)/(2a/v_x) = mv_x^2/a.\end{aligned}$$

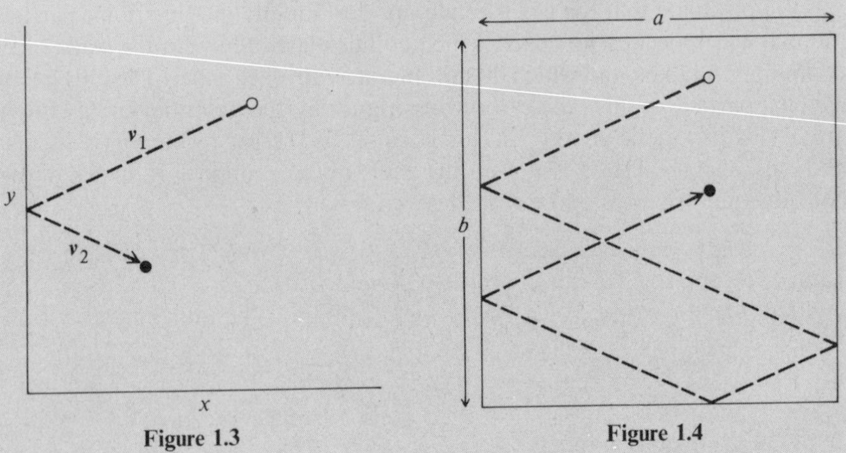

Figure 1.3

Figure 1.4

Suppose we have n moles of molecules, nN molecules (where N is Avogadro's number), with an *average* or *mean square velocity* $\overline{v_x^2}$ in the x direction. The average force on the x wall of the box due to these nN particles will then be

$$\bar{F}_x = nNm\overline{v_x^2}/a.$$

The pressure on this end of the box is given by

$$P_x = \bar{F}_x/\text{Area} = \bar{F}_x/bc = (nNm\overline{v_x^2}/a)/bc = nNm\overline{v_x^2}/abc$$
$$P_x = nNm\overline{v_x^2}/V,$$

where V is the volume of the box.

By proceeding in a similar manner we can show that

$$P_y = nNm\overline{v_y^2}/V$$

and

$$P_z = nNm\overline{v_z^2}/V.$$

But since the particles are moving randomly, $\overline{v_x^2} = \overline{v_y^2} = \overline{v_z^2}$ and therefore $P_x = P_y = P_z$. The pressure on all walls of the container is the same.

Recall that the square of the vector velocity is

$$\begin{aligned} \mathbf{v}^2 &= (\hat{\imath}v_x + \hat{\jmath}v_y + \hat{k}v_z)\cdot(\hat{\imath}v_x + \hat{\jmath}v_y + \hat{k}v_z) \\ &= v_x^2 + v_y^2 + v_z^2. \end{aligned}$$

Similarly

$$\overline{v^2} = \overline{v_x^2} + \overline{v_y^2} + \overline{v_z^2} = 3\overline{v_x^2}.$$

On the basis of our postulates, then, the pressure P on the container is deduced to be

$$P = nNm\overline{v_x^2}/V = \tfrac{1}{3}nNm\overline{v^2}/V$$

or

$$PV = \tfrac{1}{3}nNm\overline{v^2}, \tag{18}$$

where n is the number of moles of particles present, N is Avogadro's number, V is the volume of the container, and $\overline{v^2}$ is the mean square velocity, defined as the average of the squared velocities of all the particles.

The macroscopic energy E_t is then the sum of the microscopic energies ε of all the particles or

$$E_t = nN\bar{\varepsilon} = nN(\tfrac{1}{2}m\overline{v^2}), \tag{19}$$

where $\bar{\varepsilon}$ is the mean or average particle energy. Combining Eqs. (18) and (19) gives

$$PV = \tfrac{2}{3}E_t = \tfrac{2}{3}n\bar{E}_t, \tag{20}$$

where $\bar{E}_t$ is the kinetic energy per mole (molar energy). Such a monatomic ideal gas will, of course, obey the ideal gas equation. Then if our postulates are correct

$$PV = nRT = \tfrac{2}{3}n\bar{E}_t \tag{21}$$

and

$$\bar{E}_t = \tfrac{3}{2}RT. \tag{22}$$

We have now an equation which gives the kinetic energy of such a gas as a function of a state variable, the temperature. By combining Eqs. (19) and (21) we obtain

$$T = \tfrac{1}{3}(Nm\overline{v^2})/R = \tfrac{1}{3}m\overline{v^2}/k, \tag{23}$$

where k is *Boltzmann's constant*, the gas constant per molecule, with a value of 1.38044×10^{-16} erg deg^{-1}. Thus the macroscopic temperature is directly related to the motion of the microscopic particles. This important concept will be useful in dealing with more complex systems as well.

Can we now relate this energy E_t with our thermodynamic or internal energy, $\Delta E = Q + W$? Let us consider a change from a state in which the volume of one mole of a monatomic ideal gas is V_1 and the temperature is T_1 to one in which these state variables are V_1 and T_2. For this constant-volume change in state (Equation 11)

$$\Delta E = Q + W = Q_V = \int_{T_1}^{T_2} C_V \, dT = C_V(T_2 - T_1).$$

The change in the energy E_t (Eq. 22) is

$$\Delta E_t = \tfrac{3}{2}RT_2 - \tfrac{3}{2}RT_1 = \tfrac{3}{2}R(T_2 - T_1).$$

It is clear that the change in the total kinetic energy of a monatomic ideal gas is the same as the change in the thermodynamic internal energy *if* the heat capacity at constant volume is equal to $\tfrac{3}{2}R$. Experimentally C_V for helium (a monatomic gas) is found to be 2.983 cal mole^{-1} deg^{-1} or $\tfrac{3}{2}R$. The behavior expected from our postulated model of a monatomic gas seems to be in excellent agreement with experiment.

1.13. A Microscopic Interpretation of the van der Waals Gas

We can also use the ideas of the kinetic theory of gases to interpret the van der Waals equation.

In our derivation of the kinetic theory we made two assumptions which could not be true for a real gas. We assumed that a molecule is a point, taking up no space itself, and that a molecule is not influenced by the nearness of other molecules. The van der Waals equation substitutes the pressure actually felt by the molecules for the measured pressure in the ideal gas equation; for the total volume of the container it substitutes the empty volume actually available to the molecules.

If b is the volume taken up by a mole of gas molecules themselves, then the space available to a given molecule in which to move is not V, the volume of the container, but $(V - bn)$ with b in units of liters mole^{-1}.

When molecules are attracted to each other they experience a force which affects their motion. The pressure actually exerted by the molecules should include a correction term which will account for the added compressive force due to this mutual attraction. The effective pressure will then be $(P_{ext} + p')$ where P is the external pressure and p' is this self-exerted "cohesion pressure." The p' term is proportional to the square of the density. The squared term arises from the fact that when the density is doubled, both the number of attracting molecules and the number of attracted molecules are doubled. Since density is proportional to n/V, it follows that $p' = a(n/V)^2$ and the corrected pressure term is $P + a(n/V)^2$ with the proportionality constant a having units of atm-liter^2 mole^{-2}.

Replacing P with $(P + an^2/V^2)$ and V with $(V - bn)$ changes the ideal gas equation to the van der Waals equation: $(P + an^2/V^2)(V - bn) = nRT$.

The microscopic interpretation of b can be checked qualitatively in a rather straightforward manner. Even a rudimentary acquaintance with atoms and molecules would suggest that the respective volumes of hydrogen, nitrogen, chlorine, and benzene molecules are related as $H_2 < N_2 < Cl_2 < C_6H_6$. Since b has been interpreted as the volume actually filled by one mole of molecules, the b values should be in this same order. Table 1.2 indicates that, indeed, they are.

The constant a is clearly related to the forces of interaction between molecules. Our knowledge of such forces is not sufficient for a detailed discussion of this relationship. However, we might expect the ease of liquefaction of a gas to be related to the strength of the intermolecular interactions and thus to a.

Several indicators of the ease of liquefaction are given in Table 1.2, including the boiling point and the critical temperature T_c and pressure P_c. A material with a high value of the constant a and therefore a large intermolecular interaction would be expected to be a liquid even at a relatively high temperature, that is, it would have a high boiling point. Table 1.2 indicates a reasonable degree of correlation between the values of a and the boiling point, but the agreement is not complete.

When the temperature is increased sufficiently a point is reached beyond which a liquid can no longer exist, regardless of the pressure. This temperature is known as the *critical temperature* T_c. The minimum pressure which will cause liquefication at T_c is known as the critical pressure P_c. Those materials which have strong intermolecular attractions (and thus large a values) would be expected to reach their critical states at higher values of temperatures and pressure than those with low a values. Table 1.2 indicates about the same amount of agreement as with the boiling point. Indeed, the boiling point of a liquid is usually roughly two-thirds of its critical temperature (kelvin scale).

It can be shown by using the van der Waals equation in the critical state that

$$a = \frac{27(RT_c)^2}{64P_c} \tag{24a}$$

and

$$b = RT_c/8P_c = \tfrac{1}{3}V_c. \tag{24b}$$

In fact, these equations are used to calculate most tabulated values of a and b for gases from experimental values of their critical variables. The van der Waals equation with these values of a and b agrees well with the experimentally observed behavior of the gas *near its critical point*. The values of a and b do vary with temperature, however, and the observed behavior in regions removed from the critical point may differ markedly from the van der Waals prediction using a and b values based on critical data.

1.14. Energy and Enthalpy Changes for an Ideal Gas

Our analysis of the ideal monatomic gas by the kinetic theory enabled us to derive an expression (Eq. 22) for its energy as a function of a state variable: $\bar{E} = \frac{3}{2}RT$. Recall Eq. (9a):

$$d\bar{E} = (\partial\bar{E}/\partial V)_T\, dV + (\partial\bar{E}/\partial T)_V\, dT. \tag{9a}$$

For an ideal gas $(\partial\bar{E}/\partial V)_T = 0$, the energy is a function of the temperature only and not of the volume (Eq. 22); also $(\partial\bar{E}/\partial T)_V = C_V$ (Eq. 16). Therefore for an ideal gas

$$dE = nC_V\, dT \tag{25a}$$

and

$$\Delta E = n\int_{T_1}^{T_2} C_V\, dT. \tag{25b}$$

When the volume is constant, Eqs. (25a) and (25b) are true for any substance. For an ideal gas these equations are *always* true, *even when the volume changes* during the process. For an *isothermal* (constant-temperature) process involving an ideal gas the change in energy is always zero since the integral of $C_V\,dT$ would be zero.

Similarly, by beginning with Eq. (9b) instead of (9a) we can show that

$$dH = nC_P\,dT \tag{26a}$$

and

$$\Delta H = n\int_{T_1}^{T_2} C_P\,dT. \tag{26b}$$

Equations (26a) and (26b) are *always* true for an ideal gas, *even for processes in which the pressure is not constant*; also $\Delta H = 0$ when an ideal gas undergoes an isothermal change in state.

Recall Eq. (10):

$$d\overline{H} = d\overline{E} + d(P\overline{V}). \tag{10}$$

For an ideal monatomic gas, Eqs. (25) and (26) hold and we have

$$C_P\,dT = C_V\,dT + d(RT)$$
$$(C_P - C_V)\,dT = R\,dT.$$

Integrating, assuming C_P and C_V are independent of temperature over the temperature range,

$$(C_P - C_V)\int_{T_1}^{T_2} dT = R\int_{T_1}^{T_2} dT$$
$$(C_P - C_V)\,\Delta T = R\,\Delta T$$
$$C_P - C_V = R. \tag{27}$$

Since most substances expand on warming, C_P is regularly greater than C_V. For an ideal gas it is greater by an amount R.

Example 11 Determine C_P and C_V for an ideal monatomic gas.

Answer: From Eq. (22), $\overline{E} = (\frac{3}{2})RT$ and

$$C_V = (\partial\overline{E}/\partial T)_V = d(\tfrac{3}{2}RT)/dT = (\tfrac{3}{2})R = (\tfrac{3}{2})(2) = 3 \text{ cal mole}^{-1}\text{ deg}^{-1}$$
$$C_P = C_V + R = (\tfrac{5}{2})R = 5 \text{ cal mole}^{-1}\text{ deg}^{-1}.$$

1.15. Generality of Thermodynamics

Thus far we have discussed gas processes principally, but we should emphasize that thermodynamics is general and applies to all physical and chemical systems: gases, chemical reactions, dry cell batteries, waterfalls, the human body, rubber bands. We must be particularly careful, however, to distinguish between the general equations (such as $\Delta E = Q + W$, $C_V = (\partial \bar{E}/\partial T)_V$, etc.) and those which apply only if certain conditions are met (such as $C_P - C_V = R$, which applies only to ideal gases).

Chemists, of course, are interested in gas compressions and expansions. There are other systems of interest, however, and some are of much greater importance. Let us now look briefly at one of these. The next chapter will be devoted to another.

1.16. Phase Transformations

Thermodynamics treats all changes of state. The *phase transformations* of a substance from a liquid to a gas (*vaporization*), from a solid to a liquid (*fusion*) and from a solid to a gas (*sublimation*) are obviously changes in state:

$$H_2O(l) \rightleftharpoons H_2O(g)$$

$$H_2O(s) \rightleftharpoons H_2O(l)$$

$$H_2O(s) \rightleftharpoons H_2O(g).$$

The phase of the substance is specified in the parentheses: (s) for solid, (l) for liquid, and (g) for gas. If a substance has allotropic modifications, further specification is necessary. For instance, carbon exists in the solid state as both graphite and diamond. C(s) would therefore be ambiguous; we must write instead C(graphite) or C(diamond).

Accompanying these changes in physical state are changes in the energy and enthalpy of the system. These thermodynamic quantities can be evaluated by directly measuring the heat and work involved in the process. The enthalpy of vaporization ΔH_{vap} is the change in enthalpy accompanying the change from liquid to gas; the enthalpy of fusion ΔH_{fus} for the solid-to-liquid transition; and the enthalpy of sublimation ΔH_{sub} for the solid-to-gas transformation.

Example 12 In Example 4 we considered the heat involved in metabolism, some 3000 kcal/day being generated by the average adult. One of the ways mammals get rid of excess heat from body processes is by evaporation of moisture from the skin. Assume that all such heat has to be eliminated by this vaporization process. How much water would a person have to drink to replace that lost by evaporation in 24 hr?

Answer: The evaporation of 1.0 mole of water requires 10.4 kcal of heat under these conditions (that is, $\Delta H_{\text{vap}}(H_2O) = 10.4$ kcal mole^{-1}). To use up 3000 kcal would require the evaporation of (3000 kcal)/(10.4 kcal mole^{-1}) = 288 moles, or about 5.18 liters (1.37 gal) of water if this were the only mechanism of heat loss (which, of course, it is not).

The vaporization of a liquid at its normal boiling point is a good example of a volume change occurring at constant temperature and pressure (Example 6c). For this case the equation becomes $W = -P_{ext}(V_{vap} - V_{liq})$. The calculation takes a very simple but less exact form if we (1) neglect the volume of the liquid and (2) assume the vapor is an ideal gas:

$$W = -P_{ext}(V_{vap} - V_{liq}) \cong -P_{ext} V_{vap} = -nRT.$$

For example, when a mole of water is boiled to dryness at one atmosphere, $W = -1.0 \times .082 \times 373 = -30.6$ liter-atm, as a good approximation. Here neglecting the volume of the liquid (about 19 ml) as compared to the volume of the vapor (about 30,600 ml) is clearly justified.

Phase transitions will be discussed in greater detail in Chapter 4.

Problems

1.1 The following are changes in state. List the properties of each system which are different in the initial and final states.

System	*Initial state*	*Final state*
rubber band	unstretched	stretched
cigarette	before being lit	after burning completely to ash
gas-filled balloon	at sea level	on top of Mt. Washington
cask of wine	freshly sealed	after 4 years
can of Coca Cola	unopened, in cooler	after 2 hr open in the sun
egg	fresh, in carton	hard-boiled

1.2 There are 27 numbered equations in this chapter. List those which are general, applying to all substances under all conditions. Then list those which apply only under specific conditions or only to specific substances. For the latter give the restricting conditions.

1.3 The specific heats of a number of materials are listed below. Calculate the molar heat capacity for each.

a) gold, 0.032 cal g^{-1} deg^{-1}
b) mercury, 0.033 cal g^{-1} deg^{-1}
c) rust (iron oxide), 0.148 cal g^{-1} deg^{-1}
d) sodium chloride, 0.204 cal g^{-1} deg^{-1}
e) ice (0°C), 0.492 cal g^{-1} deg^{-1}
f) dextrose ($C_6H_{12}O_6$), 0.275 cal g^{-1} deg^{-1}
g) urea (N_2H_4CO), 0.320 cal g^{-1} deg^{-1}

1.4 When food is heated in a metal pot on a stove, heat must be added to warm the pot as well as the food. Calculate the average specific heat of a pot which weighs 1 lb if 5.5 kW are needed to heat the pot from 25°C to 100°C in 1 min when it contains a quart of water. Assume 100% heat transfer from the heating unit to the pot. Is the pot made of aluminum, steel, or copper? The respective specific heats are 0.22, 0.11, and .093 cal g^{-1} deg^{-1}.

1.5 As part of an attempt to determine the molecular structure of a new alkaloid with pharmacological activity which was isolated from an African mushroom, the number of nitrogen atoms in the molecule was determined by the Dumas method. First 0.05 mole of the pure material (molecular weight = 394) was oxidized to water, carbon dioxide, and nitrogen gas. After the water was liquefied and the carbon dioxide gas removed by bubbling the gas through a potassium hydroxide solution, the N_2 gas was trapped in a container of volume 0.5 liter and its pressure measured. At 27°C the pressure exerted by the N_2 was 1860 torr. How many nitrogen atoms does each molecule of the alkaloid contain?

1.6 What is the molecular weight of an unknown gaseous pollutant if 5 g of it occupying a 1.0-liter flask exerts a pressure of 1.53 atm at 25°C?

1.7 Assuming the following gases behave ideally, calculate their molar energies and their mean square velocities $\overline{v^2}$ at 25°C and at 100°C: (a) the helium in the Goodyear blimp, (b) the neon in a Howard Johnson sign, (c) the krypton in an anti-Superman ray gun.

1.8 In Example 6 we calculated the work involved in various processes. Another important type of change in state is the *adiabatic* process, one in which no heat is exchanged with the surroundings, $Q = 0$. Show that the work involved in the adiabatic expansion of an ideal gas is

$$W = nC_V(T_2 - T_1).$$

1.9 Imagine that the closed vertical piston-and-cylinder device pictured is immersed in a constant-temperature bath at 27°C. Imagined as weightless, the piston is latched in place, with 22.8 liters of evacuated space above and 1.2 liters of ideal monatomic gas confined below at a pressure of 7.0 atm. (a) If the latch is tripped, how much work is done as the gas drives the piston to the top of the cylinder? (b) The piston is obviously not weightless; suppose it is actually made of steel 2.0 in. thick. What will the work answer be in this case? The density of steel is 7.8 g cm^{-3}. (c) Suppose expansion (b) is repeated except that the device is placed not in the bath, but inside a jacket that is a perfect heat insulator. Calculate the temperature of the gas when the expansion is complete.

(d) Now picture a similar closed piston-and-cylinder device with 22.8 liters of water filling the space below the steel piston and 1.2 liters of ideal gas confined above it. A needle valve is opened and the water leaks out *very* slowly allowing the gas to expand. Calculate the work done by the gas in the process which ends when all the water has leaked out, assuming the device is kept at a constant temperature of 27°C. (e) Suppose procedure (c) is repeated with CO_2 instead of the ideal gas. What will the work of the gas be, assuming van der Waals behavior?

1.10 In Example 6 we calculated the work in processes involving ideal gases. Similar calculations may be carried out using other equations of state. As an example, derive an equation for calculating the work in the isothermal expansion between volumes V_1 and V_2 of a gas that obeys the Bertholet equation of state. Assume $P_{\text{int}} = P_{\text{ext}}$ throughout the expansion.

1.11 In Example 9, instead of allowing the water to plunge picturesquely over the falls, warming itself in the process, we could divert it through a turbine and have it do work. If the flow of the falls is 100 liters min^{-1}, calculate the work which could be performed in 1 min and the power output in watts.

1.12 A 50-g piece of iron (specific heat = 0.108 cal g^{-1} deg^{-1}) initially at 95.0°C is immersed in 100 ml of water (specific heat = 1.0 cal g^{-1} deg^{-1}) initially at 25°C in a thermos bottle. The container therefore insulates the system from the surroundings so that the process which occurs is adiabatic: no heat is exchanged with the surroundings. Calculate the final temperature of the system.

1.13 The "inflation capacity" of a bar of gold might be defined as the amount its value shrinks with a given increase in inflation. Although an increase in inflation has a significant effect on its value, the volume of a gold bar changes only insignificantly with a change in temperature. With this in mind and using the information in Problem 1.3, calculate the ΔE and ΔH which occur when a 100-gram gold bar is heated from 20°C to 25°C.

1.14 Some plants generate enough heat by exothermic chemical reactions to melt snow and ice, allowing them to push through to open air and sunlight. If 1.0 mole of reaction in the plant produces 100 kcal of heat, how many molecules must react to melt 1 gm of ice in 24 hr? What would be the power output of a heater which produces an equivalent change, assuming total conversion of work into heat? The enthalpy of fusion of water is 79.7 cal g^{-1}.

1.15 Calculate the work and the heat exchanged with the surroundings, and the change in energy, when 5.0 moles of an ideal monatomic gas are heated from 25°C to 75°C (a) when the volume is constant and (b) when the external pressure is constant.

1.16 A sample of 100 g of pure ethanol is boiled dry on a day when the barometer reads 740 torr. For this process calculate the value of each of the three terms of the first law of thermodynamics. Take the heat of vaporization of ethanol as 9.41 kcal mole^{-1} and its boiling point at this pressure as 77.6°C.

2 / Thermochemistry

2.1. Calorimetry

Chemical reactions are a most important change of state for chemists. The initial state is the mixture of reactants; the final state, the products. Accompanying all such chemical changes in state are changes in the thermodynamic state variables such as energy and enthalpy. Chemical reactions can be carried out in such a manner that the heat and work involved can be measured and the thermodynamic quantities of interest determined. These measurements are the province of *thermochemistry*.

The direct, accurate measurement of heat requires a suitable instrument. This is the *calorimeter*, and the associated technique is *calorimetry*. Figure 2.1 is a diagrammatic representation of a simple calorimeter. The calorimeter consists of a container in which the reaction releasing (or absorbing) heat is caused to occur. The evolved heat flows into a known amount of water or other material which surrounds the reaction vessel. Knowing the heat capacity of this absorbing jacket, and measuring the rise in its temperature, we may use Eq. (4) to calculate the quantity of heat released.

Depending on the reaction being investigated, the calorimeter used may be of the constant-volume or the constant-pressure type. In the former case

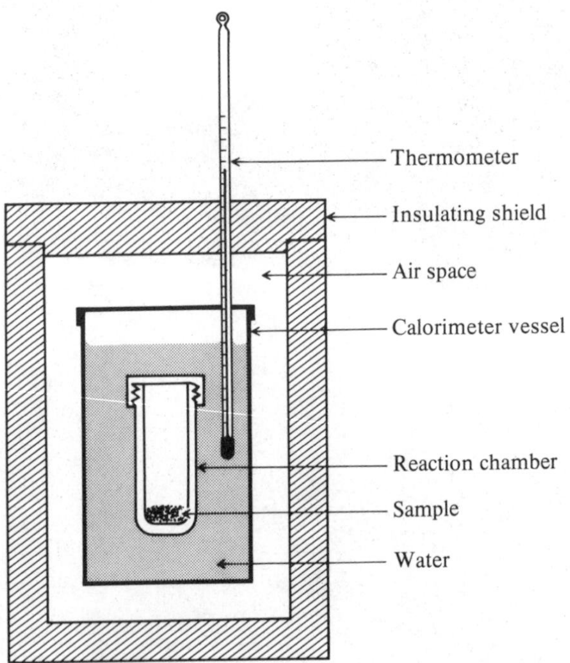

Fig. 2.1. Calorimeter (diagrammatic).

we would be measuring the energy change; in the latter, the enthalpy change. This is true since the heat which passes from the reacting materials (the system) into the calorimeter (the surroundings) in the former case is Q_V which is ΔE, but in the latter is Q_P which is ΔH. Since ΔE and ΔH can be interconverted, we may choose the more convenient calorimeter and still determine the variable we need.

Recall that $H = E + PV$ and $\Delta H = \Delta E + \Delta(PV)$. In a chemical reaction the molar volumes of the liquids and solids involved are negligibly small compared to the molar volumes of gases (see Section 1.16). When water vapor condenses into either of its other forms, for instance, the change in volume for the process is approximately equal to the negative of the volume of the vapor: $\Delta V = V_{\text{solid}} - V_{\text{vap}} \cong -V_{\text{vap}}$ and $\Delta V = V_{\text{liq}} - V_{\text{vap}} \cong -V_{\text{vap}}$. In evaluating the $\Delta(PV)$ term, we therefore regularly *consider only the gases* involved. Assuming further that the gases are ideal, we may write for an isothermal process ($\Delta T = 0$):

$$\Delta(PV) = \Delta(nRT) = (\Delta n)RT,$$

and therefore

$$\Delta H \cong \Delta E + (\Delta n)RT,$$

where Δn means moles of *gaseous* products minus moles of *gaseous* reactants. The Δn's for the following reactions are:

$$C(\text{graphite}) + O_2(g) \longrightarrow CO_2(g) \qquad \Delta n = 0$$

$$(C_{17}H_{33}COO)_3C_3H_5(l) + 3H_2(g) \longrightarrow (C_{17}H_{35}COO)_3C_3H_5(s) \qquad \Delta n = -3$$

$$H_2(g) + \tfrac{1}{2}O_2(g) \longrightarrow H_2O(l) \qquad \Delta n = -\tfrac{3}{2}$$

$$HCl(aq)^1 + H_2SO_4(l) \longrightarrow H_2SO_4(aq) + HCl(g) \qquad \Delta n = +1$$

Once determined by one of these methods, the enthalpy change accompanying a chemical reaction becomes valuable information, useful in obtaining answers to other chemical problems.

Example 13 The number of calories in margarine is to be determined for a diet list. A 1.20-g portion is weighed into a calorimeter reaction chamber which is then sealed and an excess of oxygen introduced at high pressure. The chamber is then immersed in 1400 ml of water in the outer vessel of the calorimeter, and the reaction initiated. The temperature of the water, originally 23.71°C, rises to 28.87°C.

a) Given that the heat capacity of the calorimeter itself is 450 cal deg^{-1}, calculate

[1] The state designation (aq) means "in aqueous solution."

the value to be entered in the diet table for the calories in a "small pat" (5.00 grams) of margarine.

b) The experiment as carried out in a stout sealed chamber (constant volume) would yield ΔE, the energy change. Since body processes occur at constant pressure, the enthalpy change ΔH would be more pertinent. Calculate it from the answer to (a), taking the combustion reaction as

$$C_{57}H_{108}O_6(s) + 81O_2(g) \longrightarrow 57CO_2(g) + 54H_2O(l).$$

Answer: a) The heat capacity of the water and the container is

$$(1.0 \text{ cal deg}^{-1} \text{ g}^{-1})(1400 \text{ g}) + (450 \text{ cal deg}^{-1}) = 1850 \text{ cal deg}^{-1}.$$

The heat evolved by the reaction therefore is $(28.87° - 23.71°)(1850 \text{ cal deg}^{-1}) =$ 9550 cal from a 1.20-g sample. A 5.00-g pat therefore represents $(5.00/1.20)(9550) =$ 39,800 cal. Since this heat was *released* in a *constant volume* process, we would write thermodynamically $\Delta E = -39.8$ kcal for this experiment.

b) When one mole of margarine burns, $\Delta n = 57 - 81 = -24$ moles. But, 5.00 g of margarine is only 5.00 g/888 g mole^{-1} = 0.00563 mole. Thus Δn is 0.00563 $\times (-24) = -0.135$ mole.

The calculation of ΔH at 25°C is quite straightforward.

$$\begin{aligned}\Delta H &= \Delta E + (\Delta n)RT \\ &= -39.8 \text{ kcal} + (-0.135 \text{ mole})(1.99 \text{ cal deg}^{-1} \text{ mole}^{-1}/1000 \text{ cal kcal}^{-1})298° \\ &= -39.8 - .08 = -39.9 \text{ kcal}.\end{aligned}$$

For most reactions the $(\Delta n)RT$ term is small compared to the magnitudes of ΔE and ΔH, though this case (less than one-fifth of 1%) is perhaps extreme. Note also that this combustion process would release more heat at constant pressure (ΔH) than at constant volume (ΔE). At constant pressure the system as it reacts diminishes in volume (Δn is negative). It is thus the recipient of work done upon it as the surroundings "close in." It converts this work into the extra heat observed.

In diet tables our result would probably be rounded and entered as 40 Cal, where 1 Cal or metabolic calorie is the same as 1 kcal or 1000 cal. The value as determined by calorimetry is the maximum available in metabolic oxidation. This limiting value is in general not attained in the body.

2.2. The Indirect Determination of Heats of Reaction; Hess's Law of Constant Heat Summation

There are many reactions which are not suited to calorimetric measurement. The process may be occurring in the human body or in a plant through enzyme action. The reaction may be one which does not go to completion, or it may be too slow for accurate measurement. Even if feasible, the experiment might call for reagents which are expensive, dangerous, or unavailable. Or we may simply be reluctant to devote the time and trouble which the laboratory measurement would require. Do we have an alternative? To obtain values for ΔH and ΔE without actually carrying out the direct calorimetric measurement, we must resort to an indirect method. An example will help to understand the procedure.

The complete combustion of hydrocarbons such as octane in an automobile engine yields only carbon dioxide and water:

$$C_8H_{18}(l) + (25/2)O_2(g) \xrightarrow{\Delta H_1} 8CO_2(g) + 9H_2O(l). \tag{I}$$

Note that we write the enthalpy change for the chemical process above the arrow indicating the direction of reaction. This is the enthalpy change which accompanies the reaction in which the reactants as written are converted completely into the products as written. Under constant-pressure conditions $\Delta H = Q_P$, and it is often simply called the heat of reaction. If ΔH is positive, the reaction is said to be *endothermic* (the system absorbs heat); if negative, *exothermic* (the system releases heat).

In a calorimeter with excess oxygen, the enthalpy change ΔH_1 for Reaction (I) is readily determined. But public health officials are concerned about a side-reaction which puts a dangerous pollutant, carbon monoxide, into the automobile exhaust:

$$C_8H_{18}(l) + (17/2)O_2(g) \xrightarrow{\Delta H_2} 8CO(g) + 9H_2O(l). \tag{II}$$

The enthalpy change for this reaction cannot be directly measured, since in the calorimeter Reaction (I) will always occur simultaneously to some extent. How can ΔH_2 be determined?

Enthalpy is a state function; thus ΔH depends only on the initial and final states of the system, not on the path between these two states:

$$\Delta H = H_{\text{final}} - H_{\text{initial}}. \tag{28}$$

Suppose we visualize two paths for carrying out the chemical transformation of one mole of substance A (initial state) to one mole of substance F (final state). The first path is the direct one: A → F with an accompanying enthalpy change ΔH_a which must be equal to $\overline{H}_F - \overline{H}_A$. The second path is indirect, made up of two successive steps: A → C with $\Delta H_i = \overline{H}_C - \overline{H}_A$, followed by C → F with $\Delta H_s = \overline{H}_F - \overline{H}_C$. The total enthalpy change for this second path will be $\Delta H_i + \Delta H_s$. Both paths begin with the same initial state and end with the same final state. The enthalpy change accompanying both must therefore be the same:

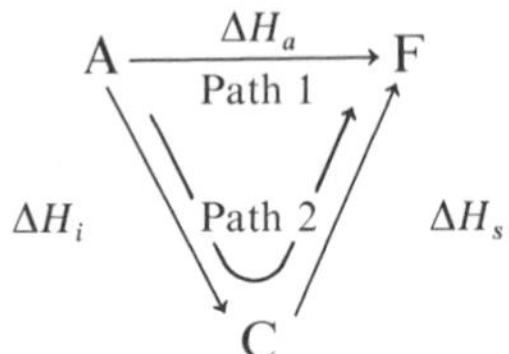

$$\Delta H_{\text{path 1}} = \Delta H_{\text{path 2}}.$$
$$\Delta H_a = \Delta H_i + \Delta H_s.$$
$$(\overline{H}_F - \overline{H}_A) = (\overline{H}_C - \overline{H}_A) + (\overline{H}_F - \overline{H}_C) = (\overline{H}_F - \overline{H}_A).$$

Therefore, if we are not able to measure the change in enthalpy for the direct path between reactants and products, we can often determine its value by measuring and adding the enthalpy changes in a sequence of known reactions which, taken together, produce the same net chemical reaction. This is *Hess's law of constant heat summation.*

Return to Reaction (II), for which we cannot measure ΔH_2. Let us see if we can construct an alternative sequence of reactions with the same initial state and the same final state as Reaction (II) and for each step of which we can measure the change in enthalpy. The key to success in this endeavor is a third reaction, one which relates carbon monoxide and carbon dioxide:

$$CO(g) + (1/2)O_2(g) \xrightarrow{\Delta H_3} CO_2(g). \tag{III}$$

We can prepare pure carbon monoxide in the laboratory, then burn it in a calorimeter, determining ΔH_3.

Reaction (II), the reaction in which we are interested, can now be carried out, on paper at least, in the following manner:

$$C_8H_{18}(l) + (25/2)O_2(g) \xrightarrow{\Delta H_1} 8CO_2(g) + 9H_2O(l) \xrightarrow{\Delta H_4} 8CO(g) + 9H_2O(l) + 4O_2(g). \tag{IV}$$

The net reaction is exactly the same as Reaction (II). Hess's law tells us that since enthalpy is a state function, its change for the reaction in which octane is oxidized to carbon monoxide and water must be the same regardless of the reaction sequence. Therefore, $\Delta H_2 = \Delta H_1 + \Delta H_4$. Even though we cannot carry out Reaction (II) in a calorimeter and measure ΔH_2 directly, we can calculate it if we can measure ΔH_1 and ΔH_4. We can measure ΔH_1 as mentioned above, but what about ΔH_4?

Reaction (IV) associated with ΔH_4 cannot, in fact, occur in a calorimeter. But notice that Reaction (IV) is the reverse of Reaction (III), for which

ΔH_3 can be measured calorimetrically.

$$\Delta H_3 = H_{\text{final}} - H_{\text{initial}} = \bar{H}_{CO_2} - (\bar{H}_{CO} + (1/2)\bar{H}_{O_2})$$
$$\Delta H_4 = (8\bar{H}_{CO} + 9\bar{H}_{H_2O} + 4\bar{H}_{O_2}) - (8\bar{H}_{CO_2} + 9\bar{H}_{H_2O})$$
$$\Delta H_4 = 8[(\bar{H}_{CO} + (1/2)\bar{H}_{O_2} - \bar{H}_{CO_2}]$$
$$\Delta H_4 = 8(-\Delta H_3).$$

Since ΔH_3 can be measured, we can determine ΔH_4. Knowing both ΔH_1 and ΔH_4, we now know ΔH_2 without carrying out the reaction: $\Delta H_2 = \Delta H_1 + \Delta H_4 = \Delta H_1 - 8\,\Delta H_3$.

2.3. Standard Enthalpies of Formation

It might be suggested that by using the individual enthalpies of $C_8H_{18}(l)$, $O_2(g)$, and $CO(g)$ we could have found ΔH_2 by using Eq. (28):

$$\Delta H_2 = H_{\text{final}} - H_{\text{initial}} = (8\bar{H}_{CO} + 9\bar{H}_{H_2O}) - (\bar{H}_{C_8H_{18}} + (17/2)\bar{H}_{O_2}).$$

In general, for any reaction the enthalpy change would be

$$\Delta H_{\text{reaction}} = H_{\text{final}} - H_{\text{initial}} = \sum_{\text{pr}} n\bar{H} - \sum_{\text{re}} n\bar{H},$$

where the n values are the stoichiometric coefficients in the reaction equation, the sums being over the products pr and reactants re respectively.

This attractive method would require a table of the absolute values of the enthalpy for all chemical substances. Each entry in such a table would represent the total enthalpy contained in one mole of that substance. This procedure is not possible since we have in general no way of determining absolute values of H, but only differences, ΔH. It is possible, however, to tabulate certain enthalpy differences that do enable the calculation of reaction enthalpies.

One type of enthalpy change often tabulated is the *standard enthalpy of formation*, ΔH_f°, defined as the enthalpy change accompanying the reaction in which one mole of a material in its designated form is created under standard conditions from its constituent elements in their most stable forms. Standard conditions for pure materials are one atmosphere pressure and some designated temperature, usually 25°C (298°K).

For instance, the standard enthalpy of formation of CO_2 would be the enthalpy change, under standard conditions, of the reaction

$$\text{C(graphite)} + O_2(g) \longrightarrow CO_2(g)$$
$$\Delta H_f^\circ(CO_2, g) = -94.05 \text{ kcal mole}^{-1}.$$

Hundreds of ΔH_f° values have been determined and tabulated. Table 2.1 records the values for a number of materials. The ΔH_f° for any element in its

Table 2.1. Standard Enthalpies of Formation at 298°K

Material	ΔH_f°, kcal mole^{-1}	*Material*	ΔH_f°, kcal mole^{-1}
C(g)	171.7	SO_2(g)	−70.96
C(graphite)	0.00	SO_3(g)	−94.45
C(diamond)	0.4532	H_2SO_4(l)	−193.91
CO(g)	−26.416	Br(g)	26.7
CO_2(g)	−94.052	Br_2(g)	7.34
CH_4(g)	−17.89	HBr(g)	−8.66
C_6H_6(l)	11.718	CaO(s)	−151.9
C_6H_6(g)	19.820	$CaCO_3$(s)	−228.45
C_2H_5OH(l)	−66.356	$Ca(OH)_2$(s)	−235.80
H_2O(l)	−68.317	NaOH(s)	−101.99
H_2O(g)	−57.789	NaCl(s)	−98.232
H(g)	52.1	glycine(s)	−126.33
O(g)	59.1	alanine(s)	−134.17
N(g)	112.5	aspartic acid(s)	−233.75
NH_3(g)	−11.04	fumaric acid(s)	−194.13
S(rhombic)	0.00	glucose(s)	−304.64
S(monoclinic)	0.071	urea(s)	−79.63
H_2S(g)	−4.815		

National Bureau of Standards Circular 500 (1952) and K. Burton and H. A. Krebs, *Biochemical Journal*, **54**, 94 (1953).

most stable form under standard conditions is zero by definition. Note however that ΔH_f° for any form of an element other than the most stable will not be zero; C(diamond), C(g), H(g), and S(monoclinic) in Table 2.1 are examples.

We can see how these ΔH_f° values are used by considering a certain commercially and sociologically important reaction which takes place in the presence of yeast enzymes and is too slow for reliable calorimetry:

$$\underset{\text{glucose}}{C_6H_{12}O_6(s)} \xrightarrow{\Delta H_5} \underset{\text{ethanol}}{2C_2H_5OH(l)} + 2CO_2(g). \tag{V}$$

The ΔH_f° values for the following reactions may be found in tables:

$$6C(\text{graphite}) + 6H_2(g) + 3O_2(g) \xrightarrow{\Delta H_6} C_6H_{12}O_6(s). \tag{VI}$$

$$4C(\text{graphite}) + 6H_2(g) + O_2(g) \xrightarrow{\Delta H_7} 2C_2H_5OH(l). \tag{VII}$$

$$2C(\text{graphite}) + 2O_2(g) \xrightarrow{\Delta H_8} 2CO_2(g). \tag{VIII}$$

Using the Hess procedure outlined in Section 2.2, we add Reactions (VII) and (VIII) and subtract Reaction (VI). The result is the reaction of interest, Reaction (V). By Hess's Law

$$\Delta H_5 = \Delta H_7 + \Delta H_8 - \Delta H_6 = 2\,\Delta H_f^\circ(\text{ethanol}) + 2\,\Delta H_f^\circ(\text{CO}_2) - \Delta H_f^\circ(\text{glucose})$$
$$= 2(-66.356) + 2(-94.052) - (-304.64) = -16.18 \text{ kcal mole}^{-1}.$$

In general

$$\Delta H^\circ_{\text{reaction}} = \sum_{\text{pr}} n\,\Delta H_f^\circ - \sum_{\text{re}} n\,\Delta H_f^\circ. \tag{29}$$

We can thus calculate the enthalpy change of any reaction *if* our table of ΔH_f° values includes every substance involved in that reaction.

One might well ask how the values of ΔH_f° for such materials as glucose and ethanol are determined, since we would not get glucose or ethanol by simply mixing carbon, oxygen, and hydrogen in a calorimeter. Here again we use the reasoning of Hess's Law, devising an alternative route between the constituent elements and the product, with the ΔH of each step either known or measurable.

2.4. Standard Enthalpies of Combustion

A particularly useful method for calculating the standard enthalpy of formation for organic molecules is based on the *standard enthalpy of combustion* ΔH_c°. This quantity is defined as the enthalpy change for the reaction in which one mole of a substance, in its designated form under standard conditions, is completely oxidized to products in their most stable forms under standard conditions.

The following sequence of combustion reactions is used to calculate ΔH_f° for ethanol using ΔH_c° values:

$$\text{C}_2\text{H}_5\text{OH(l)} + 3\text{O}_2\text{(g)} \xrightarrow{\Delta H_9} 2\text{CO}_2\text{(g)} + 3\text{H}_2\text{O(l)}. \tag{IX}$$

$$2\text{C(graphite)} + 2\text{O}_2\text{(g)} \xrightarrow{\Delta H_{10}} 2\text{CO}_2\text{(g)}. \tag{X}$$

$$3\text{H}_2\text{(g)} + (3/2)\text{O}_2\text{(g)} \xrightarrow{\Delta H_{11}} 3\text{H}_2\text{O(l)}. \tag{XI}$$

Here $\Delta H_9 = \Delta H_c^\circ(\text{ethanol})$, $\Delta H_{10} = 2\Delta H_c^\circ(\text{graphite}) = 2\Delta H_f^\circ(\text{CO}_2)$, and $\Delta H_{11} = 3\Delta H_c^\circ(\text{H}_2) = 3\Delta H_f^\circ(\text{H}_2\text{O})$. These standard enthalpies of combustion can be determined by calorimetric measurement. Adding Reactions (X) and (XI) and subtracting Reaction (IX), we obtain Reaction (VI):

$$2\text{C(graphite)} + (1/2)\text{O}_2\text{(g)} + 3\text{H}_2\text{(g)} \xrightarrow{\Delta H} \text{C}_2\text{H}_5\text{OH(l)}, \tag{VI}$$

where ΔH is the ΔH_f° for ethanol. We therefore add the enthalpy changes for Reactions (X) and (XI) and subtract that for Reaction (IX) to obtain ΔH_f°(ethanol):

$$\begin{aligned}\Delta H_f^\circ(\text{ethanol}) &= 2\Delta H_c^\circ(\text{graphite}) + 3\Delta H_c^\circ(\text{H}_2) - \Delta H_c^\circ(\text{ethanol}) \\ &= 2(-94.052) + 3(-68.317) - (-326.71) = -66.35 \text{ kcal.}\end{aligned}$$

The values of the enthalpy changes for reactions other than the formation of materials from their constituent elements can also be found by using ΔH_c° values in a similar manner. The ΔH_c° values for hundreds of compounds

Table 2.2. Standard Enthalpies of Combustion at 1 Atm[1,2]

Substance	ΔH_c°, kcal/mole 25°C	20°C	*Substance*	ΔH_c°, kcal/mole 25°C	20°C
O_2(g)	0.0		C_2H_2(g)		−312.0
N_2(g)	0.0		C_2H_4(g)		−331.6
CO_2(g)	0.0		C_2H_6(g)		−368.4
H_2O(l)	0.0		CS_2(l)		−246.6
C(graphite)	−94.05		ethylether(l)		−651.7
C(diamond)	−94.51		acetic acid(l)		−209.4
CH_4(g)	−212.8	−210.8	alanine(s)		−387.7
C_6H_6(l)	−781.0	−782.3	caffeine(s)		−1014.2
C_6H_6(g)	−789.1		fumaric acid(s)		−320.0
H_2(g)	−68.32		glycine(s)		−234.5
CO(g)	−67.64		methanol(l)		−170.9
C_2H_5OH(l)	−326.7	−327.6	nicotine(l)		−1427.7
D-glucose(s)	−669.6	−673.0	sucrose(s)		−1349.6
n-octane(l)		−1302.7	urea(s)		−151.6
benzoic acid(s)[3]		−771.2			

[1] *Handbook of Chemistry and Physics*, The Chemical Rubber Company.
[2] Products of combustion are CO_2(g), H_2O(l), N_2(g) and SO_2(g).
[3] Accepted international value for standardization.

have been measured and tabulated (see Table 2.2) for this purpose. In general

$$\Delta H^\circ_{\text{reaction}} = \sum_{\text{re}} n\Delta H_c^\circ - \sum_{\text{pr}} n\Delta H_c^\circ. \tag{30}$$

2.5. The Dependence of Enthalpy Changes on Temperature

It should be noted that the use of either Eq. (29) or Eq. (30) gives us ΔH°, the reaction enthalpy change under standard conditions (1 atm pressure and usually 298°K). Suppose we want to determine the enthalpy change to be expected if we run our reaction at some other temperature. Can we determine ΔH_T if we know ΔH°_{298}?

Consider a simple general reaction: Reactants → Products. We can determine ΔH° by using the tabulated values of ΔH°_f or ΔH°_c. We would like to predict ΔH_T, the enthalpy change at some temperature T other than the standard value. Two paths can be visualized:

$$\begin{array}{ccc} \text{Reactants at } T & \xrightarrow[\text{Path 1}]{\Delta H_T} & \text{Products at } T \\ \downarrow \Delta H_1 & \text{Path 2} & \uparrow \Delta H_2 \\ \text{Reactants at } T^\circ & \xrightarrow{\Delta H^\circ} & \text{Products at } T^\circ \end{array}$$

Here ΔH_1 is the heat necessary to warm (or cool) the reactants from the initial temperature T to the standard temperature T° at constant pressure. Similarly, ΔH_2 refers to the products undergoing the reverse temperature change. The respective values are:

$$\Delta H_1 = \int_T^{T^\circ} \sum_{\text{re}} (nC_P)\, dT \qquad \text{and} \qquad \Delta H_2 = \int_{T^\circ}^{T} \sum_{\text{pr}} (nC_P)\, dT.$$

Because enthalpy is a state function, the total enthalpy change for the two paths must be the same; hence

$$\Delta H_T = \Delta H_1 + \Delta H^\circ + \Delta H_2 = \int_T^{T^\circ} \sum_{\text{re}} (nC_P)\, dT + \Delta H^\circ + \int_{T^\circ}^{T} \sum_{\text{pr}} (nC_P)\, dT$$

$$\Delta H_T = \Delta H^\circ + \int_{T^\circ}^{T} \left(\sum_{\text{pr}} nC_P - \sum_{\text{re}} nC_P \right) dT. \tag{31}$$

This is known as *Kirchhoff's equation.* Since Δ means "final minus initial," this equation is often abbreviated

$$\Delta H_T = \Delta H^\circ + \int_{T^\circ}^{T} (\Delta C_P)\, dT. \tag{32}$$

Example 14 a) Calculate the enthalpy change in the formation of 1 mole of ethanol at 25°C by the reaction shown. Thermochemical data are given below each substance.

Reaction:	$C_2H_4(g)$ +	$H_2O(g)$ →	$C_2H_5OH(l)$
ΔH°_f(298°K), kcal mole^{-1}:	12.50	−57.80	−66.36
C_P, cal mole^{-1} deg^{-1}:	10.41	8.02	26.64

b) What is the ΔH when the reactants are initially at 15.0°C and the product at the end of the reaction is at 100.0°C?

Answer:

a) $\Delta H^\circ_{298} = \Delta H^\circ_f(\text{ethanol}) - [\Delta H^\circ_f(C_2H_4) + \Delta H^\circ_f(H_2O)]$

$= -66.36 - (12.50 - 57.80)$

$\Delta H^\circ_{298} = -21.06 \text{ kcal mole}^{-1}$.

b) $$\begin{array}{ccc} \text{Reactants (15°C)} & \xrightarrow{\Delta H} & \text{Products (100°C)} \\ \downarrow \Delta H_1 & & \uparrow \Delta H_2 \\ \text{Reactants (25°C)} & \xrightarrow{\Delta H^\circ_{298}} & \text{Products (25°C)} \end{array}$$

$$\Delta H_1 = \int_{288}^{298} \sum_{re} nC_P\, dT = \sum_{re} nC_P \Delta T = (10.41 + 8.02 \text{ cal deg}^{-1})(10.0 \text{ deg})$$

$$= 184.3 \text{ cal} = 0.18 \text{ kcal}.$$

$$\Delta H_2 = \int_{298}^{373} \sum_{pr} nC_P\, dT = \sum_{pr} nC_P \Delta T = (26.64 \text{ cal deg}^{-1})(75 \text{ deg})$$

$$= 1998 \text{ cal} = 2.00 \text{ kcal}.$$

$$\Delta H = \Delta H_1 + \Delta H_{298} + \Delta H_2 = 0.18 - 21.06 + 2.00 = -18.9 \text{ kcal}.$$

In the example above, the heat capacities involved were assumed to be constant. If the temperature range is wide or if high accuracy is required, this assumption will not be acceptable. However, empirical equations are available giving the heat capacities of the more familiar substances as functions of temperature. These are usually expressed as power series in T of the form $C_P = a + bT + cT^2$ or sometimes $C_P = a' + b'T + c'T^{-2}$.

Example 15 Calculate the heat of vaporization of water at body temperature (37°C). The heat capacity of liquid water may be taken as constant at 18.0 cal mole^{-1} deg^{-1}; the heat capacity of water vapor varies with temperature:

$$C_P(H_2O, g) = 7.256 + 2.298 \times 10^{-3}T + 2.83 \times 10^{-7}T^2.$$

$$H_2O(l) \longrightarrow H_2O(g). \qquad \Delta H_{373} = 9.70 \text{ kcal mole}^{-1}.$$

Answer:

$$\Delta H_{310} = \Delta H_{373} + \int_{373}^{310} [C_P(\text{pr}) - C_P(\text{re})]\, dT$$

$$= 9700 + \int_{373}^{310} (7.256 + 2.298 \times 10^{-3}T + 2.83 \times 10^{-7}T^2 - 18)\, dT$$

$$= 9700 - 10.744(310 - 373) + 2.298 \times 10^{-3}(310^2 - 373^2)/2$$

$$+ 2.83 \times 10^{-7} \times (310^3 - 373^3)/3$$

$$= 9700 + 677 - 49 - 2$$

$$= 10.3 \text{ kcal mole}^{-1} \text{ or } 570 \text{ cal g}^{-1}.$$

Agreement of this result with the experimental value (575 cal g^{-1}) is excellent, particularly since this equation for C_p as a function of temperature was constructed for steam ($T > 373°K$).

Problems

2.1 Use the data given in Table 2.2 to determine the standard enthalpy of formation for (a) diamond, (b) urea, (c) alanine.

2.2 Use the data in Table 2.1 to determine the standard enthalpy of combustion of (a) sulfur (rhombic), (b) sulfur (monoclinic), (c) alanine.

2.3 Deposits of solid calcium carbonate have been found in the thyroid glands of a very small number of patients. Exactly how this material finds its way into the thyroid is not known. One proposed mechanism involves the introduction into the body of an abnormally high concentration of $Ca(OH)_2$, which could enter the thyroid dissolved in the blood. It might then react with dissolved CO_2 to produce $CaCO_3$, which, being significantly less soluble than $Ca(OH)_2$, precipitates. Calculate ΔH and ΔE for the following reaction under standard conditions using the data in Tables 2.1 and 2.2:

$$Ca(OH)_2(s) + CO_2(g) \longrightarrow CaCO_3(s) + H_2O(l).$$

2.4 The source of much of the energy used by the human body is the citric acid cycle, a complex sequence of interrelated enzyme-catalyzed reactions. The net result of the cycle is the oxidation of acetic acid to carbon dioxide and water. (a) Calculate the change in enthalpy for this reaction carried out under standard conditions and at body temperature (37°C):

$$CH_3COOH(l) + 2O_2(g) \longrightarrow 2CO_2(g) + 2H_2O(l).$$

The specific heats of acetic acid, oxygen, carbon dioxide and water are 0.50, 0.22, 0.20, and 1.00, respectively. (b) The energy produced by the citric acid cycle is used in endothermic reactions which produce adenosine triphosphate (ATP). If $\Delta \overline{H}°$ for the reaction which produces ATP is 7000 cal $mole^{-1}$, what is the maximum number of moles of ATP which can be produced when 1.0 mole of acetic acid is oxidized? (c) Twelve moles of ATP are actually produced for each mole of acetic acid oxidized. What then is the efficiency of the cycle? (In this case the efficiency is 100 times the enthalpy change in ATP synthesis divided by the enthalpy change in the oxidation.)

2.5 The net result of the complex series of enzyme-catalyzed reactions in the body which produces urea may be represented as

$$2NH_3(g) + CO_2(g) \longrightarrow H_2N{-}\overset{\overset{\displaystyle O}{\|}}{C}{-}NH_2(s) + H_2O(l).$$

Calculate $\Delta H°$ and $\Delta E°$ for this reaction.

2.6 Instead of storing excess carbohydrates, the body converts them to fats and stores these fats along with excess ingested fats to meet future energy needs. By comparing the enthalpy of combustion *per gram* for sucrose and margarine (see Example 13) explain why the storage of fats is more efficient than the storage of carbohydrates.

2.7 The label on a package of Total shows that this breakfast cereal analyzes 10.5% protein, 80.5% carbohydrate, and 3.5% fat. (a) Taking these three as alanine, sucrose, and margarine (see Example 13) respectively, how many metabolic calories are there in a "1-ounce serving" of this dry cereal? (The label says 110.) (b) If breakfasters want to bring their morning intake of protein up to 10 g, they may do this by using how many ounces of milk (3.3% protein) with their ounce of cereal?

2.8 When operating, a certain 40-gal domestic water heater consumes 0.65 ft^3 of natural gas per minute. Assuming that the gas is pure methane delivered at 25°C and 1.0 atm, and that 80% of the heat of combustion of the gas is transferred to the water, what should be the "recovery time" of this heater? (Recovery time is the time required for producing a temperature rise of 100°F in a full tank of water.)

2.9 Carbon monoxide can be produced commercially by passing a limited supply of oxygen through a deep bed of white-hot coke: $2C + O_2 \rightarrow 2CO$. The heat generated by this exothermic reaction, however, is so great as to present problems. One solution is to mix some carbon dioxide with the entering oxygen so that the endothermic reaction $C + CO_2 \rightarrow 2CO$ will also occur. Assuming that operation at high temperature would change the respective enthalpies of the two reactions by the same percentage from the 25° values, and assuming a 10% heat loss from the furnace, in what proportions should the oxygen and carbon dioxide be mixed to just sustain the temperature?

2.10 Calculate the enthalpy of vaporization of benzene at its normal boiling point of 80.1°C using the following information:

	$\Delta H^\circ_{f,\,298^\circ K}$, kcal mole^{-1}	$\Delta H^\circ_{c,\,298^\circ K}$, kcal mole^{-1}	C_P, cal deg^{-1} mole^{-1}
$C_6H_6(l)$	11.718	−781.0	32.53
$C_6H_6(g)$	19.820	−789.1	$-0.4 + 77.6 \times 10^{-3}T - 264.3 \times 10^{-7}T^2$

2.11 Bees lower the temperature of their hives with a primitive air conditioning system. Water is brought into the hives and fanned by the wings of worker bees. The water evaporates, absorbing heat and cooling the hive. The enthalpy of vaporization of water at 100°C is 9.70 kcal mole^{-1}. Calculate the amount of heat removed by the evaporation of 1 g of water at 95°F, the temperature maintained in a typical bee hive.

2.12 An air conditioner cools the air passing through and at the same time causes condensation of surplus humidity; the heat and water are discharged to the outside. A dehumidifier by contrast is entirely inside; it collects the water in a tray, and returns the heat of condensation to the room itself. Though such a unit lowers the humidity, it raises the temperature somewhat, for two reasons. The machinery generates frictional heat, and the removed moisture releases its heat of condensation to the room.

On a very muggy day (barometer 736, temperature 95°F, relative humidity 90%) a dehumidifier is operated in a room 28 × 15 × 9 ft until half the original moisture is removed. (a) How much heat is released into the room due to moisture condensation?

(b) What volume of water must be emptied from the tray? The enthalpy of vaporization of water is 540 cal g^{-1}.

2.13 Industrial plants which synthesize ammonia commercially by catalytic union of nitrogen and hydrogen regularly operate at 400° to 450°C. Calculate the enthalpy of formation of ammonia at 427°C, using heat capacity equations of the form $C_P = a + b \times 10^{-3}T + c \times 10^5/T^2$. The respective *a*, *b*, and *c* constants for the three gases are

N_2: +6.83, +0.90, and −0.12.

H_2: +6.52, +0.78, and +0.12.

NH_3: +7.11, +6.00, and −0.37.

2.14 When hydrogen and oxygen are mixed, nothing happens. If a catalyst is introduced, however, water is produced explosively. Calculate the final temperature and the final pressure when 0.05 mole of oxygen and 0.05 mole of hydrogen are mixed at 25°C in the presence of a catalyst in a calorimeter with a constant volume of 1.0 liter. Assume the process is adiabatic. The following information and that given in Table 2.1 may be of help.

	Specific heat (constant volume), cal g^{-1} deg^{-1}
$H_2(g)$	2.44
$O_2(g)$	0.16
$H_2O(l)$	1.00
$H_2O(g)$	0.37
	$\Delta H^\circ_{vap}(H_2O) = 10.43$ kcal $mole^{-1}$ at 30°C

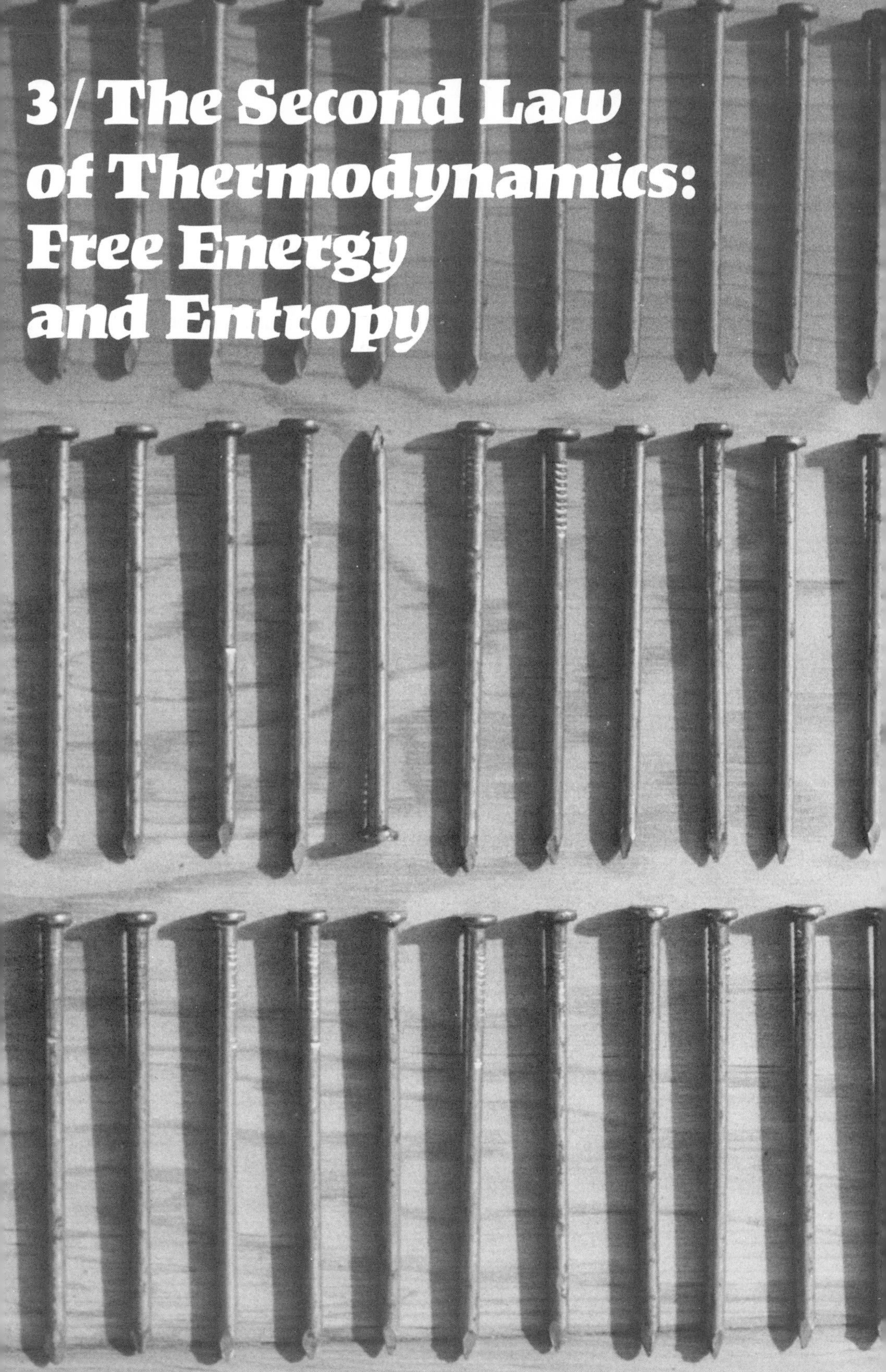

3/ The Second Law of Thermodynamics: Free Energy and Entropy

3.1. Introduction

We now have the ability to determine, by direct measurement or by calculations based on tables of standard values, the change in energy and the change in enthalpy which accompany a chemical change in state. To extract the full usefulness of thermodynamics to chemists, we must go further and examine another of its principles, the *second law of thermodynamics.*

There are a number of equivalent statements of the second law, all based on observation of what is or is not allowed to happen in the real, physical world. Often these statements refer to a cyclic device, a machine that undergoes no net change within itself as it operates (for example a gasoline or steam engine); it is purely an agent and does not "use itself up." One such statement says it is impossible for a device operating in a cycle and at constant temperature to change heat into work. If this were not true, we could conceivably design a submarine that would require no fuel. Running submerged, it would draw heat from the limitless supply in the ocean around it and use this energy to propel itself. The first law of thermodynamics ($\Delta E = Q + W$) could raise no objection to such a submarine since, with no change in its internal energy ($\Delta E = 0$), it takes in an amount of heat and completely accounts for it in work done ($Q = -W$). It is the second law that says "no" to this marvelous vessel.

For chemists, a much more fruitful statement of the second law would be one applying to the variables which we measure or calculate for our chemical systems. It should help us understand such chemical behavior as *equilibria* and the spontaneous or natural *direction* of chemical reactions. To understand the development of such a statement and be able to work with it, we must explore two new functions of state, the *free energy*, G, and the *entropy*, S. Both of these functions are defined in terms of a *reversible process.*

3.2. Reversible Processes; The Helmholtz Free Energy

A reversible process is one which can be caused to proceed in the opposite direction by making an infinitesimal change in the properties of the system at any point in its progress.

To understand this better, consider the piston-and-cylinder shown in cross section in Fig. 3.1. It confines an ideal gas and is kept at a constant temperature. In addition to atmospheric pressure the piston is held in place initially (state I) by pressure due to a pile of wheat grains. If the wheat is suddenly removed, the pressure confining the gas is decreased and the gas will immediately and spontaneously expand. This expansion will continue until the new volume is such that again $PV = nRT$ (state F). That is, it will expand

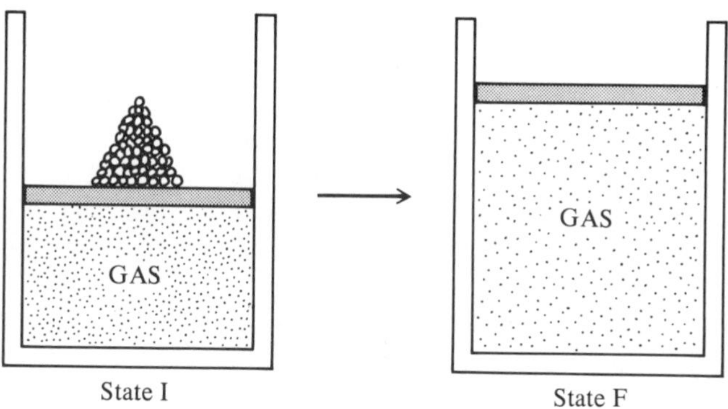

Figure 3.1

until equilibrium is reestablished and internal and external pressure are again equal. Heat will enter the cylinder from the surroundings since the process is isothermal ($\Delta T = 0$). The work which occurs in the gas expansion against the constant external (atmospheric) pressure is $-\int_I^F P_{ext}\, dV = -P\,\Delta V$. Since ΔV is positive in this expansion, the work is done *by* the gas and is therefore negative. The magnitude of this work corresponds to the area under the curve in a plot of P_{ext} versus V, the shaded portion of Fig. 3.2. Point 1 in this figure is the initial equilibrium position with the wheat in place, state I. Point 2 is the nonequilibrium state which instantaneously exists after the wheat has been removed but before the expansion begins. Point 3 is the final equilibrium position, state F. Points 1 and 3 are the only points on the curve, the only occasions during the process, when $P_{int} = P_{ext}$ and the system is in *equilibrium* with the surroundings.

Now let us suppose we have the system again in state I with the wheat in place. This time, however, we remove half the grains, allow the system to expand isothermally to equilibrium, then remove the remaining grains, and allow the system to reach equilibrium again. Figure 3.3 illustrates this sequence

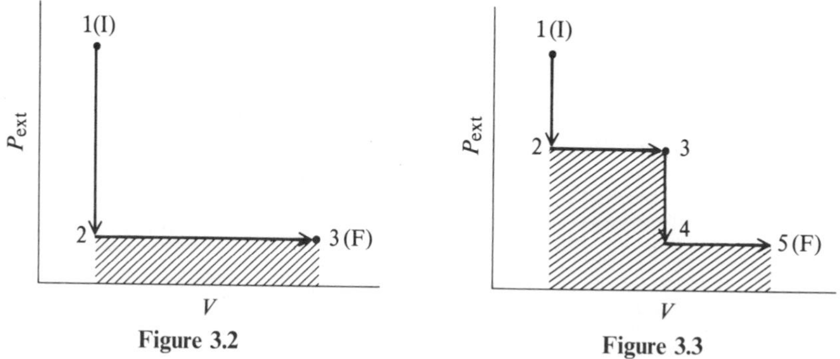

Figure 3.2

Figure 3.3

with a pressure-volume plot, the area under the curve again being equal to the magnitude of the work done. The same net change in state has occurred in the two procedures represented in Fig. 3.2 and Fig. 3.3, since the two initial states are identical and the two final states are identical. Nevertheless, the area under the curve (the work done) is greater in the second case, when we removed the wheat in two portions. Note also that Fig. 3.3 has three equilibrium states (1, 3, and 5) instead of two, and that this second process would take more time.

Figure 3.4 illustrates what happens when the wheat is removed in four batches, equilibrium being established anew after each successive removal (five equilibrium states). Figure 3.5 pictures the sequence when the removal is in eight successive batches; Fig. 3.6, when one grain at a time is removed. In each of these processes equilibrium is regained after each removal. Each equilibrium state is indicated by a point in these figures.

In Fig. 3.6 the direction of the process can be reversed at any point by adding a grain of wheat instead of removing one. But this process is still not thermodynamically reversible, since adding a grain causes more than an infinitesimal change in the external pressure. We must carry our development one step further. Suppose we grind the wheat into flour and, beginning with the flour in place on the piston (state I), take off one infinitesimal portion, a single speck of flour, at a time. This isothermal process is illustrated in Fig. 3.7. Every point on the curve is essentially an equilibrium state. At each point the process can be reversed by putting back a single speck of flour, causing an infinitesimal increase in the pressure. This corresponds to a reversible process in the thermodynamic sense. The work done on the gas in such a reversible isothermal process is called the *reversible work*, W_{rev}.

All the above processes (Figs. 3.2 through 3.7) begin at the same initial state I and end at the same final state F. Note that the *magnitude* of the work performed *by* the expanding gas on the surroundings (that is, the area under

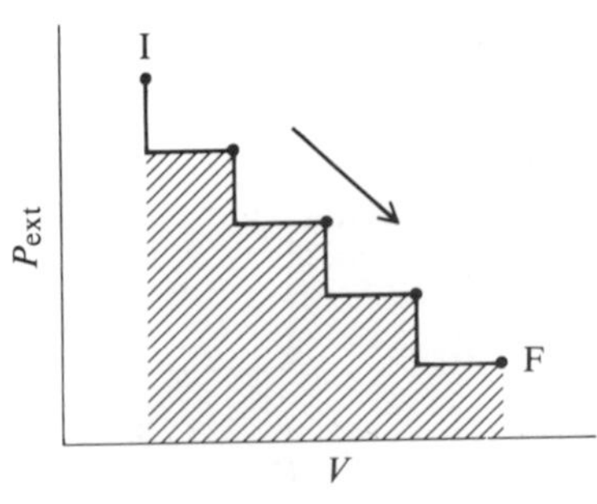

Figure 3.4

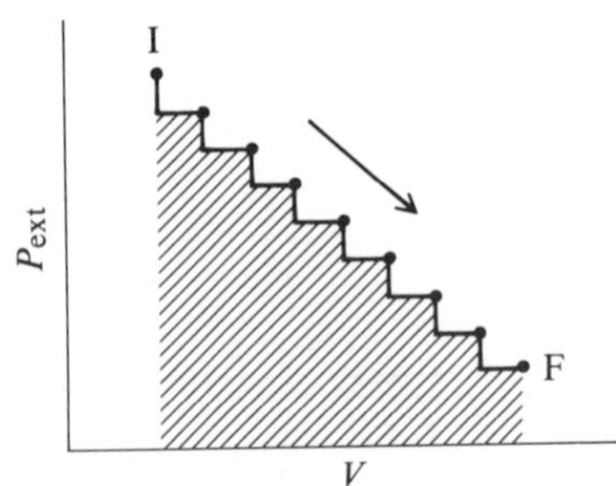

Figure 3.5

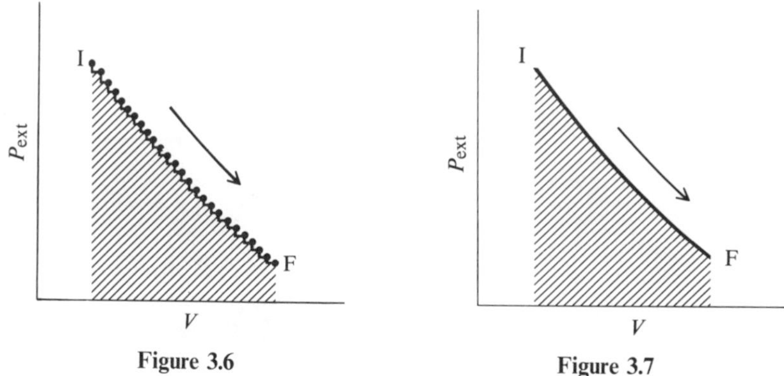

Figure 3.6

Figure 3.7

the pressure-volume curve) is greatest for the reversible process. For such an expansion process, however, the work is negative (Eq. 6); work done *on* the system is positive, work done *by* the system is negative. Taking the sign into consideration, the work done *by* the gas in the reversible process W_{rev} is *less than* (more negative than) the work done *by* the gas in the other processes.

Note also that the reversible process would take an infinitely long time and thus would not be attempted in actual practice; by contrast, the processes represented in Figs. 3.2 through 3.6 are finite and potentially realizable. We can nevertheless employ the concept of the reversible process in our thermodynamic reasoning, and can thereby calculate the thermodynamic variables which would accompany such a process.

Let us restate our results for the gas expansion in the form of the following inequality:

$$W_{rev} < W_{irrev}. \tag{33}$$

The work involved in any actual *irreversible* expansion between two given states of a system under *isothermal* conditions must always be *greater than* (less negative than) the work involved in the theoretical *reversible* expansion between the same two states.

It is of interest to consider our piston beginning in the state with no load and observe what occurs as we add wheat grains. The gas is compressed as a result of the added pressure and work is done *on* the system. The area beneath the curves again gives the magnitude of the work. In Fig. 3.8 the wheat grains are all dumped on at once; in Fig. 3.9 they are added in two batches; in Fig. 3.10, eight batches; Fig. 3.11, one grain at a time; Fig. 3.12, infinitesimally small specks of flour dust one at a time. The *minimum* amount of work which would have to be done *on* the system to obtain the desired isothermal change in state would be the work done in the reversible process illustrated in Fig. 3.12. This is the reversible work

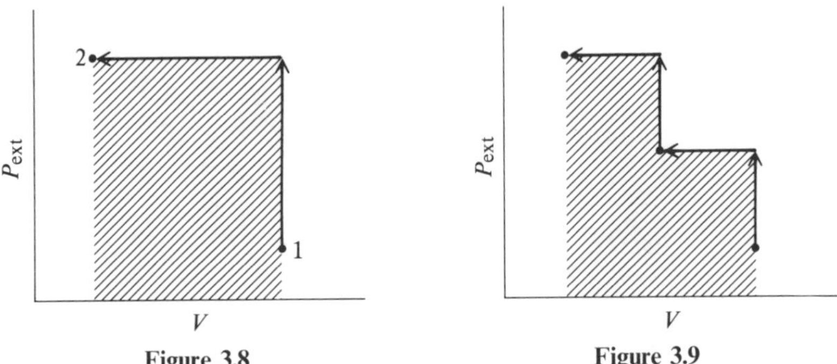

Figure 3.8 **Figure 3.9**

W_{rev} for the gas compression; it is equal in magnitude but opposite in sign to the W_{rev} of expansion (Fig. 3.7). This is true because the process illustrated in Fig. 3.12 is opposite in direction to that in Fig. 3.7, the initial and final states being reversed in the two processes.

It is clear that the work done *on* the system in the reversible compression is less than that done in any of the other compression processes; again

$$W_{rev} < W_{irrev}. \tag{33}$$

We see then that for both isothermal expansions and contractions the work done in any naturally occurring, spontaneous process must be greater than that involved in a reversible process between the same initial and final states.

Although we have only discussed the pressure-volume work involved in isothermal expansions and contractions, our results can be generalized to include all types of work (gravitational, electrical, etc.) involved in isothermal processes. We may therefore state that, in general, *the work involved in any*

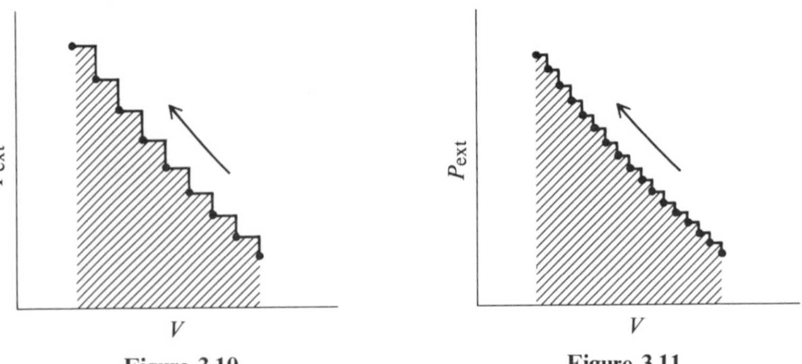

Figure 3.10 **Figure 3.11**

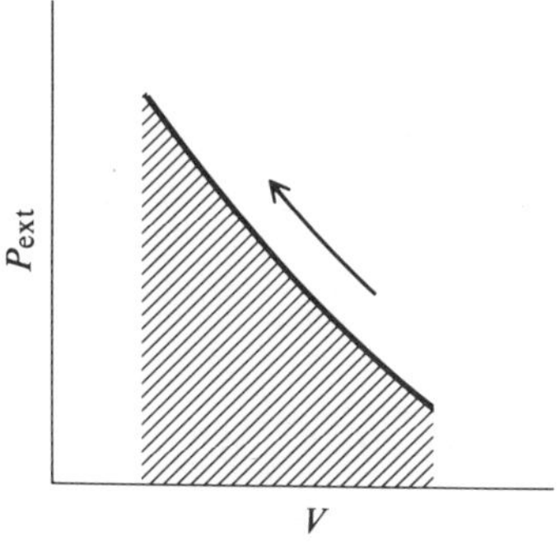

Figure 3.12

actual irreversible isothermal process is always greater than the work which would occur in a reversible process between the same two states.

Care must always be exercised in the choice of the mathematical sign of the work. When work is done *on* the system in a process, the work is *positive.* When work is done *by* the system on the surroundings, the work terms in Inequality (33) are *negative.*

There is only one limiting value of W_{rev} between two isothermal states. Thus when we specify the path (isothermal and reversible), W_{rev} corresponds to a change in a state function. This function of state is called the *Helmholtz free energy, A*:

$$\Delta A = W_{rev} \tag{34}$$

at constant temperature. Inequality (33) may be rewritten as:

$$\Delta A < W_{irrev}. \tag{33, 34}$$

$W > \Delta A$ corresponds to a naturally occurring, spontaneous isothermal process. $W = \Delta A$ for a reversible, equilibrium isothermal process. If $W < \Delta A$ for an isothermal process, it will not occur spontaneously. Equations (33) and (34) comprise a statement of the second law of thermodynamics under constant temperature conditions.

3.3. The Gibbs Free Energy

Just as we defined enthalpy to facilitate our work with chemical systems taking place under conditions of constant pressure, let us now define in a similar manner a new state function, the *Gibbs free energy, G*:

$$G = A + P_{int} V \tag{35}$$

$$dG = dA + d(P_{int} V). \tag{36}$$

Note that G bears the same relationship to A as H does to E.

For an isothermal, constant-pressure process:

$$\int_1^2 dG = \int_1^2 dA + P \int_1^2 dV$$

$$\Delta G = \Delta A + P\,\Delta V. \tag{37}$$

Recall that when a spontaneous isothermal change in state occurs, $\Delta A < W$. Thus

$$\Delta A = AG - P\,\Delta V < W. \tag{38}$$

Let us consider separately all work other than pressure-volume work W_{PV} and call it *net work*, W_{net}. This net work includes electrical work, work against gravity, etc. The total work W done on the system is then

$$W = W_{PV} + W_{\text{net}} = -\int_1^2 P\,dV + W_{\text{net}}. \tag{39}$$

For a constant-pressure process

$$W = -P\,\Delta V + W_{\text{net}},$$

and Eq. (38) becomes

$$\Delta G - P\,\Delta V < -P\,\Delta V + W_{\text{net}}$$

$$\Delta G < W_{\text{net}}. \tag{40}$$

In other words, *the net work involved in any actual irreversible change in state is always greater than the change in the Gibbs free energy when the process occurs under conditions of constant temperature and constant pressure.* If there is no work other than pressure-volume work involved in the actual process, then $W_{\text{net}} = 0$, and ΔG must be less than zero for all naturally occurring processes.

In summary, any change in state under *constant-temperature* conditions is allowed and will occur spontaneously if $W > \Delta A$ for the system. Any process involving a system whose *temperature and pressure* are constant will be spontaneous if $W_{\text{net}} > \Delta G$ ($\Delta G < 0$ if there is no work other than pressure-volume work). The relationship $W = \Delta A$ corresponds to an equilibrium process under isothermal conditions; $W_{\text{net}} = \Delta G$ indicates an equilibrium process under constant temperature and constant pressure conditions. Processes for which $W < \Delta A$ or $W_{\text{net}} < \Delta G$, under isothermal conditions and constant temperature and pressure conditions respectively, will not occur spontaneously.

We shall henceforth refer to the Gibbs free energy simply as the *free energy.*

Example 16 Three moles of an ideal gas is initially in a state in which $P_{\text{int}} = P_{\text{ext}} = 10$ atm, $T = 1000°$K. If the external pressure is suddenly reduced to 1 atm, the gas undergoes an expansion to a final equilibrium state in which $P_{int} = P_{ext} = 1$ atm, $T = 1000°K$. Prove that this expansion is spontaneous.

Answer: The work involved in this actual process is given by $-\int_1^2 P_{ext}\,dV = -(1 \text{ atm})\,\Delta V = (-1 \text{ atm}) \times (nRT/P_2 - nRT/P_1) = -221$ liter-atm.

To calculate $\Delta A = W_{rev}$ we must design a reversible process for carrying out the same change in state. We could accomplish this as we did in Section 3.2, changing the external pressure by infinitesimal amounts from the initial pressure of 10 atm to the final value of 1 atm. For this reversible isothermal expansion (see Example 6), $P_{ext} = P_{int} = nRT/V$ and

$$W_{rev} = -\int_1^2 P_{ext}\,dV = -\int_1^2 (nRT/V)\,dV = -nRT\int_1^2 dV/V$$

$$W_{rev} = -nRT \ln (V_2/V_1). \tag{41}$$

Since conditions are isothermal and the gas ideal, $P_1/P_2 = V_2/V_1$ (Boyle). Thus

$$W_{rev} = -nRT \ln (P_1/P_2) \tag{42}$$

$$W_{rev} = -3(0.0820)(1000)2.30 \log (10/1) = -566 \text{ liter-atm} = \Delta A.$$

Thus $\Delta A < W$, since $-566 < -221$, and the expansion is a spontaneous change in state.

We cannot use the ΔG criterion ($\Delta G < W_{net}$) in this example because the internal pressure is not constant.

3.4. Entropy

Although the free-energy criteria for spontaneity as we have developed them in preceding sections are quite useful, there are certain limitations. For many systems, especially chemical reactions, the determination of reversible work and net work is by no means straightforward. Also for processes which do not occur under isothermal conditions, the free-energy criteria for spontaneity do not apply. For these and other reasons, yet another state function must be developed to realize the full impact of the second law and to facilitate its application to chemical reactions. This function is the *entropy, S.*

The concept of entropy was developed by the Frenchman Sadi Carnot in 1824 in terms of an idealized, reversible cyclic heat engine. This engine is conceived as taking in a charge of heat from a hot reservoir or source, having it do the utmost possible work, then exhausting the unused residue at a lower temperature. The engine then returns to the hot reservoir for a fresh charge and repeats the cycle.

The steps in this operation may be visualized as shown in Fig. 3.13. The pressure-volume behavior of an ideal gas in such a cyclic process is illustrated in Fig. 3.14.

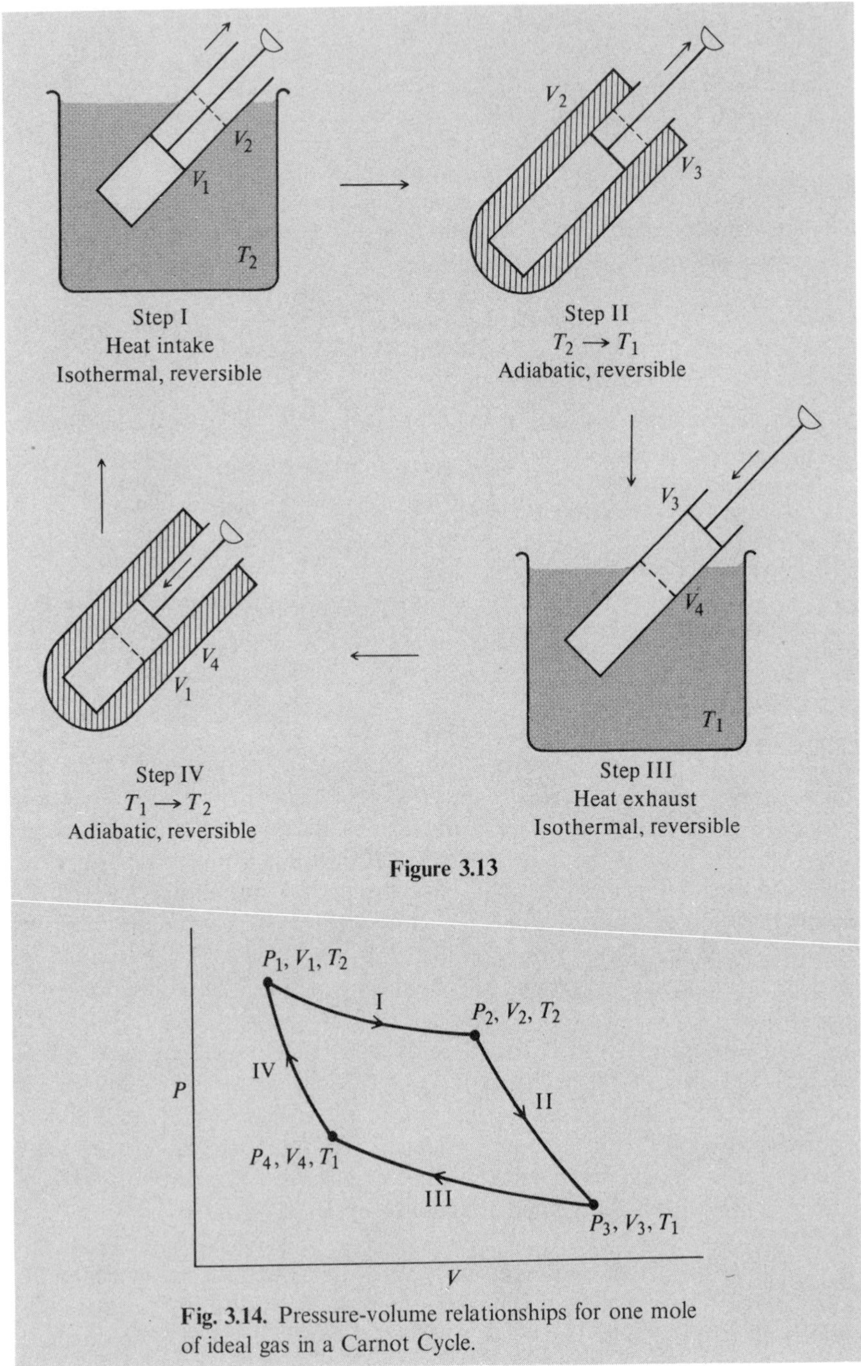

Figure 3.13

Fig. 3.14. Pressure-volume relationships for one mole of ideal gas in a Carnot Cycle.

At the beginning of the cycle, 1 mole of an ideal gas is confined in a piston-and-cylinder by an appropriate force exerted on the piston handle. The cylinder is immersed in a heat reservoir whose temperature remains constant at T_2.

In step I the piston is allowed to move back by reducing the confining force in a series of infinitesimal steps, thus meeting the requirements of reversibility (see Section 3.2). The gas expands from conditions (P_1, V_1) to (P_2, V_2) and thereby does a certain amount of work. This is accompanied by an inflow of heat from the reservoir, so that the temperature of the gas is sustained at T_2. As visualized, the process is *isothermal* and *reversible*. Since $\Delta E = 0$ for an isothermal process involving an ideal gas (Section 1.14), the work and heat involved in this first step are

$$Q_{\mathrm{I}} = -W_{\mathrm{I}} = \int P\,dV = RT_2 \ln (V_2/V_1).$$

In step II the "engine" is removed from the reservoir and slipped into a jacket which is a perfect insulator. By a similar process of infinitesimal steps, the gas is now allowed to expand further to (P_3, V_3); thus the conditions of step II are *reversible* and *adiabatic*. Since no heat can enter, the work done by the gas will be at the expense of its internal energy and the temperature falls from T_2 to some lower temperature T_1. This is indeed the purpose of step II. Now that the system (gas plus piston-and-cylinder) is at temperature T_1, the cylinder may be de-jacketed and brought into a second reservoir (the heat sink), also at T_1, with no irreversibility occurring. Since $Q_{\mathrm{II}} = 0$ and ΔE always equals $\int C_V\,dT$ for an ideal gas (Eq. 25b):

$$W_{\mathrm{II}} = \Delta E = \int_{T_2}^{T_1} C_V\,dT = C_V(T_1 - T_2).$$

In step III the gas is compressed until conditions (P_4, V_4) are reached. Because the reservoir is maintained at T_1 and the technique of infinitesimal pressure changes is again used, this compression is *isothermal* and *reversible*. Since the system is the recipient of contributed work, a resulting outflow of heat into the reservoir serves to keep the temperature of the gas from rising:

$$Q_{\mathrm{III}} = -W_{\mathrm{III}} = RT_1 \ln (V_4/V_3).$$

In step IV the cylinder at T_1 is removed, jacketed, and further compressed. This *reversible adiabatic* process is stopped when the temperature rises to T_2 and the initial conditions (P_1, V_1) have been regained. The cylinder is finally de-jacketed and returned to the heat reservoir at T_2. In step IV, $Q_{\mathrm{IV}} = 0$ and $W_{\mathrm{IV}} = C_V(T_2 - T_1)$.

The net result of this cyclic process is that our engine absorbs a certain amount of heat at a temperature T_2 and converts some of it to work, emitting heat at a lower temperature T_1. Since the process has been reversible throughout, the total amount of work W_{cycle} accomplished in one cycle is the maximum work possible for a cyclic heat engine operating between these two temperatures.

$$\begin{aligned} W_{\text{cycle}} &= W_{\text{I}} + W_{\text{II}} + W_{\text{III}} + W_{\text{IV}} \\ &= -RT_2 \ln (V_2/V_1) + C_V(T_1 - T_2) - RT_1 \ln (V_4/V_3) + C_V(T_2 - T_1) \\ &= -RT_2 \ln (V_2/V_1) + RT_1 \ln (V_3/V_4). \end{aligned} \tag{A}$$

For a reversible adiabatic process involving an ideal gas:

$$\begin{aligned} dE &= dw_{\text{rev}} \\ C_V\, dT &= -P\, dV = -(RT/V)\, dV \\ (C_V/T)\, dT &= -(R/V)\, dV. \end{aligned}$$

For the adiabatic step II:

$$\begin{aligned} C_V \int_{T_2}^{T_1} (1/T)\, dT &= -R \int_{V_2}^{V_3} (1/V)\, dV \\ C_V \ln (T_1/T_2) &= -R \ln (V_3/V_2) = R \ln (V_2/V_3). \end{aligned} \tag{B}$$

Similarly, for adiabatic step IV:

$$\begin{aligned} C_V \ln (T_2/T_1) &= -R \ln (V_1/V_4) \\ C_V \ln (T_1/T_2) &= R \ln (V_1/V_4). \end{aligned} \tag{C}$$

Combining Eqs. (B) and (C):

$$\ln (V_2/V_3) = \ln (V_1/V_4).$$

Thus

$$V_2/V_3 = V_1/V_4 \qquad \text{and} \qquad V_2/V_1 = V_3/V_4. \tag{D}$$

Returning to Eq. (A), we see that the total work is

$$\begin{aligned} W_{\text{cycle}} &= -RT_2 \ln (V_2/V_1) + RT_1 \ln (V_2/V_1) \\ &= (T_1 - T_2)R \ln (V_2/V_1). \end{aligned}$$

Since $V_2 > V_1$ and $T_2 > T_1$, W_{cycle} is negative. Therefore the gas yields this much work to the surroundings during each reversible cycle.

The efficiency of a heat engine is defined as the work performed divided by the amount of work which could have been done if the heat absorbed had been totally converted into work:

$$\begin{aligned}\text{Efficiency} &= \frac{|W_{\text{cycle}}|}{Q_{\text{I}}}\\ &= \frac{|(T_1 - T_2)R \ln (V_2/V_1)|}{RT_2 \ln (V_2/V_1)}\\ &= (T_2 - T_1)/T_2. \qquad \text{(E)}\end{aligned}$$

Notice that 100% efficiency is only approached as the lower temperature T_1 approaches absolute zero. Also note that the efficiency is zero if $T_1 = T_2$; a heat engine working in a cycle cannot convert heat into work while operating at a constant temperature (recall our marvelous submarine of Section 3.1).

The efficiency given by Eq. (E) is the maximum value for all "perfect" heat engines operating between two given temperatures. The efficiency of all real heat engines (such as steam engines) is even smaller than this due to such factors as friction and heat losses.

The cycle as described uses the transfer of heat to accomplish work. In a converse process, work may be used to accomplish the transfer of heat; this is the principle of the refrigerator, the air conditioner, and the heat pump. This process corresponds to the cycle of Figs. 3.13 and 3.14 operating in the reverse direction. In this situation the work is now the minimum work required to transfer an amount of heat absorbed at the lower temperature T_1 (such as in the interior of a refrigerator) to a region of higher temperature (such as the kitchen), where it is expelled.

Let us now return to an examination of the heat involved in the Carnot cycle. If we sum up the values of (Q/T) for the cycle we obtain a very interesting result:

$$\begin{aligned}\sum_{\text{cycle}} \frac{Q_{\text{rev}}}{T} &= \frac{Q_{\text{I}}}{T_2} + 0 + \frac{Q_{\text{III}}}{T_1} + 0\\ &= R \ln \frac{V_2}{V_1} + R \ln \frac{V_4}{V_3}. \qquad \text{(F)}\end{aligned}$$

The subscript in the expression Q_{rev} emphasizes that we are dealing with a reversible cycle. Equation (F) may be simplified by referring to Eq. (D) above, and we find that

$$\sum_{\text{cycle}} \frac{Q_{\text{rev}}}{T} = \oint \frac{dq_{\text{rev}}}{T} = 0.$$

The symbol $\oint$ is called a cyclic integral. This result may be shown to be true for all cyclic processes.

A necessary and sufficient condition for a function to be a state function is that the cyclic integral of its differential be equal to zero (see Section 1.8). Recognizing this we define a new state function, the *entropy*, S. For an infinitesimal change in state:

$$dS = dq_{\text{rev}}/T. \tag{43a}$$

For a finite process between two states 1 and 2:

$$\Delta S = \int_1^2 dq_{\text{rev}}/T. \tag{43b}$$

The change in entropy in a process is thus the sum of the heat absorbed by the system in each reversible infinitesimal step of the process divided by the temperature of that step. Thus the units of entropy are calories per degree, sometimes simply called entropy units (eu).

Now consider the cycle of Figs. 3.13 and 3.14 when its steps are *irreversible* and compare it with the reversible cycle. For the isothermal steps (I and III):

$$\Delta E_{\text{irrev}} = \Delta E_{\text{rev}}$$

$$Q_{\text{irrev}} + W_{\text{irrev}} = Q_{\text{rev}} + W_{\text{rev}}.$$

From Eq. (33):

$$W_{\text{irrev}} > W_{\text{rev}} \quad \text{and} \quad Q_{\text{irrev}} < Q_{\text{rev}}.$$

Consequently

$$\left(\frac{Q_{\text{I}}}{T_2}\right)_{\text{irrev}} < \left(\frac{Q_{\text{I}}}{T_2}\right)_{\text{rev}}$$

and

$$\left(\frac{Q_{\text{III}}}{T_1}\right)_{\text{irrev}} < \left(\frac{Q_{\text{III}}}{T_1}\right)_{\text{rev}}.$$

For the adiabatic steps (II and IV), Q/T is zero. (Adiabatic reversible processes are often referred to as *isentropic*.) Consequently

$$\oint \frac{dq_{\text{irrev}}}{T} < \oint \frac{dq_{\text{rev}}}{T} = 0. \tag{G}$$

Let us now consider the spontaneous noncyclic change in state illustrated in Fig. 3.15. For the irreversible process along the solid line we could determine $\int_1^2 dq_{\text{irrev}}/T$. The system may also be changed from state 1 to state 2 by the reversible process indicated by the dotted line in Fig. 3.15.

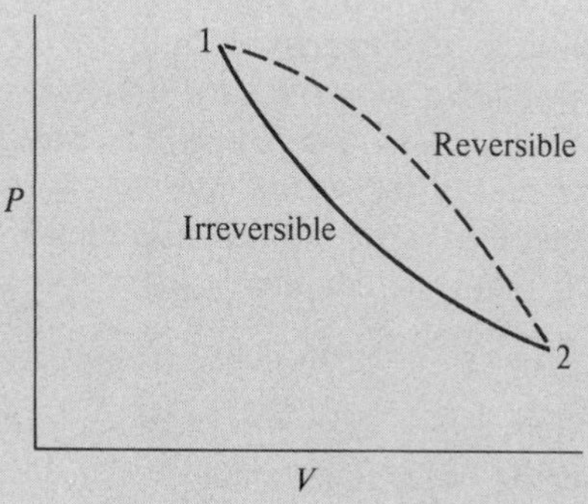

Figure 3.15

If we proceed from state 1 to state 2 along the spontaneous path and then return to state 1 via the reversible path, we have a cycle, and

$$\oint \frac{dq_{\text{irrev}}}{T} = \int_1^2 \frac{dq_{\text{irrev}}}{T} + \int_2^1 \frac{dq_{\text{rev}}}{T}.$$

Since dq_{rev}/T is an exact differential:[1]

$$\oint \frac{dq_{\text{irrev}}}{T} = \int_1^2 \frac{dq_{\text{irrev}}}{T} - \int_1^2 \frac{dq_{\text{rev}}}{T}.$$

From Eqs. (G) and (43):

$$\Delta S = \int_1^2 \frac{dq_{\text{rev}}}{T} > \int_1^2 \frac{dq_{\text{irrev}}}{T}. \tag{44}$$

The Carnot cycle treatment has led to the definition of a new state function, the *entropy*, S. For an infinitesimal change in state

$$dS = dq_{\text{rev}}/T, \tag{43a}$$

and for a finite process

$$\Delta S = \int_1^2 dq_{\text{rev}}/T. \tag{43b}$$

The difference in entropy between two states (1 and 2) is thus the sum of the heat exchanged by the system with the surroundings divided by the temperature at which the exchange would take place in a *reversible* process between the two states.

We have also seen that the maximum value of $\int_1^2 dq/T$ for a given change in state is that for a reversible process between the given initial

[1] Since dq_{irrev}/T is not an exact differential, the integral $\int_1^2 dq_{\text{irrev}}/T$ is not necessarily equal to $-\int_2^1 dq_{\text{irrev}}/T$.

and final states. Equation (44) is the mathematical statement of this fact. In other words, for a process to be possible and to occur spontaneously *the heat exchanged with the surroundings divided by the temperature at which the exchange takes place must be less than the entropy change for the same change in state.* This is a statement of the second law.

For an adiabatic process there is no heat exchanged with the surroundings; since $Q = 0$, $\int dq/T = 0$ for any actual adiabatic process. Equation (44) becomes

$$\Delta S > 0$$

for all spontaneous adiabatic processes, that is, for processes which occur in thermally isolated systems. Since the universe is considered to be isolated and no heat can be exchanged between it and any surroundings, the entropy change for all actual irreversible processes in which the universe is considered to be the system is positive. Thus the entropy of the universe is increased by every spontaneous process.

The entropy, like the free energy, is defined in terms of a *reversible* change in state. To calculate the entropy change accompanying a real process we must devise a reversible process connecting the same initial and final states. Along this reversible path the calculation of ΔS can be made unequivocally.

Example 17 Consider a thermos bottle in which a 5-g ice cube at $-5°C$ and 20 g of liquid water at 25°C are placed. For water $\Delta\bar{H}_{\text{fus}} = 1440$ cal mole^{-1}, $C_P(\text{solid}) = 8.7$ cal deg^{-1} mole^{-1} and $C_P(\text{liquid}) = 18.0$ cal deg^{-1} mole^{-1}.
a) What is the final equilibrium state of the contents of the thermos?
b) Calculate ΔS and show that this change in state is spontaneous.

Answer: a) Since the thermos is an insulating vessel, the process is adiabatic. All the heat used to raise the temperature of the ice and to melt it must come from cooling the water. Although the heat lost by one part of the system is absorbed simultaneously by the other part, the process may be visualized as occurring in several steps. First cool the liquid water to 0°C, then use the heat released in this process to heat the ice to 0°C and melt it.

The cooling of 20 g of liquid water from 25°C to 0°C releases $nC_P\,\Delta T$ calories, $(20.0 \text{ g}/18.0 \text{ g mole}^{-1})(18.0 \text{ cal deg}^{-1} \text{ mole}^{-1}) \times (25.0°) = 500$ cal.

The heating of 5 g of ice from $-5°C$ to 0°C would require $nC_P\,\Delta T = (5.0 \text{ g}/18.0 \text{ g mole}^{-1})(8.7 \text{ cal deg}^{-1} \text{ mole}^{-1})(5.0°) = 12$ cal. The melting of 5 g of ice at 0°C would require $n\,\Delta\bar{H}_{\text{fus}} = (5.0 \text{ g}/18.0 \text{ g mole}^{-1})(1440 \text{ cal mole}^{-1}) = 403$ cal. The total heat required for the heating and melting process would be 415 cal. The 500 cal released in the cooling process is more than sufficient; the "extra" heat is then used to warm the 25 g of liquid at 0°C to the final temperature:

$$500 - 415 = 85 \text{ cal} = nC_P\,\Delta T$$

$$85 \text{ cal} = (25 \text{ g}/18 \text{ g mole}^{-1})(18 \text{ cal deg}^{-1} \text{ mole}^{-1})\,\Delta T$$

$$\Delta T = 3.4°.$$

The final state is thus 25 g of liquid water at 3.4°C. In the process which has actually occurred, 20 g of liquid water cools from 25°C to 3.4°C, releasing enough heat to raise the temperature of 5 g of snow to 0°C, melt it, and raise the resulting 5 g of liquid to 3.4°C.

b) To calculate ΔS we must devise a reversible path between the initial state (5 g of ice at −5°C and 20 g of liquid at 25°C) and the final state (25 g of liquid at 3.4°C). The process we shall use is to (1) heat the ice reversibly to 0°C, (2) melt the ice reversibly at 0°C, (3) reversibly heat this 5 g of liquid to 3.4°C, and (4) cool the 20 g of liquid at 25°C to 3.4°C reversibly.

Since dq for heating or cooling n moles of a material at constant pressure is $nC_P\,dT$, the entropy change in the reversible process 1 is

$$\Delta S_1 = \int (dq_{\text{rev}}/T) = \int_{268^\circ}^{273^\circ} (nC_P\,dT/T)$$

$$= (5.0/18)(8.7)\ln(273/268) = 0.05 \text{ cal deg}^{-1}.$$

For process 2:

$$\Delta S_2 = \int (dq_{\text{rev}}/T) = n(\Delta\bar{H}_{\text{fus}}/T) = 1.48 \text{ cal deg}^{-1}.$$

For process 3:

$$\Delta S_3 = \int_{273^\circ}^{276.4^\circ} (nC_P\,dT/T) = 0.05 \text{ cal deg}^{-1}.$$

For process 4:

$$\Delta S_4 = \int_{298^\circ}^{276.4^\circ} (nC_P\,dT/T) = (20/18)(18)\ln(276.4/298)$$

$$= -1.54 \text{ cal deg}^{-1}.$$

The total $\Delta S = \Delta S_1 + \Delta S_2 + \Delta S_3 + \Delta S_4 = 0.04$ cal deg^{-1}. Since Q for the actual adiabatic process is 0, the entropy change must be greater than zero for a spontaneous process (Eq. 44); since the total entropy change found, 0.04 cal deg^{-1}, is positive, the process is indeed spontaneous.

3.5. Free Energy as a Function of Enthalpy and Entropy

As chemists we are most often interested in the criteria of spontaneity as they apply to chemical reactions. We would like to predict whether a reaction will occur spontaneously under a given set of conditions. In most cases our reactions will be run under conditions of constant temperature and constant pressure. Consider Inequality (44) under these conditions:

$$\Delta S > \frac{1}{T}\int_1^2 dq_{\text{irrev}} = \frac{Q}{T}.$$

But under conditions of constant pressure, Q is equal to ΔH and our spontaneity criterion becomes

$$T\,\Delta S > \Delta H$$

or

$$\Delta H - T\,\Delta S < 0. \tag{45}$$

The left-hand side of this inequality can be shown to equal ΔG. The following three differential equations can be obtained from Eqs. (11), (14), (34), (36), and (44):

$$dH = dE + P\,dV \qquad \text{(constant pressure)} \tag{46}$$

$$dE = dq_{\text{rev}} + dw_{\text{rev}} \qquad \text{(reversible)}$$

$$dE = T\,dS + dA \qquad \text{(constant temperature)} \tag{47}$$

$$dA = dG - P\,dV \qquad \text{(constant pressure)} \tag{48}$$

Substituting Eqs. (47) and (48) into Eq. (46) gives:

$$dH = T\,dS + dG$$

or

$$dG = dH - T\,dS,$$

and for a finite isothermal process

$$\Delta G = \Delta H - T\,\Delta S.$$

Thus for an *isothermal, constant-pressure process with no work other than pressure-volume work,*

$$\Delta G = \Delta H - T\,\Delta S < 0 \tag{49}$$

is our criterion for spontaneity.

Spontaneity therefore depends on two state functions, the entropy and the enthalpy. For a spontaneous process $\Delta H < 0$, and $\Delta S > 0$ would be favorable to spontaneity. Note that ΔH can, however, be positive (endothermic) in a spontaneous change if the $T\,\Delta S$ term is large enough to make ΔG negative. Similarly ΔS may be negative for a spontaneous process if ΔH is negative and its magnitude is great enough to overshadow the $T\,\Delta S$ term.

The same results could have been obtained simply by *defining* free energy as a composite function of enthalpy and entropy:

$$G = H - TS.$$

Under isothermal, constant-pressure conditions:

$$\Delta G = \Delta H - T\,\Delta S.$$

Employing Inequality (44) again

$$\Delta G = \Delta H - T\,\Delta S < 0. \tag{49}$$

In summary, the criteria for spontaneity which we have developed in this chapter are:

$\Delta A < W$	at constant temperature
$\Delta G < W_{net}$	at constant temperature and pressure
$\Delta S > \int_1^2 (dq/T)$	under all conditions
$\Delta S > 0$	under adiabatic conditions
$\Delta G = \Delta H - T\,\Delta S < 0$	at constant pressure and constant temperature when only pressure-volume work can occur

Under isothermal, constant-pressure conditions all these criteria must hold simultaneously. When the temperature is constant but the pressure is not, only the ΔA and ΔS criteria can be applied. When the process occurs under conditions which are not isothermal, only the ΔS criterion can be used. Notice that in each case we must compare a function involving the work or heat accompanying an *actual* process with a similar term involving the work or heat in a hypothetical *reversible* process. The last of the above criteria will be of particular importance to us in our work with chemical reactions.

3.6. The Third Law; Standard Entropy and Standard Free Energy

We have seen how to determine ΔH for a chemical reaction (Chapter 2). If we could determine ΔS for the reaction, combining ΔH and ΔS via Eq. (49) would allow us to calculate ΔG for the reaction and to predict whether or not it will be spontaneous.

Note that $\Delta H = Q_p$ can be determined directly by calorimetric measurements. For a *real* process, however, $\Delta S = \int_1^2 (dq_{rev}/T)$ must be calculated by visualizing a *reversible* process between the same initial and final states. This is rather difficult to do for a chemical reaction. Can we devise an alternative procedure?

Absolute values of E and H cannot be given at this point. The *third law of thermodynamics* does, however, allow absolute values to be determined for S.

Two statements of the third law (which can be shown to be equivalent) are:

1) A temperature of 0°K cannot be reached in a finite process from a temperature above 0°K.

2) The entropy difference between two states of the same pure ordered crystalline substance is zero at 0°K.

As a result of the third law, the entropy of all pure ordered crystalline substances can be assigned the value zero at absolute zero, 0°K. The absolute entropy of a substance at any given temperature T is then

$$\int_{0^\circ\mathrm{K}}^{T} dS = \int_{0^\circ}^{T} (dq_{\mathrm{rev}}/T)$$

$$S_T - S_{0^\circ} = S_T = \int_{0^\circ}^{T} n(C_P/T)\,dT.$$

Since C_P is not constant over such a range of temperature, the evaluation of this integral is difficult. It has been carried out, however, for many substances; values of the *standard absolute entropy* S°_{abs} for some common materials appear in Table 3.1. The term S°_{abs} is the entropy of one mole of a substance under standard conditions at some designated temperature T°, usually 298°K:

$$S^\circ_{\mathrm{abs}} = \int_{0^\circ}^{T^\circ} (C_P/T)\,dT. \tag{50}$$

Thus ΔS° for a reaction would be

$$\Delta S^\circ_{\mathrm{reaction}} = \sum_{\mathrm{pr}} nS^\circ - \sum_{\mathrm{re}} nS^\circ. \tag{51}$$

If one or more phase transitions (solid → liquid, liquid → vapor, allotropic conversions, etc.) occur between absolute zero and the standard temperature, the entropy change accompanying these transitions must be included in Eq. (50). For instance, the absolute entropy of 1.0 mole of water at 25°C would be given by

$$S^\circ_{\mathrm{abs}} = \int_{0^\circ}^{273^\circ} (C_{P,\,\mathrm{sol}}/T)\,dT + \Delta S_{\mathrm{fus}} + \int_{273^\circ}^{298^\circ} (C_{P,\,\mathrm{liq}}/T)\,dT.$$

Also available in tabular form (Table 3.1) are values of the *standard entropy of formation*, ΔS°_f, the entropy change accompanying the reaction in which 1.0 mole of the substance is formed in its designated state under standard conditions from its substituent elements in their most stable forms. Its value for a given substance is Eq. (51):

$$\Delta S^\circ_f = S^\circ_{\mathrm{abs}} - \sum_{\mathrm{elements}} nS^\circ_{\mathrm{abs}},$$

where the sum is over the absolute entropy values of the substituent elements. The tabulated values of ΔS°_f may be used to calculate ΔS° for reactions:

$$\Delta S^\circ_{\mathrm{reaction}} = \sum_{\mathrm{pr}} n\,\Delta S^\circ_f - \sum_{\mathrm{re}} n\,\Delta S^\circ_f. \tag{52}$$

By combining ΔH° and ΔS°, ΔG° for chemical reactions can be calculated:

$$\Delta G^\circ = \Delta H^\circ - T\,\Delta S^\circ. \tag{53}$$

Table 3.1. Standard Absolute Entropies, Standard Entropies of Formation, and Standard Free Energies of Formation for Selected Substances (25°C)

Substance	S°_{abs}, cal deg^{-1} mole^{-1}	ΔS°_f, cal deg^{-1} mole^{-1}	ΔG°_f, kcal mole^{-1}
H(g)	27.39	11.79	48.58
O(g)	38.47	13.97	54.99
N(g)	36.61	13.73	81.47
Br(g)	41.81	23.6	19.69
C(g)	37.76	36.40	160.85
C(graphite)	1.361	0.00	0.00
C(diamond)	0.583	−0.78	0.685
S(rhombic)	7.62	0.00	0.00
S(monoclinic)	7.78	0.16	0.023
O_2(g)	49.00	0.00	0.00
O_3(g)	56.80	−17.00	39.06
H_2(g)	31.21	0.00	0.00
N_2(g)	45.77	0.00	0.00
Cl_2(g)	53.29	0.00	0.00
Br_2(l)	36.4	0.00	0.00
Br_2(g)	58.64	22.2	0.751
H_2O(g)	45.11	−10.61	−54.64
H_2O(l)	16.72	−39.00	−56.69
CO(g)	47.30	21.44	−32.81
CO_2(g)	51.06	0.697	−94.26
CH_4(g)	44.50	−19.28	−12.14
C_2H_4(g)	52.45	−85.49	16.28
CH_3OH(g)	56.8	−31.5	−38.69
C_2H_5OH(g)	67.4	−53.5	−40.30
C_2H_5OH(l)	38.4	−82.5	−41.77
NH_3(g)	46.01	−23.69	−3.976
NO(g)	50.34	2.95	20.72
NO_2(g)	57.47	−14.43	12.39

Values of ΔG°_f are also tabulated. The *standard free energy of formation* for a chemical susbstance is defined in a manner analagous to ΔH°_f and ΔS°_f. Values for ΔG°_f may be determined using Eq. (53). Table 3.1 contains ΔG°_f values for a number of materials. Such tabulations may be used to determine ΔG° for a reaction by using the following equation:

$$\Delta G^\circ_{\text{reaction}} = \sum_{\text{pr}} n\,\Delta G^\circ_f - \sum_{\text{re}} n\,\Delta G^\circ_f. \tag{54}$$

Example 18 Various theories concerning the origin of organic life include spontaneous production of the complex molecules making up plants and animals from simpler molecules. An example of this is the following reaction which produces urea. Pertinent molar thermodynamic variables are listed under the reactants and products.

Reaction:	$CO_2(g)$ +	$2NH_3(g) \rightarrow$	$(NH_2)_2CO(s)$ +	$H_2O(l)$
S°_{abs}, cal deg^{-1} mole^{-1}:	51.06	46.01	25.00	16.72
ΔH°_f, kcal mole^{-1}:	−94.05	−11.04	−79.63	−68.32

Calculate ΔG° at 25° and comment on the spontaneity of this reaction.

Answer: Using Eq. (51), $\Delta S^\circ = (25.00 + 16.72) - (51.06 + 2(46.01)) = -101.4$ cal deg^{-1}. ΔH° is calculated using Eq. (29): $\Delta H^\circ = (-79.63 - 68.32) - (-94.05 - 2(-11.04)) = -31.82$ kcal. ΔG° is determined using Eq. (53); $\Delta G^\circ = -31.82 - (298)(-0.1014) = -1.60$ kcal.

Only pressure-volume work can occur and thus W_{net} is zero for this reaction, as it will be for most chemical reactions. Since the standard free-energy change we calculated is negative, the reaction *under standard conditions* is spontaneous ($\Delta G < 0$). Notice that $\Delta H^\circ < 0$ and $\Delta S^\circ < 0$: the enthalpy change contributes to a negative ΔG, whereas the $-T\,\Delta S$ term makes ΔG less negative.

Example 19 Assuming ΔS° and ΔH° are independent of temperature (not a particularly good assumption), what is the maximum temperature for which the reaction of Example 18 would be spontaneous?

Answer: As T increases, the $T\,\Delta S$ term becomes more important. At some temperature the $T\,\Delta S$ and ΔH terms will exactly cancel and $\Delta G = 0$. Above that temperature ΔG will be positive and the process will no longer be spontaneous; in fact, urea will spontaneously react with water to give ammonia and carbon dioxide. The maximum temperature we seek, then, is the temperature at which $\Delta G^\circ = \Delta H^\circ - T\,\Delta S^\circ = 0$ and $\Delta H^\circ = T\,\Delta S^\circ$:

$$-31.82 \text{ kcal} = T(0.1014) \text{ kcal deg}^{-1}$$
$$T = 314°\text{K}.$$

3.7. Variation of Free Energy with Temperature

The value of ΔG which we calculate using Eq. (53) or (54) is ΔG°, the change in free energy under standard conditions. What would ΔG be if the reaction is run at some other temperature? This problem may be visualized in the same manner as the variation of ΔH with temperature (Section 2.5):

$$\begin{array}{ccc} \text{Reactants at } T & \xrightarrow{\Delta G_T} & \text{Products at } T \\ \downarrow \Delta G_1 & & \uparrow \Delta G_2 \\ \text{Reactants at } T^\circ & \xrightarrow{\Delta G^\circ} & \text{Products at } T^\circ \end{array}$$

Since free energy is a state function, $\Delta G_T = \Delta G^\circ + \Delta G_1 + \Delta G_2$. Here ΔG_1 is the free-energy change for heating or cooling the reactants from temperature T to the standard temperature T°; ΔG_2 for changing the products from T° to T.

Recall that $G = H - TS = (E + PV) - TS$. Differentiating completely: $dG = dE + P\,dV + V\,dP - T\,dS - S\,dT$. But under reversible conditions with only pressure-volume work (no net work), $dE = dq + dw = T\,dS - P\,dV$ and

$$dG = V\,dP - S\,dT. \tag{55}$$

For constant pressure conditions $dP = 0$ and $dG = -S\,dT$. Thus for a process which involves changing the temperature from T_1 to T_2 at constant pressure,

$$\Delta G = -\int_{T_1}^{T_2} dG = -\int_{T_1}^{T_2} S\,dT. \tag{56}$$

Hence

$$\Delta G_1 = -\int_{T}^{T^\circ} S_{\text{re}}\,dT \text{ and } \Delta G_2 = -\int_{T^\circ}^{T} S_{\text{pr}}\,dT$$

and

$$\Delta G_T = \Delta G^\circ + \Delta G_1 + \Delta G_2 = \Delta G^\circ - \int_{T}^{T^\circ} S_{\text{re}}\,dT - \int_{T^\circ}^{T} S_{\text{pr}}\,dT$$

$$= \Delta G^\circ - \int_{T^\circ}^{T} (S_{\text{pr}} - S_{\text{re}})\,dT,$$

or

$$\Delta G_T = \Delta G^\circ - \int_{T^\circ}^{T} \Delta S\,dT. \tag{57}$$

Written in differential form this equation is

$$(\partial\,\Delta G/\partial T)_P = -\Delta S \tag{58}$$

and is known as the *Gibbs-Helmholtz* relationship. If we know ΔG at one temperature and we know ΔS for the reaction as a function of temperature, we can calculate ΔG at another temperature.

Example 20 In Example 18 we calculated ΔG° at 298°K for the following reaction:

$$CO_2(g) + 2NH_3(g) \longrightarrow (NH_2)_2CO(s) + H_2O(l).$$

Calculate ΔG at 350°K using the data given in Example 18.

Answer: In Example 18 we found the value of ΔG° to be -1.6 kcal and the value of ΔS° to be -0.101 kcal deg^{-1}. Assuming ΔS varies only a negligible amount between 298°K and 350°K, Eq. (57) becomes

$$\Delta G = \Delta G^\circ - \Delta S \int dT = \Delta G^\circ - \Delta S(T - T^\circ)$$

$$= -1.60 - (-0.101)(350 - 298)$$

$$= 3.65 \text{ kcal.}$$

3.8. Variation of Free Energy with Pressure

Suppose we wish to find ΔG at a pressure different from the standard value. Again

$$\begin{array}{ccc} \text{Reactants at } P & \xrightarrow{\Delta G_P} & \text{Products at } P \\ \downarrow \Delta G_1 & & \uparrow \Delta G_2 \\ \text{Reactants at } P^\circ & \xrightarrow{\Delta G^\circ} & \text{Products at } P^\circ \end{array}$$

Again $\Delta G_P = \Delta G^\circ + \Delta G_1 + \Delta G_2$ where ΔG_1 is the free-energy change for changing the pressure on the reactants from P to P° and ΔG_2 for changing the products from P° to P. Applying Eq. (55) at constant temperature: $dG = V\,dP$. Thus for a reversible process under isothermal conditions,

$$\Delta G = \int_{P_1}^{P_2} dG = \int_{P_1}^{P_2} V\,dP.$$

Hence

$$\Delta G_1 = \int_{P}^{P^\circ} V_{\text{re}}\,dP \text{ and } \Delta G_2 = \int_{P^\circ}^{P} V_{\text{pr}}\,dP$$

and

$$\Delta G_T = \Delta G^\circ + \Delta G_1 + \Delta G_2 = \Delta G^\circ + \int_{P}^{P^\circ} V_{\text{re}}\,dP + \int_{P^\circ}^{P} V_{\text{pr}}\,dP$$

$$= \Delta G^\circ + \int_{P^\circ}^{P} (V_{\text{pr}} - V_{\text{re}})\,dP,$$

or

$$\Delta G_P = \Delta G^\circ + \int_{P^\circ}^{P} \Delta V\,dP. \tag{59}$$

In differential form,

$$(\partial\, \Delta G/\partial P)_T = \Delta V. \tag{60}$$

If we know ΔG at one pressure and the change in volume accompanying a reaction, ΔG can be calculated at other pressures.

Example 21 Calculate ΔG for the reaction of Example 18 at a pressure of 5 atm instead of 1 atm.

Answer: If we assume that the volume of solid urea and liquid water are negligible when compared to the volume of the gaseous reactants CO_2 and NH_3, the change in volume $\Delta V = -V(\text{reactants})$. If we further assume that these gaseous materials behave ideally, the change in volume would be minus the volume occupied by 3 moles of ideal gas: $\Delta V = -nRT/P = -3RT/P$. Thus

$$\begin{aligned} \Delta G(5\ \text{atm}) &= \Delta G^\circ(1\ \text{atm}) + \int_1^5 -(3RT/P)\,dP \\ &= -1.60 - 3(1.98 \times 10^{-3})(298) \ln(5/1) \\ &= -1.60 - 2.86 = -4.46\ \text{kcal}. \end{aligned}$$

3.9. Microscopic Interpretation of Entropy

In Section 1.12 we interpreted energy as the macroscopic manifestation of molecular (microscopic) motion. Now let us consider the molecular interpretation of entropy.

Again we would need the concepts of quantum chemistry and of statistical mechanics for a complete presentation of the microscopic interpretation of entropy. At this point, however, we can gain enough insight to better understand the thermodynamics of our chemical systems in terms of molecules.

The microscopic basis of entropy is related to molecular *disorder* and to *probability*. If a process occurs spontaneously, our basic notions of probability tell us that the final state must be more probable than the initial state. The probability increases in a spontaneous process as does the entropy (isolated or adiabatic system). Entropy and probability are indeed related, but what is the explicit relationship?

As in our discussion of the kinetic theory of gases, we postulate that our system is made up of particles (atoms). In our simplified treatment we shall assume that the only energy these particles can possess is translational energy. The probability of a given macroscopic state for such a system is proportional to the number Ω of different microscopic arrangements of these particles which give that same macroscopic state. This can be better understood by considering an analogy.

Suppose we have a box with three compartments (see Fig. 3.16) and wish to place three marbles in the box. The number of ways Ω in which we could put one marble in each of the compartments would be six. This is illustrated in Fig. 3.16(a). In Fig. 3.16(b) we see that there are three different ways of putting two marbles in the first compartment and one marble in the second compartment ($\Omega = 3$). (Likewise there are three different ways of putting two in the second compartment and one in the first compartment, etc.) And there is only one way ($\Omega = 1$) of putting all three marbles in the first compartment. If we throw the three marbles randomly into the box we will therefore be twice as likely to find one marble in each of the compartments as to find two in the first compartment and one in the second. The probability of the first arrangement is twice that of the second. Similarly the probability of all three marbles falling into the first compartment is 1/6 of the probability of one marble falling into each of the three compartments.

Note that in this example the lower the value of Ω, the more "ordered" is the arrangement. By this we mean that when $\Omega = 1$ as in Fig. 3.16(c), we are certain that marble number one is in compartment one; when $\Omega = 3$ marble number one may be in either compartment one or two; and when $\Omega = 6$ it may be in any of the three boxes. The uncertainty in the positions of the particles which make up the system (that is, the *disorder* of the system) increases with the number of possible arrangements of these particles.

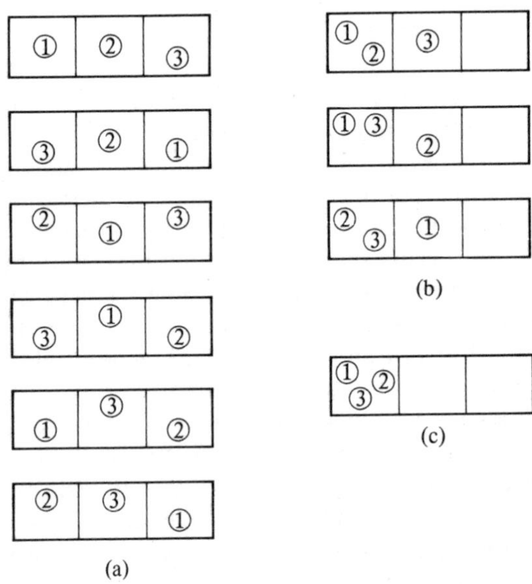

Figure 3.16

Let us carry this example further. Suppose that each compartment will hold only one marble. There will then be a *total* of six possible arrangements of the three marbles. If we wish to place four marbles in a box with four compartments, each of which will hold only one marble, there would be twenty-four possible configurations (Fig. 3.17).

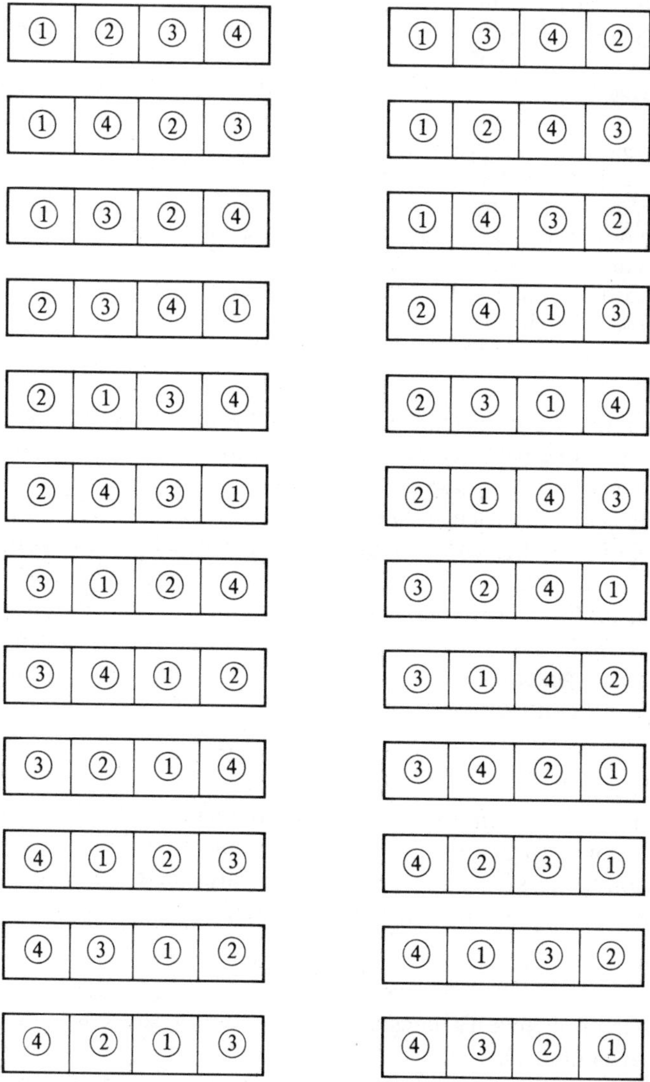

Figure 3.17

Example 22 What is the number of possible arrangements of five marbles in five compartments?

Answer: An attempt to count the possible arrangements of five marbles in five compartments in the manner of Figs. 3.16 and 3.17 would indicate that as the number of particles and compartments increases, the task of determining the number of configurations rapidly becomes extremely laborious. Luckily we can simplify the procedure.

Consider the four-marbles-in-four-compartments case. The first marble can be placed in any one of the four compartments. There are thus four possible positions for it. The second marble, however, can be placed in any one of only three compartments; the third in just one of the two remaining compartments; and the fourth has only one possible position. The total number of configurations for the four marbles is equal to the product of the possible individual arrangements for each of the marbles: $\Omega = 4 \times 3 \times 2 \times 1$. Mathematically this product is written as 4! ("four factorial").

Similarly, in the case of the five marbles in five compartments, the first has five possible positions; the second has four; the third, three; the fourth, two; and the fifth, one. The total number of possible configurations is $\Omega = 5! = 5 \times 4 \times 3 \times 2 \times 1 = 120$.

In general, for n particles in n compartments where each compartment will hold only one particle, there will be $n!$ possible configurations.

Another important case is that in which there are n particles to place in m compartments (each of which holds only one particle) when n is less than m. Consider five marbles placed in a box with eight compartments. The first marble can go into any of eight positions; there are seven choices for the second marble; six for the third; five for the fourth; and four for the fifth. There are thus $8 \times 7 \times 6 \times 5 \times 4$ configurations. This can be written as $8! \div 3!$. In general, with n particles and m compartments, there will be $m! \div (m - n)!$ configurations, or

$$\Omega = m!/(m - n)! \tag{61}$$

Let us now return to our attempt to relate probability and entropy. We have pointed out that the probability (and therefore Ω) increases in a spontaneous process; we know that entropy also increases in such a process. Boltzmann postulated that the two are related by

$$S = k \ln \Omega, \tag{62}$$

where k is Boltzmann's constant (equal to the gas constant divided by Avogadro's number).

If the entropy of one system is S_1 and of a second system is S_2, the total entropy of the two systems would be the sum of the individual entropies, or $S = S_1 + S_2$. This is true since entropy is an extensive state variable. If the probability of the first system is proportional to Ω_1 and of the second system to Ω_2, probability theory requires that the total value of Ω be $\Omega_1\Omega_2$. The Boltzmann postulate is consistent with these observations since $S_1 = k \ln \Omega_1$, $S_2 = k \ln \Omega_2$ and

$$S = S_1 + S_2 = k \ln \Omega_1 + k \ln \Omega_2 = k \ln \Omega_1\Omega_2 = k \ln \Omega.$$

In a change of state from an initial state I to a final state F, the change in entropy is related to the change in probability by Boltzmann's postulate as

$$\Delta S = S_{\mathrm{F}} - S_{\mathrm{I}} = k \ln \Omega_{\mathrm{F}} - k \ln \Omega_{\mathrm{I}} = k \ln \left(\frac{\Omega_{\mathrm{F}}}{\Omega_{\mathrm{I}}}\right). \tag{63}$$

We know how to determine ΔS by thermodynamic calculation. In order to test the accuracy of Boltzmann's postulate, we should also determine ΔS from the probabilities of initial and final states and compare the two values.

Consider the isothermal expansion of 1.0 mole of an ideal gas from volume V_1 to volume V_2. From our thermodynamic equations

$$\begin{aligned}\Delta S &= \int_1^2 \frac{dq_{\mathrm{rev}}}{T} \\ &= \frac{1}{T}\int_1^2 (dE - dw_{\mathrm{rev}}) = \frac{1}{T}\left(\Delta E - \int_1^2 dw_{\mathrm{rev}}\right).\end{aligned}$$

For an ideal gas, $\Delta E = 0$ for any isothermal process.

$$\Delta S = -\frac{1}{T}\int_1^2 dw = \frac{1}{T}\int_1^2 P\,dV = \int_1^2 \frac{R}{V}\,dV = R \ln \frac{V_2}{V_1} \tag{64}$$

Now consider the calculation of the change in entropy using the Boltzmann postulate and Eq. (63). To use this approach, we need to know Ω_1 and Ω_2, the number of different ways of arranging the 1.0 mole (Avogadro's number N) of gas particles of the system before and after expansion.

Suppose we mentally divide up the total volume V_1 into tiny compartments, each of which has volume v, just large enough to contain only one particle. There would be $V_1/v = n_1$ of these compartments. From Eq. (61),

$$\Omega_1 = n_1!/(n_1 - N)! \tag{65}$$

When n_1 is very large, $n_1!$ is virtually impossible to evaluate. Stirling's approximation

$$\ln n! \simeq n \ln n - n \tag{66}$$

is, however, very accurate when n is large. Therefore

$$\begin{aligned}\ln \Omega_1 &= (n_1 \ln n_1 - n_1) - [(n_1 - N) \ln (n_1 - N) - (n_1 - N)] \\ &= n_1 \ln n_1 - (n_1 - N) \ln (n_1 - N) - N.\end{aligned}$$

Similarly we divide V_2 into compartments of volume v. There will be $n_2 = V_2/v$ of these and

$$\ln \Omega_2 = n_2 \ln n_2 - (n_2 - N) \ln (n_2 - N) - N.$$

From Eq. (63)

$$\begin{aligned}\Delta S &= k \ln (\Omega_2/\Omega_1) = k[\ln \Omega_2 - \ln \Omega_1] \\ &= k[n_2 \ln n_2 - (n_2 - N) \ln (n_2 - N) - N - n_1 \ln n_1 \\ &\quad + (n_1 - N) \ln (n_1 - N) + N)] \\ &= k[n_2 \ln n_2 - n_2 \ln (n_2 - N) + N \ln (n_2 - N) - n_1 \ln n_1 \\ &\quad + n_1 \ln (n_1 - N) - N \ln (n_1 - N)].\end{aligned}$$

Since we are dealing with a dilute gas, n_1 and n_2 are much larger than N; we may write $(n_1 - N) \simeq n_1$ and $(n_2 - N) \simeq n_2$. Then

$$\Delta S = k[N \ln n_2 - N \ln n_1] = Nk \ln (n_2/n_1) = R \ln (n_2/n_1).$$

Recall that $n_1 = V_1/v$ and $n_2 = V_2/v$:

$$\Delta S = R \ln [(V_2/v)/V_1/v)] = R \ln \frac{V_2}{V_1}. \tag{67}$$

Equation (64), calculated for the isothermal expansion of an ideal gas using only macroscopic thermodynamic state variables, and Eq. (67), obtained via Boltzmann's postulate and the postulate of a gas made up of particles, are identical. Often this visualization in terms of probability and disorder will be of primary importance in the successful application of the concept of entropy to chemical problems.

It should be noted that since $\Delta S > 0$ for a spontaneous change in an adiabatic system, all natural processes in such systems proceed in that manner for which $\Omega_F > \Omega_I$ (Eq. 63), that is, from a state of higher order to one of less order. This is the source of the statement that since the universe is an adiabatic system its disorder is constantly increasing, approaching total chaos.

3.10. Microscopic Interpretation of Free Energy

We have seen that free energy can be written as a function of enthalpy and entropy:

$$G = H - TS$$

$$\Delta G = \Delta H - T\,\Delta S. \tag{49}$$

For an isothermal, constant-pressure process involving only pressure-volume work, $\Delta G < 0$ for a natural, spontaneous process. If $\Delta S = 0$, $\Delta H < 0$ for a spontaneous change. Recalling the relationship between enthalpy and energy and the microscopic interpretation of energy, the natural direction is a decrease in energy, a slowing down of molecular motion. If $\Delta H = 0$, $\Delta S > 0$ for a spontaneous change; the natural direction is an increase in the disorder of the system.

In cases in which neither ΔS nor ΔH is zero, a spontaneous natural process may be one of three types. It may proceed to a state of lower energy and higher disorder; it may proceed to a state of higher energy but higher disorder, in which case $T\,\Delta S$ must be greater than ΔH; or it may proceed to a state of greater order and lower energy, in which case ΔH must be large enough in a negative sense to offset the $T\,\Delta S$ term.

The free energy may therefore be interpreted microscopically as a combination of molecular motion and of molecular disorder. The interplay of these two determines the natural direction of a molecular process.

Problems

3.1 Note that ΔS_f° and ΔG_f° for such substances as $O_2(g)$, $H_2(g)$, S(rhombic) and C(graphite) are given as zero in Table 3.1. Explain why.

3.2 (a) Using the data of Table 3.1, calculate ΔS°, ΔG°, and ΔH° for the transformation of 1 mole of ethanol from liquid to vapor under standard conditions. (b) Which form is more stable under standard conditions?

3.3 (a) Calculate ΔS° and ΔG° for our sociologically important reaction of Chapter 2:

$$\underset{\text{glucose}}{C_6H_{12}O_6(s)} \longrightarrow \underset{\text{ethanol}}{2C_2H_5OH(l)} + 2CO_2(g).$$

(b) Use both the entropy criterion and the free-energy criterion to show that this reaction is spontaneous. (This is not an adiabatic process.)

3.4 On a morning when the thermometer reads 50°F, a motorist checks his E-78 tires to 26 lb at his neighborhood service station in preparation for a trip. On the highway when the tire temperature builds up to 120°F, one of the tires blows out like a gunshot. Assuming ideal gas behavior, the C_v of air as 5.05 cal mole^{-1} deg^{-1} at this temperature, and an inside volume of 45 liters for the tire, calculate for the escaping air (a) the entropy change undergone and (b) the work done.

3.5 In Example 14 we dealt with a reaction which appeared to be a reasonable method for producing ethanol:

$$C_2H_4(g) + H_2O(g) \longrightarrow C_2H_5OH(l).$$

Is this in fact a feasible reaction under standard conditions?

3.6 The following reaction is a typical example of the net results of the photosynthetic production of carbohydrates in plants:

$$6CO_2(g) + 6H_2O(l) \longrightarrow 6O_2(g) + \underset{\text{glucose}}{C_6H_{12}O_6(s)}.$$

(a) Calculate the free-energy change, the entropy change, the enthalpy change, and the energy change for this reaction under standard conditions. (b) Is this reaction an allowed process under standard conditions?

3.7 The allotropic forms of a substance may have quite different physical properties. Graphite and diamond certainly do not appear to be composed of the same element, but they are both pure carbon. (a) Using the values for ΔH_f° from Table 2.1 and those for ΔS_f° from Table 3.1, calculate ΔG° for the transition

$$\text{C(graphite)} \longrightarrow \text{C(diamond)}.$$

Confirm this value using the values of ΔG_f° from Table 3.1. (b) Which allotrope of carbon is more stable under standard conditions? Why does the less stable form not change quickly and spontaneously to the more stable form? (This last question will be discussed in depth in the kinetics section of this text.) (c) For obvious reasons, it would be nice to be able to make the form of carbon which is less stable from the more stable form. Since $\Delta G^\circ > 0$ for this process, however, it is forbidden under standard conditions. Can ΔG be reversed so that $\Delta G < 0$ for this conversion by (1) increasing the temperature at constant pressure? (2) decreasing the temperature at constant pressure? (3) increasing the pressure at constant temperature? The density of diamond is 3.51 g cm^{-3}; for graphite, 2.26 g cm^{-3}. (4) decreasing the pressure at constant temperature? Explain your answer to each. If your answer is "yes" to any of the above, calculate the temperature or pressure at which the less stable form becomes the more stable.

3.8 In Problem 2.3, we discussed the following reaction as a possible mechanism for this precipitation of $CaCO_3$ in the thyroid gland:

$$Ca(OH)_2(s) + CO_2(g) \longrightarrow CaCO_3(s) + H_2O(l).$$

(a) Calculate ΔG° for this reaction; ΔG_f° for $Ca(OH)_2$ is -214.33. (b) Is the reaction spontaneous under standard conditions? (c) What is ΔG at 37°C and 0.1 atm?

3.9 In Problem 2.14 we calculated the final temperature when water is formed from 0.05 mole of oxygen and 0.05 mole of hydrogen, mixed at 25°C in the presence of a catalyst in a constant-volume, 1-liter calorimeter. (a) Assuming the process is adiabatic, calculate ΔH, ΔS, and ΔG. (b) Why cannot the ΔG criterion for spontaneity be used? (c) Show that the process is indeed spontaneous by using the ΔS criterion.

3.10 In terms of the microscopic order-disorder interpretation of entropy, explain (a) why the standard entropy of formation of most molecules is negative. (b) why the standard entropies of formation of CO and CO_2 are positive. (c) why the entropy of formation of large molecules (such as glucose) is more negative than that for smaller molecules (such as alanine). (d) why ΔS° for vaporization, for sublimation and for melting is positive. (e) whether a system consisting of two gases at the same temperature and pressure, but separated by a partition which divides the container in half, will undergo an entropy change when the partition is withdrawn. (f) why ΔS° is negative when crystals form from a saturated solution. (g) why, when a pousse-café consisting of 6 alcoholic layers of different colors each floated successively onto the next denser beneath, is allowed to stand, there will be an increase in entropy (of the drink, that is).

3.11 If a single coin is flipped, what is the probability of the head on the coin being up? If two coins are flipped, what is the probability of obtaining two heads? When four are flipped what is the probability of four heads? Obtain a general relation for the probability of *n* heads when *n* coins are flipped.

3.12 The following two reactions occur in biological oxidations, the first in the oxidation of lactic acid and the second in the oxidation of succinic acid:

$$\underset{\text{pyruvic acid}}{CH_3COCOOH} \xrightarrow{\text{enzyme}} \underset{\text{acetaldehyde}}{CH_3CHO} + CO_2$$

$$\underset{\text{fumaric acid}}{HO\overset{\overset{O}{\|}}{C}(CH)_2\overset{\overset{O}{\|}}{C}OH} + H_2O \xrightarrow{\text{enzyme}} \underset{\text{malic acid}}{HO\overset{\overset{O}{\|}}{C}CH_2\overset{\overset{HO}{|}}{C}H\overset{\overset{O}{\|}}{C}OH}$$

Predict which of these reactions will have a positive entropy change and which will have a negative entropy change.

3.13 A fellow offers you a dice game at even money. "You throw this pair of dice," he says, "and the 1's and 2's will work for me, while the 3's, 4's, 5's, and 6's work for you. Every time a 1 or a 2 appears, you pay me; every time *neither* a 1 nor a 2 appears, I pay you." Moral considerations aside, should you take him up on his offer?

4 / Phase Transitions

4.1. Components and Phases

We considered phase transitions briefly in Section 1.16. We now have developed the ideas of thermodynamics sufficiently to undertake a more complete discussion of these phenomena.

The *number of components* is defined as the least number of chemical species necessary to describe the composition of a system. Equivalently, it is the number of chemical species whose concentrations can be independently varied in a system. For instance, the system composed of hydrogen and oxygen gas in equilibrium with steam at an elevated temperature may at first appear to be a three-component (H_2, O_2, H_2O) system. But since the equilibrium constant K equals $[H_2O]^2/[H_2]^2[O_2]$ (where the brackets signify concentration), specifying the concentration of any two of the materials automatically fixes the third. Accordingly, the number of components is two rather than three.

A *phase* is defined as a physically distinct region. The $H_2 + O_2 \rightarrow H_2O$ system described above is a two-component, one-phase (gas) system. Ice in a glass of water is a one-component (H_2O), two-phase (solid and liquid) system.

4.2. The Clapeyron Equation

Consider two pure phases of the same substance in equilibrium (a one-component, two-phase system), for instance $H_2O(\text{liquid}) \rightleftarrows H_2O(\text{vapor})$. If the molar free energy of the liquid at equilibrium were greater than that of the vapor ($\overline{G}_{\text{liq}} > \overline{G}_{\text{vap}}$) the free-energy change for the phase transition liquid $\rightarrow$ vapor would be negative ($\Delta G_{\text{vapn}} = G(\text{vap}) - G(\text{liq}) < 0$), and liquid would change spontaneously to vapor. Conversely if $\overline{G}(\text{liq}) < \overline{G}(\text{vap})$ at equilibrium, ΔG for the transition vapor $\rightarrow$ liquid ($G(\text{liq}) - G(\text{vap})$) would then be negative, and vapor would change spontaneously to liquid. But at equilibrium the relative amounts of liquid and vapor are not changing; the liquid phase is not changing spontaneously to vapor nor is the vapor condensing spontaneously to liquid. Consequently at equilibrium the molar free energy of the liquid and that of the vapor must be equal: $\overline{G}(\text{liq}) = \overline{G}(\text{vap})$; and thus the molar free energy of vaporization $\Delta\overline{G}_{\text{vapn}} = \overline{G}(\text{vap}) - \overline{G}(\text{liq})$ is 0 at equilibrium.

If we now change the temperature of this phase equilibrium by an infinitesimally small amount, we find that the pressure must also change if equilibrium is to be maintained. The system was in equilibrium before the change, then finds a new equilibrium afterward. Originally the molar free energy of the liquid and the vapor were identical; afterward they must again be equal, but at a new value. It follows that the two free energies must change by the same amount, or

$$d\overline{G}(\text{vap}) = d\overline{G}(\text{liq}).$$

For each phase of this one-component system we can write, from Eq. (55),

$$d\overline{G} = \overline{V}\, dP - \overline{S}\, dT, \tag{68}$$

where $\overline{G}$, $\overline{S}$, and $\overline{V}$ are molar quantities. Substituting this into the previous equation, we may write

$$\overline{V}(\text{vap})\, dP - \overline{S}(\text{vap})\, dT = \overline{V}(\text{liq})\, dP - \overline{S}(\text{liq})\, dT$$
$$[\overline{V}(\text{vap}) - \overline{V}(\text{liq})]\, dP = [\overline{S}(\text{vap}) - \overline{S}(\text{liq})]\, dT$$
$$dP/dT = [\overline{S}(\text{vap}) - \overline{S}(\text{liq})]/[\overline{V}(\text{vap}) - \overline{V}(\text{liq})]$$
$$dP/dT = \Delta\overline{S}/\Delta\overline{V}. \tag{69}$$

This relation is known as the *Clapeyron equation.* It tells us how much the pressure must change to maintain equilibrium between two phases if we change the temperature, and vice versa. Although we derived it specifically for liquid-vapor equilibria, the Clapeyron equation applies exactly to all phase transition equilibria.

Example 23 Show that for the equilibria liquid$\rightleftharpoons$vapor and solid$\rightleftharpoons$vapor, the Clapeyron equation can be written in the form

$$d(\ln P)/dT = \Delta\overline{H}/RT^2, \tag{70}$$

known as the *Clausius-Clapeyron* equation.

Answer: Since for a phase equilibrium $\Delta\overline{G} = 0 = \Delta\overline{H} - T\,\Delta\overline{S}$ then $\Delta\overline{S} = \Delta\overline{H}/T$. The molar volume of a liquid or solid is negligible when compared with the molar volume of its vapor (see Section 1.16). Therefore for equilibria involving vapor, $\Delta\overline{V}$ is approximately equal to $\overline{V}_{\text{vap}}$, which in turn is approximately predicted by the ideal gas equation as RT/P.

Making these substitutions in the Clapeyron equation, we obtain

$$dP/dT = \frac{\Delta\overline{H}}{T}\Big/\frac{RT}{P} = P\,\Delta\overline{H}/RT^2,$$

and since $dP/P = d(\ln P)$,

$$d(\ln P)/dT = \Delta\overline{H}/RT^2. \tag{70}$$

Table 4.1 contains values of $\Delta\overline{H}_{\text{trans}}$, $\Delta\overline{S}_{\text{trans}}$, and the temperature at which the transition equilibria exist at 1 atm pressure.

Integrating Eq. (70) between limits, assuming $\Delta\overline{H}$ to be temperature independent,

$$\int_{P_1}^{P_2} d(\ln P) = (\Delta\overline{H}/R)\int_{T_1}^{T_2} T^{-2}\,dT$$

$$\ln(P_2/P_1) = (\Delta\overline{H}/R)(1/T_1 - 1/T_2). \tag{71}$$

We can use this equation to predict the equilibrium vapor pressure P_2 of a pure liquid at a desired temperature T_2 from its known ethalpy or latent heat of vaporization $\Delta\overline{H}_{\text{vapn}}$ and its known vapor pressure P_1 at some other temperature T_1 (for example, its normal boiling point, where $P = 1$ atm). Equation

Table 4.1. Selected Values of Enthalpy and Entropy Changes and Normal Transition Temperatures at 1 Atm

Vaporization (liquid → gas):

Substance	$\Delta\overline{H}_{\text{vapn}}$, kcal mole^{-1}	T_b, °C	$\Delta\overline{S}_{\text{vapn}}$, cal mole^{-1} deg^{-1}
O_2	1.630	−182.97	18.07
CH_4	1.955	−161.49	17.51
H_2O	9.717	100.0	26.04
NH_3	5.581	−33.43	23.28
CCl_4	7.17	76.7	20.3
$CHCl_3$	7.02	61.2	20.8
$(CH_3)_2CO$(acetone)	7.22	56.2	21.90
C_6H_6(benzene)	7.36	80.1	20.83
$(C_2H_5)_2O$(ether)	6.21	34.5	20.18

Fusion (solid → liquid):

Substance	$\Delta\overline{H}_{\text{fus}}$, kcal mole^{-1}	T_m, °C	$\Delta\overline{S}_{\text{fus}}$, cal mole^{-1} deg^{-1}
H_2O	1.436	0.0	5.26
NH_3	1.351	−77.76	6.91
CCl_4	0.6	−22.9	2.4
$CHCl_3$	2.2	−63.5	10.5
Fe	3.6	1535	2.0

National Bureau of Standards Circular 500, (1952) and Norbert A. Lange, *Handbook of Chemistry*, 10th ed., New York: McGraw-Hill (1967).

(71) also applies to the vapor pressure of solids; in that case ΔH is the latent heat or enthalpy of sublimation. Note further that, whereas the Clapeyron equation (69) was exact, the integrated Clausius-Clapeyron (71) is only a very good approximation, due to the three inexact assumptions made in its derivation.

Equation (71) is not applicable (but Eq. 69 is) to equilibria between two condensed (that is, nongaseous) phases, such as solid $\rightleftarrows$ liquid in fusion, or $\text{solid}_1 \rightleftarrows \text{solid}_2$ in allotropic transition.

Example 24 Mercury, either ingested or inhaled, is a serious heavy-metal poison, ultimately producing (among other effects) irreversible damage to the kidneys. Continued breathing of air containing as little as 0.1 mg per cubic meter is considered the threshold limit. Suppose that spilled mercury, lingering in floor cracks and inaccessible areas after clean-up, introduces only 10% of its equilibrium vapor pressure into the air. Calculate whether, at room temperature of 25°C, this exceeds the acceptable limit. Take the boiling point of mercury as 357°C and its average enthalpy of vaporization as 14.56 kcal mole^{-1}.

Answer: Substituting in Eq. (71) to get the vapor pressure of mercury at 25°C gives

$$2.30 \log \left(\frac{760 \text{ torr}}{p_1}\right) = \frac{14560 \text{ cal}}{1.99 \text{ cal deg}^{-1} \text{ mole}^{-1}} \left(\frac{1}{298^\circ} - \frac{1}{630^\circ}\right)$$

$$\log 760 - \log p_1 = \frac{14560}{2.30 \times 1.99} (33.56 - 15.87) \times 10^{-4}$$

$$\log p_1 = 2.881 - (3180)(17.69 \times 10^{-4})$$

$$\log p_1 = -2.744 = -3 + 0.256$$

$$p_1 = \text{antilog}\,(-3 + 0.256) = 0.00180 \text{ torr at equilibrium.}$$

The ideal gas equation $PV = (W/M)RT$ will then give the weight W of mercury in a cubic meter of air saturated with this monatomic vapor:

$$W = \frac{MPV}{RT} = \frac{(201 \text{ g mole}^{-1})(0.00180 \text{ torr})(1000 \text{ liters m}^{-3})}{(760 \text{ torr atm}^{-1})(0.0821 \text{ liter-atm deg}^{-1} \text{ mole}^{-1})(298 \text{ deg})}$$

$$W = 0.0195 \text{ g} = 19.5 \text{ mg m}^{-3} \text{ at equilibrium.}$$

Thus if, because of ventilation, the exposed mercury establishes only 10% of its equilibrium vapor pressure in the room, the resulting 1.95 mg m^{-3} is still about 20 times the tolerance limit.

Equation (71) can also be used conversely to determine latent heats. Most tabulated heats of vaporization are in fact the result of measuring vapor pressures at successive temperatures, then solving for ΔH either graphically or analytically. The basis for the graphical method is obtained by integrating Eq. (70) completely and converting to ordinary logs:

$$(\log P) = \frac{-\Delta H}{2.3R}\left(\frac{1}{T}\right) + \frac{C}{2.3},$$

where C is the integration constant. Semi-log paper is then used to plot P on the logarithmic ordinate against the reciprocal of the kelvin temperature on the abscissa. Ideally a straight line will result; from the above equation its slope is $-\Delta H/2.3R$. Therefore

$$\Delta H = -2.3R \times \text{slope}.$$

Example 25 A chemical handbook gives the vapor pressure of toluene at various temperatures, as shown in the two columns on the left below. Use the graphical method to find the heat of vaporization of toluene in this temperature range.

Vapor pressure, mm Hg	*Centigrade temperature,* °C	*Kelvin temperature,* °K	*Reciprocal kelvin temperature,* $10^4/T$
10	6.4°C	279.4°K	35.79
60	40.3	313.3	31.92
200	69.5	342.5	29.20
600	102.5	375.5	26.63
1200	127.5	400.5	24.97

Answer: Proceeding systematically, we fill in the last two columns. We then use three-cycle semi-log paper to plot the first column on the logarithmic ordinate against the last column on the abscissa. After locating the five data points, we draw the *best straight line* through and among these points (see Fig. 4.1).

The numerical slope, $(y_2 - y_1)/(x_2 - x_1)$, is best obtained by choosing a whole log cycle as the interval, thus making $y_2 - y_1 = 1.0$. The values of x_2 and x_1 are then read from the graph (dotted lines) as 25.47×10^{-4} and 30.68×10^{-4}, respectively. The slope is therefore $1/(25.47 - 30.68) \times 10^{-4} = 10^4/(-5.21) = -1920$. The latent heat of vaporization of toluene is now determined:

$$\Delta H = -(2.30R)(\text{slope}) = -(2.30)(1.99)(-1920) = 8789 \text{ cal mole}^{-1}.$$

(The handbook value is 8402 cal mole^{-1}. Our $+4.6\%$ discrepancy is principally due to experimental errors, approximations involved in deriving the equation in the form in which we used it, variation of ΔH with temperature, and errors in visually estimating the best straight line.)

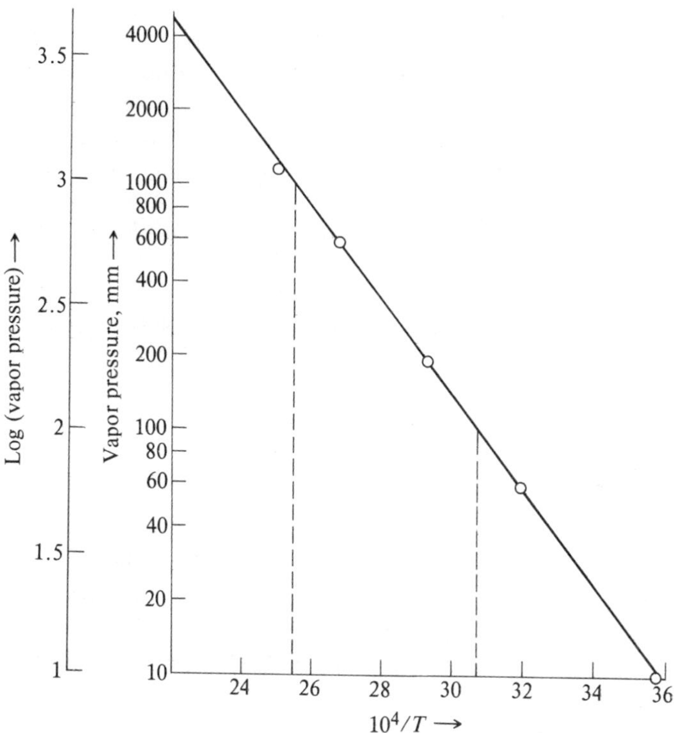

Fig. 4.1. Toluene.

4.3. The Molecular View of Phase Transitions

Our microscopic view of energy (Section 1.12) and of entropy (Section 3.9) can also be used to interpret phase transitions on the molecular level.

The molecules of the vapor phase are moving randomly within the container. There is great disorder. The motion is rapid and the attractive forces between molecules are not great enough to affect significantly their relative motion. As the temperature is lowered, the molecular translational energy decreases until a point is reached where the attractive forces between molecules are able to cause a coalescence into the liquid phase. The molecules are still moving about; they still possess translational energy, but they are now much closer to each other. Thus the volume of the liquid phase is smaller than that of the vapor phase. Moreover the now-effective intermolecular forces cause the molecules to arrange themselves into ordered groups. This grouping is transitory and is constantly changing, but at any given instant there is considerably more order in the liquid phase than there was in the vapor. Another way of saying this is that the entropy (disorder) of the material decreases when liquefaction occurs: $\overline{S}(\text{g}) > \overline{S}(\text{l})$, $\Delta\overline{S}_{\text{vapn}} > 0$; $\overline{H}(\text{g}) > \overline{H}(\text{l})$, $\Delta\overline{H}_{\text{vapn}} > 0$.

If the temperature is lowered further, the translational energy decreases further until a second point is reached at which the attractive forces between molecules succeed in stopping the translational motion completely, and the molecules arrange themselves in fixed, almost totally three-dimensionally ordered patterns relative to each other. A decrease in volume usually (but not always) accompanies this solidification as the molecules pack more tightly together. A decrease of entropy also occurs due to the increased order: $\overline{S}(\text{l}) > \overline{S}(\text{s})$, $\Delta\overline{S}_{\text{fus}} > 0$; $\overline{H}(\text{l}) > \overline{H}(\text{s})$, $\Delta\overline{H}_{\text{fus}} > 0$.

Similar logic may be applied to the process of sublimation: $\Delta\overline{S}_{\text{sub}} > 0$, $\Delta\overline{H}_{\text{sub}} > 0$.

4.4. Phase Diagrams

If we plot the equilibrium pressure against the equilibrium temperature for a liquid ⇄ vapor phase equilibrium, Eq. (69) indicates that the slope of the curve will be $\Delta\overline{S}/\Delta\overline{V}$. Figure 4.2 is such a plot. Each point on the solid line corresponds to an equilibrium state. We learned in Section 4.3 that both $\Delta\overline{V}$ and $\Delta\overline{S}$ for such a transition are positive. The slope of this curve $(\Delta\overline{S}/\Delta\overline{V})$ will therefore be positive.

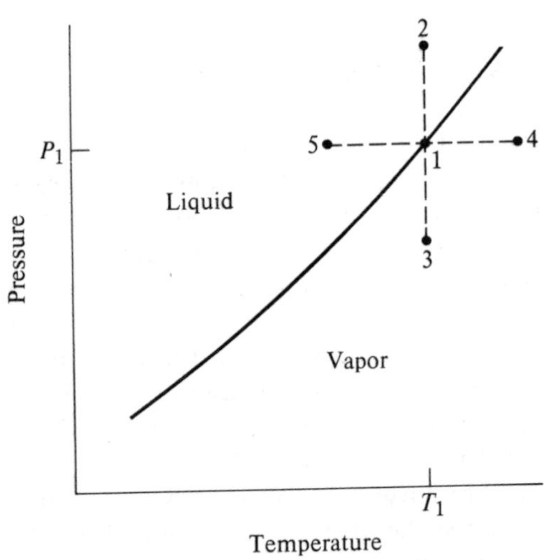

Figure 4.2

At any given temperature T_1, if we increase the pressure above the equilibrium pressure P_1 (from point 1 to point 2 in Fig. 4.2), we no longer have an equilibrium state; we move off the equilibrium line. The change in pressure will cause the free energy of the vapor and that of the liquid to change by different amounts. This can be demonstrated by recalling

$$d\overline{G} = \overline{V}\,dP - \overline{S}\,dT, \tag{68}$$

from which we obtain under constant-temperature conditions

$$\left(\frac{\partial \overline{G}}{\partial P}\right)_T = \overline{V}. \tag{72}$$

Thus $(\partial \overline{G}_{\text{vap}}/\partial P)_T = \overline{V}_{\text{vap}}$ and $(\partial \overline{G}_{\text{liq}}/\partial P)_T = \overline{V}_{\text{liq}}$. Since $\overline{V}_{\text{vap}} > \overline{V}_{\text{liq}}$,

$$(\partial \overline{G}_{\text{vap}}/\partial P)_T > (\partial \overline{G}_{\text{liq}}/\partial P)_T,$$

and for a given change in pressure the free energy of the vapor will increase by a greater amount than the free energy of the liquid:

$$d\overline{G}_{\text{vap}} > d\overline{G}_{\text{liq}}.$$

Since the molar free energies of the two phases were equal in the equilibrium state ($\overline{G}_{\text{vap}_1} = \overline{G}_{\text{liq}_1}$), an increase in pressure at constant temperature will result in

$$\overline{G}_{\text{vap}_2} = \overline{G}_{\text{vap}_1} + d\overline{G}_{\text{vap}} > \overline{G}_{\text{liq}_1} + d\overline{G}_{\text{liq}} = \overline{G}_{\text{liq}_2}.$$

With the molar free energy of the vapor now greater than the molar free energy of the liquid, ΔG for the phase transition vapor → liquid is negative ($\overline{G}_{\text{liq}_2} - \overline{G}_{\text{vap}_2} < 0$): the material will spontaneously change from vapor to liquid if isothermal conditions are maintained.

Alternatively, if the pressure is lowered (from point 1 to point 3 in Fig. 4.2), dP is negative. This requires $d\overline{G}$ also to be negative in Eq. (72) since $\overline{V}$ is always positive. Consequently the value of $d\overline{G}_{\text{vap}}$ is now less than (more negative than) that of $d\overline{G}_{\text{liq}}$, and $\overline{G}_{\text{vap}}$ is now less than $\overline{G}_{\text{liq}}$. Thus ΔG for the process liquid → vapor is negative, and the process is spontaneous: all the material becomes vapor if isothermal conditions are maintained.

A similar argument based on Eq. (68) at constant pressure,

$$\left(\frac{\partial \overline{G}}{\partial T}\right)_P = -\overline{S}, \tag{73}$$

demonstrates that since $S(\text{g}) > S(\text{l})$ an increase in temperature at a given pressure (point 1 to point 4 in Fig. 4.2) will destroy the phase equilibrium and cause all the material to become vapor, while a decrease in temperature (point 1 to point 5 in Fig. 4.2) will result in liquid.

Similar plots can be made for the solid ⇄ liquid and solid ⇄ vapor equilibria for any given material. If we plot all three phase equilibria on a single pressure-versus-temperature graph, we obtain the *phase diagram* for the material (Fig. 4.3). Point 1 is the *triple point* of the material. At this point all three phases are *simultaneously* in equilibrium: $\overline{G}_{\text{liq}} = \overline{G}_{\text{vap}} = \overline{G}_{\text{sol}}$. There is only one such point for any one-component system in three phases. Point 2 lies on the liquid-vapor equilibrium line: $\overline{G}_{\text{liq}} = \overline{G}_{\text{vap}} < \overline{G}_{\text{sol}}$. Point 3 is on the liquid-solid equilibrium line: $\overline{G}_{\text{liq}} = \overline{G}_{\text{sol}} < \overline{G}_{\text{vap}}$. Point 4 is on the solid-vapor equilibrium line: $\overline{G}_{\text{sol}} = \overline{G}_{\text{vap}} < \overline{G}_{\text{liq}}$. At point 5, $\overline{G}_{\text{vap}} < \overline{G}_{\text{liq}}$ and $\overline{G}_{\text{vap}} < \overline{G}_{\text{sol}}$; point 6: $\overline{G}_{\text{liq}} < \overline{G}_{\text{vap}}$ and $\overline{G}_{\text{sol}}$; point 7: $\overline{G}_{\text{sol}} < \overline{G}_{\text{vap}}$ and $\overline{G}_{\text{liq}}$. If we had a solid or a vapor held at a temperature and pressure corresponding to point 6 it would spontaneously liquefy; a liquid or solid held at point 5 would spontaneously vaporize; a liquid or vapor at point 7 would spontaneously solidify.

All three of the equilibrium P-versus-T lines generally have positive slopes since $\Delta\overline{V}$ and $\Delta\overline{S}$ are both usually positive for liquid → vapor, solid → vapor, and solid → liquid. Water is an exception. Liquid water is more dense than ice at equilibrium pressures and temperatures, and thus $\Delta V < 0$ for solid → liquid. For water the slope $\Delta\overline{S}/\Delta\overline{V}$ is therefore negative for the melting/freezing process. The phase diagram for water is given in Fig. 4.4.

Example 26 An ice skater glides smoothly over the ice because there is a thin layer of liquid water between the skate blades and the solid ice. This layer of liquid results from the melting of the ice caused by friction between the skate blade on the ice *and* from the lowering of the melting point of ice with the increase of pressure due to the skater's weight on the skate blades, which are designed so that only a very small area is in contact with the ice at any instant.

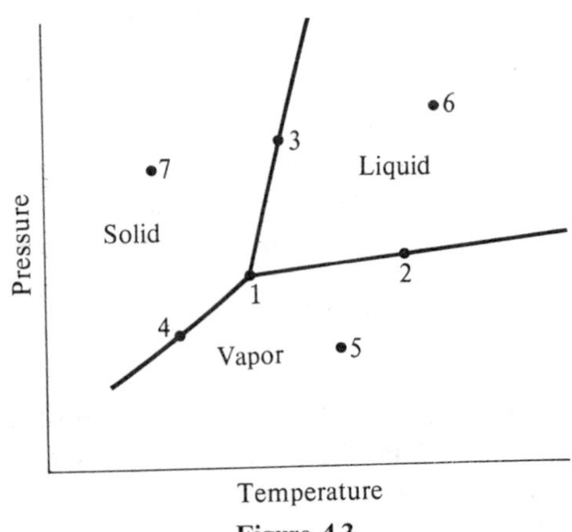

Figure 4.3

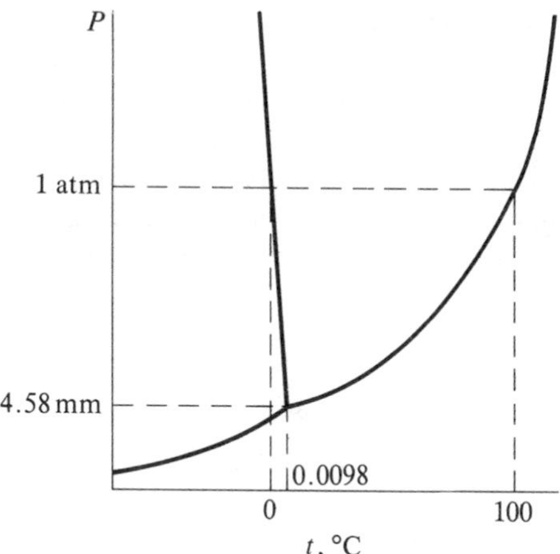

Fig. 4.4. Phase diagram for water. Adapted from G. W. Castellan, *Physical Chemistry*, 2nd ed. (1971), Addison-Wesley, Reading, Mass.

If we assume that the latter is totally responsible for the melting, what would be the melting point of ice under the pressure of a 150-lb man (68.0 kg) on skates whose contact area is 10^{-2} cm^2? The density of the liquid at 0°C is 1.000 g cm^{-3}; the density of the solid at 0°C is 0.917 g cm^{-3}; $\Delta\overline{H}_{\text{fus}} = 1.44$ kcal mole^{-1}.

Answer: The pressure in addition to normal atmospheric pressure would be (mass) × (acceleration due to gravity)/(area) = (68000 g)(981 cm sec^{-2})/(10^{-2} cm^2) = 6.67×10^9 dyne cm^{-2} = 6580 atm. The molar volume of the liquid would be the molecular weight ÷ the density = 18×10^{-3} liter. Similarly, for the solid, $\overline{V} = 19.63 \times 10^{-3}$ liter. Hence $\Delta\overline{V} = -1.63 \times 10^{-3}$ for solid → liquid.

$$dP/dT = \Delta\overline{S}/\Delta\overline{V} = \Delta\overline{H}/T\,\Delta\overline{V}$$

$$\int_{1\text{ atm}}^{6581\text{ atm}} (\Delta\overline{V}/\Delta\overline{H})\,dP = \int_{273}^{T} (1/T)\,dT$$

$$(\Delta\overline{V}/\Delta\overline{H})\,\Delta P = \ln(T/273) = 2.30\log(T/273)$$

$$\log(T/273) = (1/2.30)(-1.63 \times 10^{-3}\text{ liter})(6580\text{ atm})$$

$$\times (24.2 \times 10^{-3}\text{ kcal liter}^{-1}\text{ atm}^{-1})/(1.44\text{ kcal})$$

$$= -0.078 = -1 + 0.922$$

$$T/273 = 0.835$$

$$T = 228°\text{K}.$$

4.5. Phase Diagrams for Allotropic Materials

An added complication occurs when a substance has more than one solid form. A good example of this phenomenon is sulfur, which possesses two crystalline forms: monoclinic and rhombic. The phase diagram for sulfur is given in Fig. 4.5. Each of the solid lines is an equilibrium line. The line separating the monoclinic and rhombic regions contains the points of equilibrium S(rhombic) ⇄ S(monoclinic).

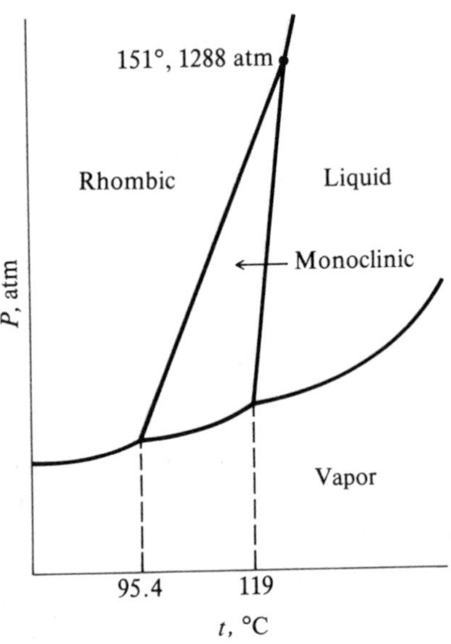

Fig. 4.5. Phase diagram for sulfur. Adapted from G. W. Castellan, *Physical Chemistry*, 2nd ed. (1971), Addison-Wesley, Reading, Mass.

4.6. Phase Correlations

In the phase diagram of water (Fig. 4.4), the liquid ⇄ vapor equilibrium line is a plot of the vapor pressure of water as a function of temperature. The vapor pressure of pure water is predetermined by its temperature alone; that is, for each temperature there is only one vapor pressure. Such a system is said to be *univariant*, or to have one *degree of freedom*. This means that just one variable of state may be manipulated (within reasonable limits), and all other variables must adjust their values to conform. A similar situation exists for the solid ⇄ vapor equilibrium line which records the vapor pressure of ice, and for the liquid ⇄ solid equilibrium line which pictures the influence of pressure on the melting of ice.

At the triple point, however, with all three phases present, the system is fixed, or *invariant*; the temperature must be 273.16°K and the pressure 4.58 torr, corresponding to the coordinates of the triple point. If on this system we should now impose an arbitrary change in either of these values, one of the three phases will disappear. For example, if we raise the temperature slightly we lose the solid phase; raise the pressure and we lose the vapor phase; lower the pressure (or the temperature) and we lose the liquid phase.

In line with these observations we may define an invariant system as one in which no variable can be changed without an accompanying loss of a phase. The *number of degrees of freedom* $\mathscr{F}$ of a system denotes the number of independent variables that must be fixed before the system becomes fixed or invariant. If we reflect on the observed behavior of water, we might hit upon the relation that $\mathscr{P} + \mathscr{F} = 3$, where $\mathscr{P}$ is the number of phases present at equilibrium. This is also true for all other one-component systems. If liquid water alone is present, then $\mathscr{P} = 1$ and $\mathscr{F} = 2$, meaning that with two degrees of freedom we may change both the temperature and the pressure, within bounds, without producing a phase change. Since any system must exist in at least one phase, then $\mathscr{F} = 2$ represents the greatest possible degree of freedom for such a system.

Suppose we now complicate matters by bringing a second component into the system, say by dissolving some glucose in the water. This addition of a solute introduces concentration as a new variable, since the proportions we use may be manipulated. It also increases the number of conceivable phases. Moreover, we may add to the water both some glucose and some alcohol. Clearly, predicting the variance and the phases of multi-component systems can become very complex.

4.7. The Phase Rule

A generalization of great usefulness in such phase equilibria is known as the *phase rule*. It may be written very simply as

$$\mathscr{P} + \mathscr{F} = \mathscr{C} + 2,$$

where $\mathscr{C}$ is the number of components in the system (see Section 4.1). Obviously pure water is a one-component system, resulting in our previous relation that $\mathscr{P} + \mathscr{F} = \mathscr{C} + 2 = 3$. The glucose solution, however, is a two-component system, for which $\mathscr{P} + \mathscr{F} = 4$. Here (since there must be at least one phase present) the maximum degrees of freedom would be three: temperature, pressure, and composition. On the other hand an invariant system would allow four phases in simultaneous equilibrium, as a maximum. Taking the general case of A and B as the two components, five phases can be envisioned: solid A, solid B, liquid A, liquid B, and vapor. (There can be but one vapor phase, since all gases blend with each other homogeneously.) The phase rule says, however, that these five phases cannot coexist; only four. An example of such a four-phase invariant system is the familiar ice-and-salt freezing mixture. Here the four phases are solid salt, solid ice, saturated brine, and vapor. Regardless of their relative amounts, if all four are present at equilibrium the system is "fixed": temperature $-21.13°C$, pressure 0.69 torr, and brine 23.3% salt.

In addition to the types of equilibria illustrated, the phase rule is of great utility in understanding the formation and properties of alloys, the behavior of azeotropes, the crystallization and vapor pressure of salt hydrates, and many other multi-phase systems. It is invaluable to geologists in interpreting the layers of salt beds, formed when prehistoric seas reached saturation and dried up.

Problems

4.1 Give the number of components in each of the following systems and identify the phases present: (a) one ice cube in a glass of water, (b) ten ice cubes in a glass of water, (c) ten ice cubes in a glass of sweetened water, (d) one ice cube in a glass of sweetened water with surplus sugar crystals in the bottom, (e) water and octane (no air) in a stoppered flask, (f) a freshly charged seltzer bottle, (g) the pousse-café of Problem 3.10(g).

4.2 The latent heat of vaporization of ammonia at its normal boiling point ($-33.3°C$) is 327.4 cal g^{-1}. Calculate the molar entropy of vaporization at this temperature and pressure.

4.3 Refrigerators and air conditioners work in a cycle; a motor-driven pump causes a refrigerant substance to evaporate inside the structure, picking up enthalpy of vaporization as it does so, then to condense outside the structure, releasing its enthalpy to the surroundings. An ordinary home refrigerator may contain as refrigerant 150 g of Freon-12, CCl_2F_2, which has an enthalpy of vaporization of 40.0 cal g^{-1}. If the ice-maker unit in such a refrigerator takes in water at 60°F and ejects ice cubes at 20°F, what will be the minimum number of cycles the Freon charge must make to turn out 3.0 lb of ice?

4.4 Calculate (a) the work done when 75 liters of steam at 100°C and 400 torr is isothermally compressed by a slow-moving piston until the vapor is completely liquefied. Calculate also (b) ΔH, (c) ΔS, and (d) ΔG for this process.

4.5 A liquid boils when its vapor pressure is equal to atmospheric pressure. Below this temperature the vapor pressure (which is a direct measure of the ability of molecules to escape into the vapor phase) is less than the opposing atmospheric pressure. Above the boiling point the vapor pressure of the liquid is greater than atmospheric pressure, and the liquid spontaneously vaporizes. The normal boiling point of a liquid is defined as the temperature at which its vapor pressure equals the atmospheric pressure at sea level (1 atm or 760 torr, the pressure exerted by a column of mercury 760 mm high). What is the boiling point of water at LeConte Lodge, located 6000 ft above sea level on top of Mt. LeConte in the Smoky Mountain National Park, where atmospheric pressure is 659 torr?

4.6 A simple and readily available technique for producing sub-freezing temperatures in small-scale experiments (such as determining the freezing point of milk samples to test their water content) is to use a water aspirator to draw a stream of air through liquid ether causing the ether to vaporize, absorbing heat from the surroundings. Suppose a device using this method is loaded at 25°C with 100 ml of ether, in which is suspended a test tube containing 12.0 ml of water to be frozen. Assuming perfect heat transfer, how much liquid ether will remain at the time the water is just completely frozen? In a blank run with the test tube empty, 6.0 ml of ether was evaporated in chilling the vessel and ether from 25° to 0°C. The enthalpy of vaporization of ether may be taken as 6.22 kcal mole^{-1}, its molecular weight as 74.12 g mole^{-1}, and its density as 0.708 g ml^{-1}.

4.7 In our discussion of Fig. 4.3, we stated that a liquid or vapor held at point 7 (the temperature and pressure is kept constant at the values of point 7) would spontaneously solidify. Suppose, however, 10 g of water were supercooled to -10°C and *then* allowed to freeze adiabatically. (a) What would the final state of this system be? The specific heat of ice is 0.5 cal deg^{-1} g^{-1}. (b) Calculate ΔH and ΔS for this process and prove that the process is spontaneous.

4.8 It is observed that the entropy of vaporization for many substances at their normal boiling points is approximately 21 cal deg^{-1} mole^{-1} (Trouton's rule). Deviations from this value may be explained by attributing to the molecules in the liquid phase more or less order than is the case for "normal" molecules for which $\Delta S_{\text{vapn}} = \Delta H_{\text{vapn}}/T = 21$. For water, ΔS_{vapn} is 26.04 cal deg^{-1} mole^{-1}. Is the liquid phase of water more ordered or less ordered than those substances which obey Trouton's rule? (This effect is due to *hydrogen bonding*, to be discussed in Chapter 17.)

4.9 Use Trouton's rule to predict the latent heat of vaporization of formaldehyde whose normal boiling point is -21°C.

4.10 (a) The equilibrium vapor pressure of ice at -5°C is 3.01 torr. A 10-g cube of ice is placed in a 10-liter container which has been evacuated (that is, a vacuum exists in the container). If the container and its contents are kept at -5°C, how much of the ice cube will evaporate? (b) The equilibrium vapor pressure of liquid water at 50°C is 92.5 torr. If 10 ml of water is placed in an evacuated 10-liter container, how much of the liquid will evaporate if the container and its contents are kept at 50°C? (c) Trace the course of these two processes on a phase diagram of water (Fig. 4.4).

4.11 Compact stoves which burn butane gas are becoming popular with backpackers. The butane is supplied from a canister of liquid butane. (a) What is the minimum pressure exerted on the butane in the container to keep it a liquid at 25°C if the normal boiling point of butane is -0.5°C and the latent heat of vaporization is 91.5 cal g^{-1}?

(b) At what temperature would the butane stove become useless to a backpacker on top of Mt. LeConte (see Problem 4.5)? In other words, below what temperature would butane not spontaneously vaporize under 659 torr pressure?

4.12 (a) Show that the Clapeyron equation, when applied to fusion in the near vicinity of 1.0 atm, predicts the influence of pressure on the melting point as

$$\frac{\Delta T}{\Delta P} = \frac{T(\bar{V}_{\text{liq}} - \bar{V}_{\text{sol}})}{\Delta H_f}.$$

Predict $\Delta T/\Delta P$, the change in melting point per atmosphere of applied pressure (b) for water, and (c) for some other substance of your choice. Do the same (d) for the rhombic-monoclinic transition temperature of sulfur, whose normal value is 95.6°C and for which $\Delta H = 2.52$ cal g^{-1}.

Pertinent density values are

H_2O(l)	1.0 g cm^{-3}
H_2O(s)	0.92 g cm^{-3}
S(rhombic)	2.07 g cm^{-3}
S(monoclinic)	1.96 g cm^{-3}

4.13 An advanced student is redistilling some high-purity toluene on a day when the barometer reads 743 torr. If the handbook gives the normal boiling point as 110.6°C, what boiling temperature may be expected? For toluene, $\Delta\bar{H}_{\text{vapn}} = 8000$ cal mole^{-1}.

4.14 Four forms of sulfur are well known: rhombic, monoclinic, liquid, and vapor. All four forms cannot, however, coexist in a stable system. Why?

4.15 A closed desiccator held at a fixed temperature and initially evacuated contains a mass of copper sulfate crystals which partially effloresce according to the equation

$$CuSO_4 \cdot 5H_2O(s) \rightleftharpoons CuSO_4 \cdot 3H_2O(s) + 2H_2O(g).$$

Under the circumstances described (a) Will the pressure inside be constant and predictable? (b) Which, if either, of the two solids will disappear on long standing in an open container? Why?

4.16 For each of the following equilibrium systems decide the number of components and the number of degrees of freedom, and identify which variables of state the latter might conveniently be. (a) A closed 1-liter evacuated flask into which has been introduced 250 ml of water and 250 ml of benzene. Each liquid is appreciably soluble in the other. (b) A small amount of iodine is added to the system in (a). Iodine is somewhat soluble in both liquids, but more so in the benzene layer. (c) Further iodine is added to the system in (b) until some iodine crystals remain undissolved. (d) The temperature of the system in (c) is lowered, causing half the benzene to freeze. (e) Some ice cubes are added to the system in (d).

See also Problems 1.14, 1.16, 2.10, 2.11, 2.12, 3.2, 3.7, 3.10(d) and 3.10(f).

5 / Chemical Potential; Reactions in the Gas Phase

5.1. Definition

Before we can extend our thermodynamic treatment to chemical applications of even greater interest, we need to introduce a most important intensive variable, the *chemical potential* μ or *partial molar* (or *molal*) *free energy*.

Extensive state functions (E, H, G, S, etc.) are functions of the amount of material present. At any instant during the course of a reaction $A \rightarrow F$, the total free energy of the system is

$$G_{\text{system}} = n_A \overline{G}_A + n_F \overline{G}_F,$$

where n is the number of moles of a given substance present at that instant and $\overline{G}$ is the molar free energy of that substance under the conditions prevailing in the reacting system at that instant. For example, in the gas phase reaction in which carbon monoxide is removed from the air by water vapor:

$$CO(g) + H_2O(g) \longrightarrow CO_2(g) + H_2(g),$$

the total free energy at a particular instant during the course of the reaction will be

$$G = n_{CO}\overline{G}_{CO} + n_{H_2O}\overline{G}_{H_2O} + n_{CO_2}\overline{G}_{CO_2} + n_{H_2}\overline{G}_{H_2}.$$

Our expression for the total free energy of the system should therefore explicitly include the amounts of reactants and products in addition to the pressure and temperature:

$$G_{\text{system}} = f(P, T, n_A, n_F),$$

and similarly for other functions of state. For our example,

$$G = f(P, T, n_{CO}, n_{H_2O}, n_{CO_2}, n_{H_2}).$$

Since G is a function of state we can write the exact differential

$$dG = (\partial G/\partial P)_{T,\, n_A,\, n_F}\, dP + (\partial G/\partial T)_{P,\, n_A,\, n_F}\, dT + (\partial G/\partial n_A)_{T,\, P,\, n_F}\, dn_A + (\partial G/\partial n_F)_{T,\, P,\, n_A}\, dn_F. \qquad (74)$$

If the amounts of A and F remain constant, $-dn_A = dn_F = 0$ and Eq. (74) reduces to Eq. (68).

The *chemical potential* $\mu(A)$ of a substance A is defined as

$$\mu(A) = (\partial G/\partial n_A)_{T,\, P,\, n_j}, \qquad (75)$$

where the subscript n_j means that the amounts of all substances other than substance A are held constant. Thus for the reaction mentioned above,

$$\mu(CO) = (\partial G/\partial n_{CO})_{T,\, P,\, n_{H_2O},\, n_{CO_2},\, n_{H_2}},$$

$$\mu(H_2O) = (\partial G/\partial n_{H_2O})_{T,\, P,\, n_{CO},\, n_{CO_2},\, n_{H_2}}, \text{ etc.}$$

If T and P are held constant during the reaction, Eq. (74) becomes

$$dG = \mu(\mathrm{A})\, dn_{\mathrm{A}} + \mu(\mathrm{F})\, dn_{\mathrm{F}}. \tag{76}$$

For our CO-scavenging reaction,

$$dG = \mu(\mathrm{CO})\, dn_{\mathrm{CO}} + \mu(\mathrm{H_2O})\, dn_{\mathrm{H_2O}} + \mu(\mathrm{CO_2})\, dn_{\mathrm{CO_2}} + \mu(\mathrm{H_2})\, dn_{\mathrm{H_2}}.$$

Similarly, for a general reaction $a\mathrm{A} + b\mathrm{B} \to f\mathrm{F} + g\mathrm{G}$, at any instant ($T$ and P constant)

$$dG = \mu(\mathrm{A})\, dn_{\mathrm{A}} + \mu(\mathrm{B})\, dn_{\mathrm{B}} + \mu(\mathrm{F})\, dn_{\mathrm{F}} + \mu(\mathrm{G})\, dn_{\mathrm{G}}. \tag{77}$$

5.2. General Relationship among Chemical Potentials under Equilibrium Conditions

Suppose that in our general reaction $a\mathrm{A} + b\mathrm{B} \to f\mathrm{F} + g\mathrm{G}$, an infinitesimal amount of rightward reaction, dn, occurs. Because of the reaction stoichiometry $a\, dn$ moles of A and $b\, dn$ moles of B disappear while $f\, dn$ moles of F and $g\, dn$ moles of G appear. Then

$$\begin{aligned} dn_{\mathrm{A}} &= -a\, dn \\ dn_{\mathrm{B}} &= -b\, dn \\ dn_{\mathrm{F}} &= f\, dn \\ dn_{\mathrm{G}} &= g\, dn, \end{aligned}$$

and Eq. (77) becomes

$$dG = -a\mu(\mathrm{A})\, dn - b\mu(\mathrm{B})\, dn + f\mu(\mathrm{F})\, dn + g\mu(\mathrm{G})\, dn. \tag{78}$$

At equilibrium if no net work is involved $dG = 0$ and

$$-a\mu(\mathrm{A})\, dn - b\mu(\mathrm{B})\, dn + f\mu(\mathrm{F})\, dn + g\mu(\mathrm{G})\, dn = 0$$

$$a\mu(\mathrm{A}) + b\mu(\mathrm{B}) = f\mu(\mathrm{F}) + g\mu(\mathrm{G}). \tag{79}$$

Thus a criterion for equilibrium for a chemical reaction under conditions of constant temperature and pressure and with only pressure-volume work permitted (no net work) is that the sum of the stoichiometrically weighted chemical potentials of the products is equal to the sum of the stoichiometrically weighted chemical potentials of the reactants:

$$\sum_{\mathrm{re}} c\mu = \sum_{\mathrm{pr}} c\mu, \tag{80}$$

where c is the stoichiometric coefficient of the substance in the reaction equation. When our carbon monoxide reaction ($\mathrm{CO} + \mathrm{H_2O} \to \mathrm{CO_2} + \mathrm{H_2}$) reaches equilibrium

$$\mu(\mathrm{CO}) + \mu(\mathrm{H_2O}) = \mu(\mathrm{CO_2}) + \mu(\mathrm{H_2}).$$

5.3. General Relationship among Chemical Potentials under Nonequilibrium Conditions

Now let us look at the situation under conditions other than equilibrium. At constant temperature and pressure Eq. (78) holds for our general reaction. For a spontaneous process in which only pressure-volume work is involved $\Delta G < 0$ and

$$-a\mu(\mathrm{A})\,dn - b\mu(\mathrm{B})\,dn + f\mu(\mathrm{F})\,dn + g\mu(\mathrm{G})\,dn < 0$$

or

$$[f\mu(\mathrm{F}) + g\mu(\mathrm{G})]\,dn < [a\mu(\mathrm{A}) + b\mu(\mathrm{B})]\,dn,$$

which can also be written as

$$\{[f\mu(\mathrm{F}) + g\mu(\mathrm{G})] - [a\mu(\mathrm{A}) + b\mu(\mathrm{B})]\}\,dn < 0. \tag{81}$$

Employing our previous notation this inequality becomes

$$\left(\sum_{\mathrm{pr}} c\mu - \sum_{\mathrm{re}} c\mu\right) dn < 0. \tag{82}$$

There are two situations which would satisfy this inequality:

(1) If the sum of the stoichiometrically weighted chemical potentials of the reactants is greater than the corresponding sum for the products, the value within the parentheses in Eq. (82) would be negative. This would require dn to be positive for a spontaneous reaction to occur. A positive dn means that the reaction proceeds rightward, creating products (F and G) and consuming reactants (A and B). This process will continue until the weighted sums of the chemical potentials of the products and of the reactants are equal, that is, until equilibrium is reached (Eq. 80).

(2) If the sum of the stoichiometrically weighted chemical potentials of the products is greater than the corresponding sum for the reactants, the value within the parentheses of Eq. (82) will be positive, requiring dn to be negative. A negative dn means that the reaction proceeds spontaneously leftward, using up products (G, F) and generating reactants (A, B).[1] Again the reaction will continue until equilibrium is reached, that is, until the two weighted chemical potential sums are equal.

These two cases can be generalized by stating that *when a chemical reaction proceeds spontaneously the system moves to a state of lower chemical potential.*

---

[1] To avoid possible confusion, keep in mind that the terms "reactant" and "product" always refer to the reaction *as written*: Reactants → Products. In case 1 above, the reaction proceeds as written. In case 2, however, it proceeds spontaneously in the opposite direction: Reactants ← Products.

When $\sum_{\text{pr}} c\mu < \sum_{\text{re}} c\mu$, the reaction proceeds spontaneously rightward. When $\sum_{\text{pr}} c\mu > \sum_{\text{re}} c\mu$, the reaction proceeds spontaneously leftward. When $\sum_{\text{pr}} c\mu = \sum_{\text{re}} c\mu$, the reaction is at equilibrium.

If $\mu(CO) + \mu(H_2O) > \mu(CO_2) + \mu(H_2)$, the reaction $CO + H_2O \rightarrow CO_2 + H_2$ occurs spontaneously. If $\mu(CO) + \mu(H_2O) < \mu(CO_2) + \mu(H_2)$, however, the opposite reaction is spontaneous: $CO_2 + H_2 \rightarrow CO + H_2O$.

5.4. Chemical Potential for a Pure Substance

In order to use these relationships we must develop expressions for the chemical potentials for various kinds of substances in various situations.

For one component in a single phase, the chemical potential equals the molar free energy:

$$\mu(i) = (\partial G/\partial n_i)_{T,\,P,\,n_j} = (\partial G/\partial n_i)_{T,\,P} = \overline{G}(i), \tag{83}$$

since there is only one material present.

Although the treatment of phase transitions of pure substances in the previous chapter could have employed the chemical potential rather than the molar free energy, the use of the chemical potential concept adds little to our understanding of such systems.

5.5. Chemical Potential of a Gas in a Mixture of Gases

Now consider a gas mixture. Since there is more than one substance present we no longer have a pure phase. If we assume that the gases are all ideal, however, they must obey *Dalton's law of partial pressures*: each gas behaves as if it were the only gas present—as if it were a pure gas at a pressure equal to its *partial pressure* in the mixture, $p_i = n_i RT/V$, where V is the total volume of the container and n_i the number of moles of gas i. The total pressure P is equal to the sum of the partial pressures of all the gases present: $P = \sum p_i$. Since the molecules of ideal gases do not interact, the presence of other gases does not affect the properties (such as free energy) of a particular ideal gas. The free energy of an ideal gas in a mixture is thus equal to the number of moles of that gas present multiplied by its molar free energy, just as it would be if there were no other gases present. Hence for an ideal gas, even when it is not the only species in the gas phase, its chemical potential is equal to its molar free energy:

$$\mu(i) = \overline{G}(i). \tag{84}$$

The amount of a particular material present in a gaseous reaction varies during the course of the reaction and, since p_i is proportional to n_i, so does the partial pressure of that substance. Recall (Eq. 72) that $\overline{G}$ is a function of pressure: $(\partial\overline{G}/\partial P)_T = \overline{V}$. Since the pressure of a gas in a mixture is p_i, this equation may be rewritten as

$$[\partial\overline{G}(i)/\partial p_i]_T = \overline{V}_i. \tag{85}$$

Therefore if the temperature is constant

$$d\overline{G}(i) = d\mu(i) = \overline{V}_i \, dp_i$$

Integrating from state 1 to state 2:

$$\int_{\mu_1}^{\mu_2} d\mu(i) = \int_{p_1}^{p_2} \overline{V}_i \, dp_i$$

$$\mu_2(i) - \mu_1(i) = \int_{p_1}^{p_2} (RT/p_i) \, dp_i$$

$$\mu_2(i) = \mu_1(i) + RT \ln (p_{i_2}/p_{i_1}). \tag{86}$$

If state 1 is the *standard state* of the gas, $p_{i_1} = 1$ atm and

$$\mu(i) = \mu^\circ(i) + RT \ln p_i, \tag{87}$$

where $\mu^\circ(i)$ is the standard chemical potential of gaseous substance i, its chemical potential when its partial pressure is 1 atm.

5.6. Gas Phase Reactions

If we assume that all the participants in the reaction $CO + H_2O \rightarrow CO_2 + H_2$ behave as ideal gases, equations similar to Eqs. (86) and (87) may be written for them:

$$\mu(CO) = \mu^\circ(CO) + RT \ln p_{CO}$$
$$\mu(H_2O) = \mu^\circ(H_2O) + RT \ln p_{H_2O}$$
$$\mu(CO_2) = \mu^\circ(CO_2) + RT \ln p_{CO_2}$$
$$\mu(H_2) = \mu^\circ(H_2) + RT \ln p_{H_2}.$$

Now if state 2 is an equilibrium state

$$\mu_2(CO) + \mu_2(H_2O) = \mu_2(CO_2) + \mu_2(H_2),$$

and using Eq. (86),

$$\mu_1(CO) + RT \ln [(p_{CO})_2/(p_{CO})_1] + \mu_1(H_2O) + RT \ln [(p_{H_2O})_2/(p_{H_2O})_1]$$
$$= \mu_1(CO_2) + RT \ln [(p_{CO_2})_2/(p_{CO_2})_1] + \mu_1(H_2) + RT \ln [(p_{H_2})_2/(p_{H_2})_1].$$

This last equation can be rewritten as

$$[\mu_1(CO_2) + \mu_1(H_2)] - [\mu_1(CO) + \mu_1(H_2O)] = -RT\{\ln [(p_{CO_2})_2/(p_{CO_2})_1] + \ln [(p_{H_2})_2/(p_{H_2})_1] - \ln [(p_{CO})_2/(p_{CO})_1] - \ln [(p_{H_2O})_2/(p_{H_2O})_1]\}$$

Since $\mu_i = \overline{G}_i$ for an ideal gas,

$$\overline{G}_1(CO_2) + \overline{G}_1(H_2) - \overline{G}_1(CO) - \overline{G}_1(H_2O) = -RT[\ln (p_{CO_2})_2 + \ln (p_{H_2})_2 - \ln (p_{CO})_2 - \ln (p_{H_2O})_2] + RT[\ln (p_{CO_2})_1 + \ln (p_{H_2})_1 - \ln (p_{CO})_1 - \ln (p_{H_2O})_1],$$

and

$$\Delta G_1 = -RT \ln [p_{CO_2} p_{H_2}/p_{CO} p_{H_2O}]_2 + RT \ln [p_{CO_2} p_{H_2}/p_{CO} p_{H_2O}]_1.$$

Since state 2 is an equilibrium state the ratio of pressures in this state is equal to the equilibrium constant K_p for this reaction. The ratio of pressures in state 1 will be simply abbreviated M. Therefore

$$\Delta G_1 = -RT \ln K_p + RT \ln M. \tag{88}$$

Equation (88) will hold for any gas phase reaction involving ideal gases. For the general reaction $aA + bB \rightarrow fF + gG$,

$$K_p = [p_F^f p_G^g / p_A^a p_B^b]_{equil}. \tag{89}$$

M for the general case is the same stoichiometrically weighted ratio of partial pressures except under nonequilibrium conditions.

Example 27 Calculate the free energy change for the reaction of carbon monoxide with water vapor:

$$CO(g) + H_2O(g) \longrightarrow CO_2(g) + H_2(g)$$

when the partial pressures of CO and H_2O are both 0.10 atm, those for CO_2 and H_2 are 0.20 atm and 0.40 atm respectively, and the temperature is 25°C. Also determine what the final partial pressure of each of the four gases will be if they are mixed at the above partial pressures and allowed to react to equilibrium. The equilibrium constant for this reaction is 3.2×10^3.

Answer: The partial pressure ratio $(p_{CO_2} p_{H_2}/p_{CO} p_{H_2O})$ is always found to be 3.2×10^3 at equilibrium at 25°C regardless of the initial partial pressures. M for this case is $(0.20)(0.40)/(0.10)^2 = 8.0$ and $K_p = 3.2 \times 10^3$. Thus using Eq. (88):

$$\begin{aligned}\Delta G &= -RT[\ln (3.2 \times 10^3) - \ln (8.0)] \\ &= -2.3(1.98 \text{ cal deg}^{-1} \text{ mole}^{-1})(298°) \log (3.2 \times 10^3/8.0) \\ &= -3.54 \text{ kcal mole}^{-1}.\end{aligned}$$

Since $\Delta G < 0$ the reaction proceeds spontaneously *under these conditions*, that is, if the partial pressures of CO, H_2O, CO_2, and H_2 are maintained at 0.1, 0.1, 0.2 and 0.4 respectively.

If, however, the gases have these partial pressures initially but neither is CO and H_2O added nor CO_2 and H_2 removed, the partial pressures of CO and H_2O decrease and those of CO_2 and H_2 increase. Consequently M increases as the reaction proceeds and ΔG increases (becomes less negative). When $M = K_p$, $\Delta G = 0$ and no further change occurs.

When y moles of CO disappear in this reaction y moles of H_2O are also consumed and y moles of both CO_2 and H_2 appear. Assuming ideal gas behavior, n_i is proportional to p_i, and y moles is proportional to x atm. The final equilibrium partial pressure ratio is:

$$K_p = (p_{CO_2} p_{H_2} / p_{CO} p_{H_2O})_{equil}$$

$$3.2 \times 10^3 = (0.2 + x)(0.4 + x)/(0.10 - x)(0.10 - x)$$

$$x = 0.093 \text{ atm.}$$

The final partial pressures are: $p_{CO_2} = 0.29$ atm, $p_{H_2} = 0.49$ atm, $p_{CO} = 0.007$ atm, and $p_{H_2O} = 0.007$ atm.

Suppose we choose the conditions of state 1 of Eq. (88) to be standard conditions. In this case $p_{A_1} = p_{B_1} = \cdots = 1.0$ atm, $M_1 = 1$ and $\ln M = 0$. ΔG_1 under these conditions is of course the standard free energy change ΔG°. Equation (88) becomes

$$\Delta G^\circ = -RT \ln K_p. \tag{90}$$

Equation 90 is a most important relationship. It says that the innate drive of a reaction under standard conditions, its ΔG°, may be learned from the magnitude of the reaction's equilibrium constant. Also we have seen that ΔG° can be determined from tables. From this calculated value of ΔG° for a reaction its equilibrium constant can be calculated. Thus we can now predict the relative amounts of reactants and products at equilibrium *without carrying out the reaction.*

Example 28 Calculate ΔG° for the carbon monoxide reaction of Example 27 from the equilibrium constant and also from the data of Table 3.1.

Answer:

$$\begin{aligned} \Delta G^\circ &= -RT \ln K_p \\ &= -(1.99 \text{ cal deg}^{-1} \text{ mole}^{-1})(298^\circ) \ln (3.2 \times 10^3) \\ &= -4.76 \text{ kcal.} \end{aligned}$$

From Table 3.1:

$$\begin{aligned} \Delta G^\circ &= \overline{G}_f^\circ(CO_2) + \overline{G}_f^\circ(H_2) - \overline{G}_f^\circ(CO) - \overline{G}_f^\circ(H_2O) \\ &= -94.26 + 0 + 32.81 + 56.69 \\ &= -4.76 \text{ kcal.} \end{aligned}$$

In general, substituting Eq. (90) into Eq. (88) gives

$$\Delta G = \Delta G^\circ + RT \ln M. \tag{91}$$

Note that this equation can be written as

$$\Delta G = RT \ln (M/K).$$

When $M < K$, $\Delta G < 0$ and the process reactants → products occurs spontaneously until $M = K$, at which point $\Delta G = 0$; equilibrium is reached. When $M > K$, then $\Delta G > 0$ and the reverse process, products → reactants, is spontaneous until $M = K$.

Example 29 A common atmospheric pollutant, sulfur dioxide (SO_2), arises principally from the burning of sulfur-containing fuels. In the atmosphere it is oxidized to sulfur trioxide (SO_3) which then combines with water to form sulfuric acid (H_2SO_4), which damages library volumes, limestone buildings, and human lungs, among other things.

a) Calculate the equilibrium constant at 25°C for the reaction which produces SO_3 from SO_2.
b) If 1 liter of air contains 0.008 mole O_2, 2×10^{-7} mole SO_2, and 2×10^{-9} mole SO_3, would sulfur trioxide be formed or would it decompose spontaneously?

Reaction:	$SO_2(g) + \frac{1}{2}O_2(g)$	$\longrightarrow$	$SO_3(g)$
ΔG_f°, kcal mole^{-1}	-71.79 $\quad$ 0.0		-88.52

Answer: a) $-RT \ln K = \Delta G^\circ = -88.52 + 71.79 - 0.0 = -16.73$ kcal

$$\ln K = 16.73/(2 \times 10^{-3})(298) = 27.9$$

$$K \simeq 10^{12}.$$

b) $M = p_{SO_3}/(p_{SO_2} p_{O_2}^{1/2}) = (nRT/V)_{SO_3}/(nRT/V)_{SO_2}(nRT/V)_{O_2}^{1/2}$

$M = (2 \times 10^{-9} RT/V)/(2 \times 10^{-7} RT/V)(8 \times 10^{-3} RT/V)^{1/2}$

$M = 0.11(V/RT)^{1/2} = 2.26 \times 10^{-4}$.

Since $M < K$, $\Delta G < O$ and SO_3 would spontaneously form.

5.7. Real Gases; Fugacity

Strictly speaking, Eqs. (86) and (87) define the chemical potential for an ideal gas only. All the thermodynamic relationships derived using this definition therefore apply only to ideal gases and not to real gases. This includes Eqs. (89) and (90).

As has been pointed out previously, we can assume that in most of the cases we deal with no great error will result from assuming that real gases behave like ideal gases. But there are instances when such an assumption can be damaging, such as at low temperature or high pressure or when great accuracy is desired.

In dealing with real gases we could return to the point in our derivation where we assumed that our gas was ideal and instead of the ideal gas equation insert a more accurate equation of state such as that of van der Waals. The equations which we would then derive would be more accurate but more unwieldly. Rather than following this procedure, scientists do an interesting thing: to preserve the simplicity of the equations derived for ideal gases they define for real gases a new variable, the *fugacity* f, such that in every equation in which a partial pressure p_i appears, the fugacity f_i is used instead. For a real gas the equation $\mu(i) = \mu^\circ(i) + RT \ln p_i$ is only approximately true; if written

$$\mu(i) = \mu^\circ(i) + RT \ln f_i \tag{92}$$

it becomes exactly true *by definition*. Thus pressure is the *measurable* property but fugacity is the *determining* property, thermodynamically speaking. When the real gas behaves ideally, the fugacity and the pressure are identical, but as the situation becomes less and less ideal (as when the gas is compressed) the observed pressure strays farther and farther from the fugacity. Thus fugacity is the thermodynamically *effective* partial pressure of a gas, expressed in atmospheres. The observed partial pressure p lies between f and nRT/V.

The standard state of a real gas is $f_i = 1$ atm. The *reference state* for a gas is zero pressure; the gas approaches ideal behavior as the pressure approaches the reference state:

$$\lim_{p_i \to 0} f_i = p_i. \tag{93}$$

The *fugacity coefficient* γ_i is defined as

$$\gamma_i = f_i/p_i, \tag{94}$$

and $\lim_{p_i \to 0} \gamma_i = 1$.

If a detailed equation of state can be found for a gas or if the details of its behavior can be determined, its fugacity can be calculated. Recall Eq. (85): at constant temperature $d\mu = \overline{V}\,dp$. This is true for all gases, real or ideal. We can write

$$d\mu_{\text{ideal}} = \overline{V}_{\text{ideal}}\,dp = (RT/p)\,dp$$
$$d\mu_{\text{real}} = \overline{V}_{\text{real}}\,dp.$$

$\overline{V}_{\text{ideal}}$ is calculated from the observed temperature and pressure using the ideal gas law. $\overline{V}_{\text{real}}$ is measured directly or calculated using the real equation of state. Thus

$$d(\mu_{\text{real}} - \mu_{\text{ideal}}) = (\overline{V}_{\text{real}} - RT/p)\,dp.$$

Integrating from $p = 0$ where $\mu_{\text{real}} = \mu_{\text{ideal}}$ (where $p = f$) to p, the pressure of interest:

$$\mu_{\text{real}} - \mu_{\text{ideal}} = \int_0^p [\overline{V}_{\text{real}} - (RT/p)]\, dp.$$

Since $\mu_{\text{real}} = \mu^\circ + RT \ln f$ and $\mu_{\text{ideal}} = \mu^\circ + RT \ln p$,

$$\ln f = \ln p + \int_0^p [(\overline{V}_{\text{real}}/RT) - 1/p]\, dp. \tag{95}$$

In this text we shall continue to assume that we may use ideal gas equations when dealing with real gases. But it is well to remember that such is not always the case and to be prepared to take nonideality into account.

5.8. Other Partial Molar Quantities

We have to this point confined our attention to partial molar free energy or chemical potential. However, partial molar functions can be written corresponding to any extensive state variable. If F is an extensive function of state such as E, H, S, A, G, V, the partial molar value of F for a substance i in a mixture is defined as

$$\overline{\mathscr{F}}(i) = (dF/dn_i)_{T,\, p,\, n_j}. \tag{96}$$

Many of the thermodynamic equations we have previously used for pure substances have their partial molar counterparts. For instance:

$$\overline{\mathscr{H}}(i) = \overline{\mathscr{E}}(i) + p_i \overline{\mathscr{V}}(i)$$
$$\mu(i) = \overline{\mathscr{A}}(i) + p_i \overline{\mathscr{V}}(i)$$
$$d\mu(i) = \overline{\mathscr{V}}(i)\, dp_i - \overline{\mathscr{S}}(i)\, dT,$$

where $\overline{\mathscr{H}}$, $\overline{\mathscr{E}}$, $\overline{\mathscr{V}}$, $\overline{\mathscr{A}}$, and $\overline{\mathscr{S}}$ are partial molar enthalpy, partial molar energy, partial molar volume, partial molar Helmholtz free energy, and partial molar entropy, respectively.

Problems

5.1 (a) Prove that K_p has no units. (b) Although K_p is unitless, its value in many cases does depend on the pressure units employed. For the following reaction calculate K_p using (1) pressures in atm and (2) pressures in torr:

$$SO_2Cl_2(g) \longrightarrow SO_2(g) + Cl_2(g)$$

equilibrium partial pressures: SO_2Cl_2: 0.25 atm
SO_2: 0.75 atm
Cl_2: 0.75 atm

(c) Calculate ΔG° for the above reaction (1) when the partial pressures are given in atm and (2) when partial pressures are in torr. What are the standard state partial pressures in each case? (d) Calculate ΔG for this reaction using Eq. (91) when the partial pressures are:

SO_2Cl_2: 1 atm
SO_2: 2 atm
Cl_2: 1 atm

Use both ΔG° values determined in (c) and show that if we are consistent in the units used for the partial pressures in K_p and in Q, ΔG is the same in both cases (as it must be, of course).

5.2 For which of the following equilibria will the value of K_p depend on the units used for the partial pressures?

a) $H_2(g) + Cl_2(g) \rightleftarrows 2$
b) $H_2(g) + \frac{1}{2}O_2(g) \rightleftarrows H_2O(g)$
c) $NH_3(g) + \frac{5}{4}O_2(g) \rightleftarrows NO(g) + \frac{3}{2}H_2O(g)$ (Ostwald process)
d) $NOBr(g) \rightleftarrows NO(g) + \frac{1}{2}Br_2(g)$
e) $CO(g) + Cl_2(g) \rightleftarrows COCl_2(g)$ (phosgene)
f) $N_2(g) + O_2(g) \rightleftarrows 2NO(g)$ (arc process)
g) $2N_2(g) + O_2(g) \rightleftarrows 2N_2O(g)$ (laughing gas)
h) $C_2H_2(g) + HCl(g) \rightleftarrows C_2H_3Cl(g)$ (vinyl chloride which polymerizes to form plastics)

5.3 What will occur in each of the gas equilibria of the preceding problem if the equilibrium mixture of gases is compressed?

5.4 Consider the all-gaseous reactions (I) $CO + \frac{1}{2}O_2 \rightarrow CO_2$ and (II) $H_2O \rightarrow H_2 + \frac{1}{2}O_2$. Show that, when these two reactions are combined to give the reaction cited in Section 5.1, $CO + H_2O \rightarrow CO_2 + H_2$, it follows that the resultant $\Delta G^\circ = \Delta G_{\mathrm{I}}^\circ + \Delta G_{\mathrm{II}}^\circ$ and that $K = K_{\mathrm{I}} K_{\mathrm{II}}$.

5.5 Reaction II of Problem 5.4 might be written in the following ways:

1. $H_2O(g) \rightleftarrows H_2(g) + \frac{1}{2}O_2(g)$
2. $2H_2O(g) \rightleftarrows 2H_2(g) + O_2(g)$

What is the relation of ΔG_1° to ΔG_2°? Of K_1 to K_2?

5.6 An automobile engine warmed up and idling has in its exhaust carbon dioxide 13.5% (by volume), carbon monoxide 3.0%, water vapor 5.0%, and hydrogen 0.0%. If a sample of this exhaust is confined at 25°C and 1 atm and the only reaction occurring is $CO(g) + H_2O(g) \rightarrow CO_2(g) + H_2(g)$, what will be the partial pressure of CO when equilibrium is reached?

5.7 Consider the following all-gaseous reaction, for which the values appearing beneath it represent the standard free energy of formation in kcal mole^{-1} of each substance respectively.

$$\underset{-17.89}{CH_4} + \underset{-25.50}{CCl_4} \longrightarrow \underset{-21.00}{2CH_2Cl_2}$$

(a) If 1.0 liter each of CH_4 and CCl_4 are mixed with 2.0 liters of CH_2Cl_2 (all volumes measured at 1.0 atm pressure) and allowed to react at 25°C and 1.0 atm, will association or dissociation occur? (b) What will be the mole percent of each species when equilibrium is reached under these conditions?

5.8 Realizing that the decomposition of steam into hydrogen and oxygen is very slight even at elevated temperatures, show that, as a good approximation, the fraction decomposed is inversely proportional to the cube root of the applied pressure.

5.9 At 1.0 atm pressure, carbon dioxide is quoted as being "2% decomposed at 2000°C" into carbon monoxide and oxygen. (a) Calculate K_p for the dissociation reaction at 2000°C. (b) If the system is compressed to a pressure of 5.0 atm, what does the 2% figure become?

5.10 Calculate the percentage dissociation of carbon dioxide in Problem 5.9 if, instead of compressing as in (b), the 5.0-atm total pressure is achieved by forcing in (a) carbon dioxide, or (b) carbon monoxide, or (c) oxygen, or (d) helium, keeping the total volume constant in each case.

5.11 Consider again the equilibrium of Problem 5.9(a) above. Suppose that instead of adding gases as in 5.10(a), (b), (c), and (d) until we five-fold the pressure at constant volume, we choose to add each gas until the volume is five times the initial volume, keeping the external pressure constant in each case. In which direction, if at all, will the previous 2% equilibrium figure change in each of the four experiments—adding (a) CO_2, (b) CO, (c) O_2, and (d) He?

5.12 The free-energy change in a reaction always has a finite value. (a) Show that it is therefore impossible for a reaction to "go to completion," that is, for all of any reactant to be changed to product. (b) Calculate the number of molecules of reactants present at equilibrium at 1.0 atm and 25°C when 1 mole each of CO and Cl_2 are mixed, producing the poison gas phosgene. (Phosgene was used extensively during World War I as an anti-personnel gas.)

Reaction:	CO(g)	+ Cl_2(g)	$\longrightarrow$	$COCl_2$(g)
ΔG_f°, kcal mole^{-1}	−32.81	0		−50.31

5.13 If 0.30 mole of pure phosgene is introduced into a 12-liter evacuated vessel at 25°C, what total pressure will be established inside at equilibrium?

5.14 Some pure phosgene is introduced into a cylinder fitted with a pressure gauge. If the piston is moved until it is observed that the sample is 3% dissociated at 25°C, what will the pressure gauge read?

5.15 We have seen in previous chapters that even though $\Delta G^\circ > 0$ for a process (such as graphite → diamond), it could be made spontaneous by varying the pressure and/or the temperature since free energy is a function of both pressure and temperature. We now see that a reaction whose $\Delta G^\circ > 0$ may be made spontaneous by changing the concentrations (partial pressures) of the reacting species. (a) The equilibrium constant for the reaction

$$\mathrm{NOBr(g)} \longrightarrow \mathrm{NO(g)} + \tfrac{1}{2}\mathrm{Br_2(g)}$$

is 0.15 at 25°C (partial pressures in atmospheres). Calculate ΔG°. (b) If the initial partial pressure of NOBr is 1 atm and the initial partial pressures of NO and Br_2 are equal, what is the largest value which the latter may have for the reaction to be spontaneous as written?

5.16 Suppose 1.0 mole of NOBr gas is placed in a 1-liter container at 25°C and allowed to react as in Problem 5.2(d). Plot ΔG versus the partial pressure of NO as the reaction proceeds. The equilibrium constant with pressures in atmospheres is 0.15.

6 / Chemical Reactions in Solution

6.1. Definitions

Any work with chemical reactions will almost inevitably involve solutions. A *solution* is a one-phase system with two or more components. Solutions may be gas-phase, liquid-phase or solid-phase. Of these types, liquid solutions are the most familiar and the most useful in the laboratory.

If in a solution of two substances A and B the number of moles of A is greater than that of B, then A is said to be the *solvent* and B the *solute*. This distinction is somewhat arbitrary and sometimes leads to difficulty (when one of the molecular weights is unknown, for instance). In most cases, however, the distinction is convenient and helpful. If the number of moles of the solvent is much greater than that of the solute, the solution is said to be *dilute*. Much of the chemist's work is done with dilute solutions.

6.2. Raoult's Law and Henry's Law

It is found experimentally that a volatile solvent in a dilute liquid solution obeys *Raoult's law*: the partial pressure of solvent vapor, that is, its *vapor pressure* p_A, above a solution is equal to the mole fraction solvent present times the vapor pressure of the solvent when pure, or

$$p_A = X_A p_A^\circ . \tag{97}$$

If in a certain solution, for example, the solvent molecules comprise nine-tenths of those present ($X_A = 0.90$), the vapor pressure of solvent from that solution will be nine-tenths of "full value", or nine-tenths the vapor pressure characteristic of the pure solvent at that temperature.

It is found experimentally that a volatile *solute* in a dilute liquid solution obeys *Henry's law*: the solubility X_B of a gas in a liquid is proportional to the partial pressure of the gas p_B. Alternatively stated, the partial pressure of solute vapor above the solution (its vapor pressure) is proportional to the mole fraction of the solute in the solution:

$$p_B = X_B K_{AB} . \tag{98}$$

The proportionality constant is not p_B° here but a constant which depends on the solution in question. Normally expressed in atmospheres, it is an inverse measure of the solubility of the gas.

Although Raoult's law and Henry's law are only obeyed exactly by solvents and solutes in extremely dilute solutions, they do provide useful approximate predictions for solutions which are somewhat less dilute.

Example 30 Through their gill systems, fish breathe the dissolved air present in water. Though the concentration of air in water is not great, show that the air a fish breathes is richer in oxygen percentagewise than our own. The partial pressure of oxygen in air is 0.20 atm and that of nitrogen is 0.80 atm at 20°C. The Henry's law constant for oxygen is 4.1×10^4 atm; for nitrogen it is 8.3×10^4 atm.

Answer: The solubility of O_2 and of N_2 in water in terms of mole fraction is given by Eq. (98):

$$\begin{aligned} X_{O_2} &= p_{O_2}/K_{O_2,\,H_2O} \\ &= 0.20\ \text{atm}/4.1 \times 10^4\ \text{atm} \\ &= 4.9 \times 10^{-6} \\ X_{N_2} &= p_{N_2}/K_{N_2,\,H_2O} \\ &= 0.80\ \text{atm}/8.3 \times 10^4\ \text{atm} \\ &= 9.6 \times 10^{-6}. \end{aligned}$$

The concentration of N_2 is greater than that of O_2, but only about twice as great. The dissolved air a fish breathes is thus about $\frac{1}{3}$ oxygen, whereas our air is only $\frac{1}{5}$ oxygen.

6.3. Chemical Potential of Solvent and Solute

When the *solvent* of a liquid-phase solution is in equilibrium with its vapor (Eq. 80):

$$\mu_{\text{A, solvent}} = \mu_{\text{A, vap}}.$$

If we assume the vapor of the solvent to be ideal we may use the chemical potential expression for an ideal vapor (Eq. 87):

$$\mu_{\text{A, solvent}} = \mu^\circ_{\text{A, vap}} + RT \ln p_{\text{A, vap}}.$$

If now we use the experimental equation for the solvent in dilute solution, Raoult's law (Eq. 93), and simplify we obtain the following:

$$\begin{aligned} \mu_{\text{A, solvent}} &= \mu^\circ_{\text{A, vap}} + RT \ln (X_{\text{A, soln}}\, p^\circ_{\text{A}}) \\ &= \mu^\circ_{\text{A, vap}} + RT \ln p^\circ_{\text{A}} + RT \ln X_{\text{A, soln}}. \end{aligned}$$

The partial pressure of the vapor in equilibrium with pure liquid A, p°_{A}, is a constant characteristic of the substance. We therefore combine the first two terms on the right into one constant term and call it the *standard chemical potential* of the *solvent*, $\mu^\circ_{\text{A, solvent}}$:

$$\mu^\circ_{\text{A, solvent}} \equiv \mu^\circ_{\text{A, vap}} + RT \ln p^\circ_{\text{A}}. \tag{99}$$

Making this substitution in the previous equation, we obtain the expression for the chemical potential of the *solvent* in a dilute solution:

$$\mu_{A,\,\text{solvent}} = \mu^{\circ}_{A,\,\text{solvent}} + RT \ln X_{A,\,\text{soln}} . \tag{100}$$

For the *solute* in a dilute solution in equilibrium with its vapor:

$$\mu_{B,\,\text{solute}} = \mu_{B,\,\text{vapor}} .$$

Assuming the vapor to be ideal,

$$\mu_{B,\,\text{solute}} = \mu^{\circ}_{B,\,\text{vap}} + RT \ln p_{B,\,\text{vap}} .$$

Substituting the experimental expression for the solute in a dilute liquid solution (Henry's Law, Eq. 94),

$$\begin{aligned}\mu_{B,\,\text{solute}} &= \mu^{\circ}_{B,\,\text{vap}} + RT \ln (X_{B,\,\text{soln}} K_{A,\,B}) \\ &= \mu^{\circ}_{B,\,\text{vap}} + RT \ln K_{A,\,B} + RT \ln X_{B,\,\text{soln}} .\end{aligned}$$

Again $K_{A,\,B}$ is a constant characteristic of the solution and we can combine the first two terms on the right:

$$\mu^{\circ}_{B,\,\text{solute}} \equiv \mu^{\circ}_{B,\,\text{vap}} + RT \ln K_{A,\,B} . \tag{101}$$

Making this substitution we obtain an equation for the chemical potential of the *solute* in a dilute solution:

$$\mu_{B,\,\text{solute}} = \mu^{\circ}_{B,\,\text{solute}} + RT \ln X_{B,\,\text{soln}} . \tag{102}$$

Equations (100) and (102) have much the same appearance, but the difference between $\mu^{\circ}_{A,\,\text{solvent}}$ and $\mu^{\circ}_{B,\,\text{solute}}$ should always be remembered (Eqs. 99 and 101).

We now have equations for the chemical potentials of substances in liquid-phase solutions in terms of the mole fraction. The chemical potential of the solute, however, is more often expressed in terms of *molality* instead of mole fraction.

In a dilute solution of solvent A and solute B, $n_A \gg n_B$, and mole fraction approaches mole ratio:

$$X_B = \frac{n_B}{n_A + n_B} \simeq n_B/n_A . \tag{103}$$

Molality is defined as the number of moles of solute per kilogram of solvent. By this definition, a solution of molality m_B contains m_B moles of solute and 1000 g (or $1000/M_A$ moles) of solvent. Its mole fraction solute may therefore be formulated

$$X_B \simeq \frac{n_B}{n_A} = \frac{m_B}{1000/M_A} = \frac{M_A}{1000} m_B . \tag{104}$$

If *dilute*, the conversion factor for molality to mole fraction is $M_A/1000$. If dilute *and aqueous*, the factor becomes 0.018.

Substituting Eq. (104) into Eq. (102), we have

$$\mu_{B,\,solute} = \mu^{\circ}_{B,\,solute} + RT \ln (m_B M_A/1000)$$
$$= \mu^{\circ}_{B,\,solute} + RT \ln (M_A/1000) + RT \ln (m_B).$$

The first two terms on the right are constants and can be combined into a single term, the *standard chemical potential* of the *solute*, $\mu^{*}_{B,\,solute}$:

$$\mu^{*}_{B,\,solute} \equiv \mu^{\circ}_{B,\,solute} + RT \ln (M_A/1000). \qquad (105)$$

Substituting, we obtain an expression for the chemical potential of a solute in terms of its molality:

$$\mu_{B,\,solute} = \mu^{*}_{B,\,solute} + RT \ln (m_B). \qquad (106)$$

From Eq. (100), the standard chemical potential of a solvent μ°_A is the chemical potential of the pure liquid, $X_A = 1$. The *standard state* for a *solvent* is thus the *pure liquid*.

From Eq. (106), the standard chemical potential of the solute μ^{*}_B is the chemical potential of the solute when it is at unit molality, or $m_B = 1$. The *standard state* for the *solute* is then *unit molality*.

6.4. Reactions in Liquid-Phase Solutions

Now consider a reaction occuring in solution with all reactants and products as solutes at low concentration:

$$bB + cC \longrightarrow fF + gG.$$

At equilibrium (Eq. 80)

$$b\mu_B + c\mu_C = f\mu_F + g\mu_G$$

Employing Eq. (106) for each of the reactants and products,

$$b(\mu^{*}_B + RT \ln m_B) + c(\mu^{*}_C + RT \ln m_C)$$
$$= f(\mu^{*}_F + RT \ln m_F) + g(\mu^{*}_G + RT \ln m_G)$$
$$f\mu^{*}_F + g\mu^{*}_G - b\mu^{*}_B - c\mu^{*}_C = -RT[f \ln (m_F) + g \ln (m_G) - b \ln (m_B) - c \ln (m_C)].$$

The left side of the above equation is the sum of the standard chemical potentials of the products minus the sum of the standard chemical potentials of the reactants. This is the change in the free energy which would accompany this reaction if carried out under standard conditions (all products and reactants with concentration of 1.0 molal.) The terms on the left of the equation are therefore equal to ΔG°:

$$\Delta G^\circ = -RT \ln (m_F^f m_G^g / m_B^b m_C^c). \tag{107}$$

The term within parentheses is simply the *equilibrium constant* in terms of molalities, K_m:

$$\Delta G^\circ = -RT \ln K_m. \tag{108}$$

Equation (108) is a very important relationship. The comments concerning Eq. (88) also apply here.

Again, in general,

$$\Delta G = \Delta G^\circ + RT \ln M_m, \tag{109}$$

where M_m is the ratio of the molalities of products and reactants under nonequilibrium conditions:

$$M_m = (m_F^f m_G^g / m_B^b m_C^c).$$

For example, for the following reaction in which barbital (a sedative used in sleeping pills) is formed,

$$(NH_2)_2CO \text{ (0.1 molal)} + (C_2H_5)_2C(CO_2C_2H_5)_2 \text{ (0.1 molal)} \longrightarrow$$

urea malonate

H_5C_2 C_2H_5 O O HN NH O (0.1 molal)

barbital

$$M_m = (m_{\text{barbital}})/[(m_{\text{urea}})(m_{\text{malonate}})] = 0.1/(0.1)(0.1) = 10.$$

6.5. Calculation of ΔG° for Solution Reactions

Eq. (54) can be used to yield ΔG° for a reaction in solution:

$$\Delta G^\circ_{\text{reaction}} = \sum_{\text{pr}} n\,\Delta G_f^\circ - \sum_{\text{re}} n\,\Delta G_f^\circ.$$

However, ΔG_f° is now the standard free-energy change in the process in which one mole of the substance is created at the standard concentration of one molal from its constituent elements in their most stable forms under standard conditions. (As before, the standard condition for the pure reacting substances will be 1.0 atm pressure and a designated temperature.)

For example the $\Delta G^\circ_{f,\text{solution}}$ for methanol is the change in free energy accompanying the process:

$$\tfrac{1}{2}O_2(g) + C(\text{graphite}) + 2H_2(g) \xrightarrow{\Delta G_1} CH_3OH(l) \xrightarrow{\Delta G_2} CH_3OH(\text{soln, 1 molal}).$$

The ΔG°_f we are interested in is $\Delta G_1 + \Delta G_2$. The value of ΔG_1 is obtained from the tabulations of ΔG°_f previously discussed; ΔG_2 from calorimetric measurement. Table 6.1 contains values for $\Delta G^\circ_{f,\text{soln}}$ for various materials in aqueous solution.

Table 6.1. Standard Free Energies and Enthalpies of Formation of Selected Substances in Aqueous Solution

Substances	ΔG°_f, kcal mole^{-1}	ΔH°_f, kcal/mole^{-1}
HCl	−31.350	−40.023
H_2S	−6.54	−9.4
H_2SO_4	−177.34	−216.90
NH_3	−6.37	−19.32
CO_2	−92.31	−98.69
$PbCl_2$	−68.51	−79.65
$Hg(OH)_2$	−66.0	—
$Ca(OH)_2$	−207.37	−239.68
NaOH	−100.184	−112.236
NaCl	−93.939	−97.302
$CaCl_2$	−194.88	−209.82
acetic acid	−95.48	—
acetate$^-$ + H^+	−88.99	—
L-alanine	−88.75	—
ethanol	−43.39	—
fumaric acid	−154.67	—
fumarate^{-2} + $2H^+$	−144.41	—
α-D-glucose	−219.38	—
succinic acid	−178.39	—
succinate^{-2} + $2H^+$	−164.97	—
NH_3	−6.37	—
$NH_4^+ + OH^-$	−19.00	—
$HCO_3^- + H^+$	−140.29	—
oxaloacetate^{-2} + $2H^+$	−190.53	—
citrate^{-3} + $3H^+$	−273.90	—
malate^{-2} + $2H^+$	−201.98	—
pyruvate$^-$ + H^+	−113.32	—

National Bureau of Standards Circular 500 (1952) and K. Burton and H. A. Krebs, *Biochemical Journal*, **54**, 94 (1953).

Example 31 Reasoning similar to that above can be applied to derive a more general equation relating ΔG° and the equilibrium constant K, in which the partial pressures of gases and the molalities of solutes appear in the same expression. In Example 29 we mentioned that the gaseous pollutant SO_3 reacts with water to form sulfuric acid. What would the partial pressure of SO_3 be above a 1.0 molal aqueous solution of H_2SO_4 at 25°C?

$$\begin{array}{lllll} \text{Reaction:} & SO_3(g) + H_2O(l) & \xrightarrow{\Delta G_1} & H_2SO_4(\text{aq, 1 molal}) & \text{(I)} \\ \Delta G_f^\circ,\ \text{kcal/mole} & -88.5 \quad -56.7 & & -177.3 & \end{array}$$

Answer: $\Delta G_f^\circ(H_2SO_4$, 1 molal) is the sum of the free-energy changes in the following two reactions:

$$H_2(g) + S(s) + 2O_2(g) \xrightarrow{\Delta G_2} H_2SO_4(l) \qquad \text{(II)}$$

$$H_2SO_4(l) \xrightarrow{\Delta G_3} H_2SO_4(\text{aq, 1 molal}) \qquad \text{(III)}$$

The amount of water present in Reaction (III) is just enough to dilute 1.0 mole of H_2SO_4 to 1.0 molal. Accordingly, $\Delta G_1^\circ = \Delta G_f^\circ(H_2SO_4,\ \text{aq}) - \Delta G_f^\circ(SO_3) - \Delta G_f^\circ(H_2O) = -177.3 + 88.5 + 56.7 = -32.1$ kcal. Since $\Delta G_1^\circ = -RT \ln K$, then $K = 2.2 \times 10^{23}$; and in this case $K = m_{H_2SO_4}/p_{SO_3}$. When $m_{H_2SO_4} = 1.00$,

$$p_{SO_3} = 1.00/2.25 \times 10^{23} = 4.4 \times 10^{-24} \text{ atm.}$$

The concentration of water does not appear in the equilibrium constant expression since its standard state is pure water ($X = 1$ for solvent) and we assume it to be essentially pure in this dilute solution.

6.6. Nondilute Solutions; Activity

The accuracy of thermodynamic expressions for solutes such as Eq. (107) is dependent upon the correctness of our expression for the chemical potential of the solute (Eq. 106). A review of the approximations made in the derivation of this equation gives convincing evidence that it can only be correct in very dilute solutions. In fact, significant deviations from behavior predicted by thermodynamic relations based on Eq. (106) are observed for many solutes at concentrations well below the standard state of unit molality.

Solutions in which the solvent and solutes all exactly obey Eqs. (100) and (106) respectively are often called *ideal solutions*. Divergencies from these equations which occur in more concentrated solutions are then spoken of as departures from ideality.

Nonideal gases were treated in Chapter 5 (Section 5.7) by introducing a new variable, the fugacity, in order to retain the simplicity of the ideal gas functions and equations. In the same way we now introduce a new variable in order to retain the simplicity of our solution equations and concepts. We define the *activity* a_i of a substance in solution as that function which, if substituted for concentration in any of our thermodynamic relationships, will give a resulting equation which agrees perfectly with the observed behavior of the solution. By definition Eq. (106) becomes

$$\mu_i = \mu_i^\circ + RT \ln a_i . \tag{110}$$

The activity can be thought of as the corrected or effective concentration of a substance in solution.

The equilibrium constant for the general solution reaction $aA + bB \rightarrow fF + gG$ would be *exactly*

$$K = a_F^f a_G^g / a_A^a a_B^b . \tag{111}$$

The *standard state* of a component of a solution is $a_i = 1$; its *reference state* (see Section 5.7) is infinite dilution. As the solution becomes more dilute the solvent and the solutes behave more "ideally," Raoult's and Henry's laws are followed more exactly, and the activity is more nearly equal to the concentration. Stated mathematically for a solute

$$\lim_{m_i \rightarrow 0} a_i = m_i .$$

The *activity coefficient* for a solute is defined as

$$\gamma_i = a_i / m_i . \tag{112}$$

Departures from ideality observed for all but the most dilute solutions are primarily a result of intermolecular interactions between the molecules of the solution. Such interactions will be discussed in Chapter 17.

Activities and activity coefficients must be experimentally determined. Colligative properties and electrochemistry (to be discussed in Chapters 7 and 8) can be used for this purpose.

We shall continue to assume that the solutions we deal with in this text are ideal unless otherwise noted. It should be remembered that for all but the most dilute situations this assumption can lead to significant error, and we should be prepared to determine and use activities instead of molalities if highly accurate results are required.

6.7. Molarity

In many applications concentrations appear in units of molarity c, moles per liter of solution. If the solution is sufficiently dilute for us to assume that the volume of the solution is essentially the same as that of the solvent, molality may be converted to molarity by multiplying by the density ρ of the solvent:

$$c = \rho m. \tag{113}$$

Since the density of water is 1 g ml^{-1}, it follows that molarity is equal to molality for dilute aqueous solutions.

Problems

6.1 When divers descend to great depths in the ocean, the pressure on their bodies increases due to the weight of the water above them. The human body can continue to function under a pressure of approximately 6 atm. (a) At what depth in fresh water would a diver encounter such a pressure? (b) The amount of air absorbed by the blood in the lungs depends on the solubility of the gases of which air is composed, principally N_2 (~80%) and O_2 (~20%). The solubility of nitrogen is predicted by Henry's law. What is the partial pressure of N_2 in air when the barometric pressure is 1 atm? The Henry's law constant for nitrogen in water is 8.3×10^4 atm. (c) Plot the solubility of N_2 in the blood (approximate as water) from sea level to the depth at which the pressure on the diver is 6 atm. (d) When divers surface too quickly the pressure decreases abruptly and the solubility of the gases in their blood therefore decreases. Some of the previously dissolved gas comes out of solution and forms bubbles in the blood stream. These bubbles obstruct blood circulation, and a painful and sometimes fatal condition known as "the bends" results. The body of an adult contains approximately 6 liters of blood. Calculate the volume of N_2 gas which would be released into the circulatory system as bubbles if the pressure were suddenly reduced from 5 atm to 2 atm. Assume the diver is being supplied air of normal constitution (80% N_2 and 20% O_2).

6.2 When the antacid Tums reaches the stomach, one of its principal active ingredients, $CaCO_3$, neutralizes stomach acid in the following reaction:

$$CaCO_3(s) + 2HCl(aq) \longrightarrow CaCl_2(aq) + CO_2(g) + H_2O(l)$$

(a) Suppose a second Tums tablet is swallowed. How would this affect the equilibrium state which has been established with the first tablet? (b) If instead of taking an additional tablet, part of the CO_2 gas produced by the first tablet is eliminated in some uncouth manner, how would this affect the equilibrium? (c) Calculate ΔG° and the equilibrium constant for this reaction.

6.3 Derive Eq. (109). See the derivation of Eq. (90) in Chapter 5.

6.4 As pointed out in Example 31, no concentration terms for pure solids or liquids appear in the equilibrium constant. Such terms also do not appear in the nonequilibrium ratio M when calculating free energies (Eq. 109). Prove this (see Problem 6.3) by deriving

$$\Delta G = \Delta G^\circ + RT \ln M$$

for the following reaction in which the much-maligned pesticide DDT may be produced:

$$\underset{\text{chlorobenzene}}{2C_6H_5Cl(l)} + \underset{\text{chloral hydrate}}{(HO)_2CHCCl_3(aq,\ 1.0\ \text{molal})} \longrightarrow \underset{\text{DDT}}{(C_6H_4Cl)_2CHCCl_3(s)} + 2H_2O(l)$$

(Chlorobenzene and DDT are insoluble in water.)

6.5 Give the concentration ratios of the equilibrium constant for each of the following reactions. (a) Neutralization of stomach acid by the antacid Gelusil:

$$Al(OH)_3(s) + 3HCl(aq) \longrightarrow AlCl_3(aq) + 3H_2O(l).$$

(b) The action of peroxide as a bleaching agent:

$$H_2O_2(aq) \longrightarrow H_2O(l) + \tfrac{1}{2}O_2(g).$$

(c) The production of the artificial sweetener saccharin:

$$\text{C}_6\text{H}_4(\text{COOH})(\text{SO}_2\text{NH}_2)(aq) \longrightarrow \text{C}_6\text{H}_4(\text{C{=}O})(\text{SO}_2)\text{NH}\ (aq) + H_2O(l)$$

6.6 From the data of Table 6.1 calculate the standard free-energy change and the equilibrium constant for the following ionizations:

a) acetic acid (aq) $\rightarrow$ acetate$^-$(aq) + H^+(aq)
b) $CO_2(aq) + H_2O(l) \rightarrow HCO_3^-(aq) + H^+(aq)$
c) fumaric acid (aq) $\rightarrow$ fumarate^{-2}(aq) + $2H^+$(aq)

6.7 Salicylic acid is a pain reliever and a precursor of aspirin (acetylsalicylic acid). The equilibrium constant for the ionization of its acidic proton is 10^{-3}:

$$\text{C}_6\text{H}_5\text{COOH}\ (aq) \longrightarrow \text{C}_6\text{H}_5\text{COO}^-\ (aq) + H^+(aq)$$

(a) Calculate $\Delta G°$. (b) Calculate ΔG when 0.2 g salicylic acid is placed in the stomach. Assume 0.5 liter of solution with a pH of 4.7. (c) When equilibrium is reached in the ionization of (b), what percentage of the salicylic acid will have ionized?

6.8 The following reactions are associated with the citric acid cycle, the major energy-yielding set of chemical reactions of the human body. (a) For each, calculate $\Delta G°$ and K. (b) Calculate ΔG for each reaction when conditions are more nearly physiological: pH = 7 and all other concentrations are equal to 0.01 molal.

1) oxaloacetate^{-2} + acetate$^-$ $\longrightarrow$ citrate^{-3}
2) fumarate^{-2} + H_2O $\longrightarrow$ malate^{-2}
3) oxaloacetate^{-2} + H_2O $\longrightarrow$ pyruvate$^-$ + HCO_3^-

6.9 For a solvent (and in some cases for solutes as well) it is convenient to define activity in terms of mole fraction. Instead of being equal to the molality in an infinitely dilute solution, the activity becomes equal to the mole fraction. The activity coefficient is in this case $\gamma_j = a_j/X_j$ and $\lim_{X_j \to 1} \gamma_j = 1$. For a solution which is not extremely dilute, Eq. (100) should be written

$$\mu_{A,\text{ solvent}} = \mu^{\circ}_{A,\text{ solvent}} + RT \ln a_{\text{solvent}}.$$

When a volatile solvent is in equilibrium with its vapor,

$$\mu_{A,\text{ solvent}} = \mu_{A,\text{ vapor}} = \mu^{\circ}_{A,\text{ vapor}} + RT \ln p_{A,\text{ vapor}}$$

(assuming ideal gas behavior for the vapor). (a) Prove that $\gamma_A = p_A/X_A p^{\circ}_A$. (b) At 25°C the vapor pressure of pure water is 23.8 torr. When 20 g of KOH is added to 100 g of water the vapor pressure is lowered to 20.7 torr. Calculate the activity coefficient and the activity of the water in this solution.

7/Solution Phenomena: Solubility and Colligative Properties

7.1. Introduction

In the previous chapter we discussed chemical reactions occurring in solution. There are other solution phenomena with which the chemist must deal and concerning which thermodynamics can provide valuable insight. In this chapter we shall be particularly interested in the solubility of materials and in the so-called colligative properties, all of which depend on thermodynamic functions.

There are four phenomena which are collectively called *colligative properties*. These are the lowering of the vapor pressure of the solvent when a solute is introduced, the lowering of the solvent's freezing point by solute, the elevation of the solvent's boiling point by solute, and the osmotic pressure of a solution. Four general observations characterize these colligative properties: (1) They are properties of the solvent, which (2) are modified in value by the introduction of a solute; the extent of this modification (3) is proportional to the concentration of added solute molecules, but (4) is independent of their nature.

7.2. Lowering of Vapor Pressure

When a solute is introduced into a solvent, the vapor pressure of the solvent is lowered, the change in vapor pressure being proportional to the mole fraction of solute. This relationship is summarized in Raoult's Law (Eq. 97):

$$p_{\text{solvent}} = X_{\text{solvent}} p^{\circ}_{\text{solvent}},$$

where p_{solvent} is the solvent vapor pressure above the solution and $p^{\circ}_{\text{solvent}}$ is the vapor pressure of the pure solvent, both measured at the temperature of the experiment. We can write this equation in terms of the solute mole fraction:

$$p_{\text{solvent}} = (1 - X_{\text{solute}})p^{\circ}_{\text{solvent}}$$

$$p_{\text{solvent}} - p^{\circ}_{\text{solvent}} = -X_{\text{solute}} p^{\circ}_{\text{solvent}}$$

$$\Delta p = (p^{\circ}_{\text{solvent}} - p_{\text{solvent}}) = X_{\text{solute}} p^{\circ}_{\text{solvent}}. \tag{114}$$

We have previously shown (Eq. 104) for a dilute solution that $X_{\text{solute}} \simeq (M_{\text{solvent}}/1000)m_{\text{solute}}$, where m is the molality and M the molecular weight. Substituting this expression into Eq. (114) we obtain the lowering of the solvent vapor pressure above the solution as a function of the molality of solute present:

$$\Delta p = (Mp^{\circ}/1000)_{\text{solvent}} m_{\text{solute}} \tag{115}$$

$$= K_{vp} m_{\text{solute}}. \tag{116}$$

Note that K_{vp} depends only on the solvent.

Example 32 Use of diethyl ether as an anesthetic for operations under primitive conditions near the equator is difficult since ether's normal boiling point is 34.6°C (94.3°F). One possible method of circumventing this difficulty is the addition of a nonvolatile solute to lower the vapor pressure. If the vapor pressure of ether is 921 torr at 40°C (104°F) what would be the minimum solute molality necessary to keep ether from boiling at this temperature?

Answer: Since ether will boil at sea level whenever its vapor pressure exceeds 1 atm (760 torr), there must be at least enough solute present to lower the vapor pressure to a value just less than 760 torr. Using Eq. (115),

$$(921 - 760)\text{torr} = [(74 \text{ g mole}^{-1})(921 \text{ torr})/1000]m$$

$$m = 161/68.2 = 2.4 \text{ molal.}$$

7.3. Solubility, Experimental Results

Let us perform an experiment. Suppose we mix 4.0 moles of component A and 1.0 mole of component B and heat the mixture until complete dissolution occurs; the resulting liquid solution has a mole fraction $X_A = 0.80$ and $X_B = 0.20$. Now cool the mixture by withdrawing heat at a steady rate. Figure 7.1(a) pictures the time-temperature sequence we would observe, Fig. 7.2(a) the temperature-composition behavior. Point 1 is the initial solution with temperature T_1 and concentration $X_A = 0.80$ and $X_B = 0.20$. The concentration of the solution remains unchanged as the temperature falls until point 2 is reached. At point 2, solid A begins to separate from the solution. The solution is now *saturated*; the substance A in solution is in equilibrium with solid A, and would remain in equilibrium if the temperature were held constant. The solubility of A is a function of temperature, however, and as the temperature is progressively lowered the solubility continues to decrease. A continues to precipitate and X_A in solution continues to diminish (and X_B increases) until point 3 is reached.

At point 3 crystals begin to appear which are visibly different. This solid is found to contain A and B in exactly the same proportions as the surrounding liquid solution at point 3. Consequently there is no further change in composition of the liquid phase as more of this mixed solid appears, and the temperature holds at T_3 until solidification is complete (point 4). T_3 is the *eutectic temperature*.

Continued withdrawal of heat after solidification is complete simply lowers the temperature of this solid mass along the final leg of the Fig. 7.1(a) curve. The overall composition of this solid is, of course, the same as the composition of the initial solution.

Suppose we now repeat the experiment, starting this time with a solution of less extreme proportions, say $X_A = 0.60$. Beginning at the same original temperature, the curves of Figs. 7.1(b) and 7.2(b) picture the new observations.

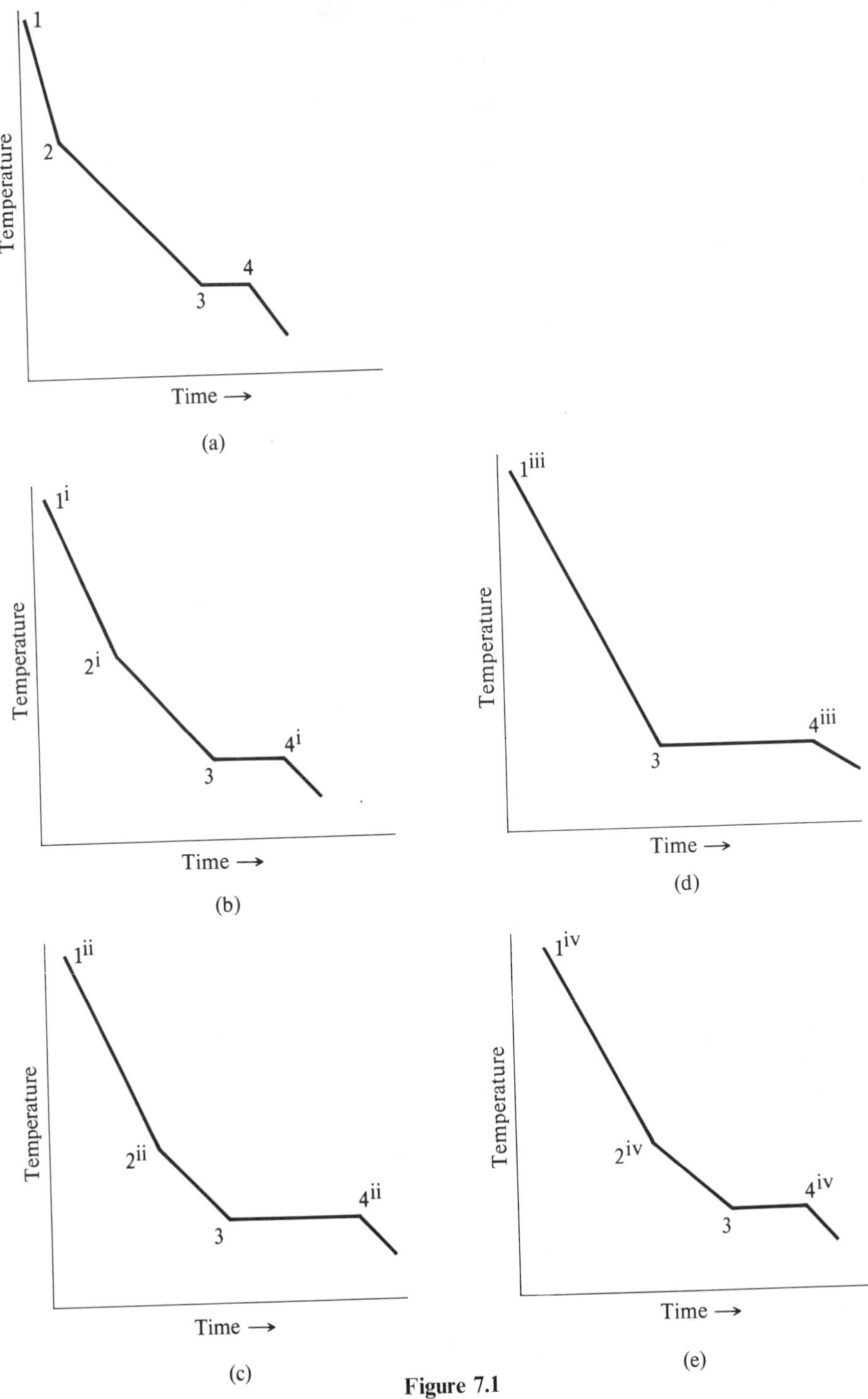
1
2
3
4
Temperature
Time →
(a)
1^i
2^i
3
4^i
Temperature
Time →
(b)
1^iii
3
4^iii
Temperature
Time →
(d)
1^ii
2^ii
3
4^ii
Temperature
Time →
(c)
1^iv
2^iv
3
4^iv
Temperature
Time →
(e)

Figure 7.1

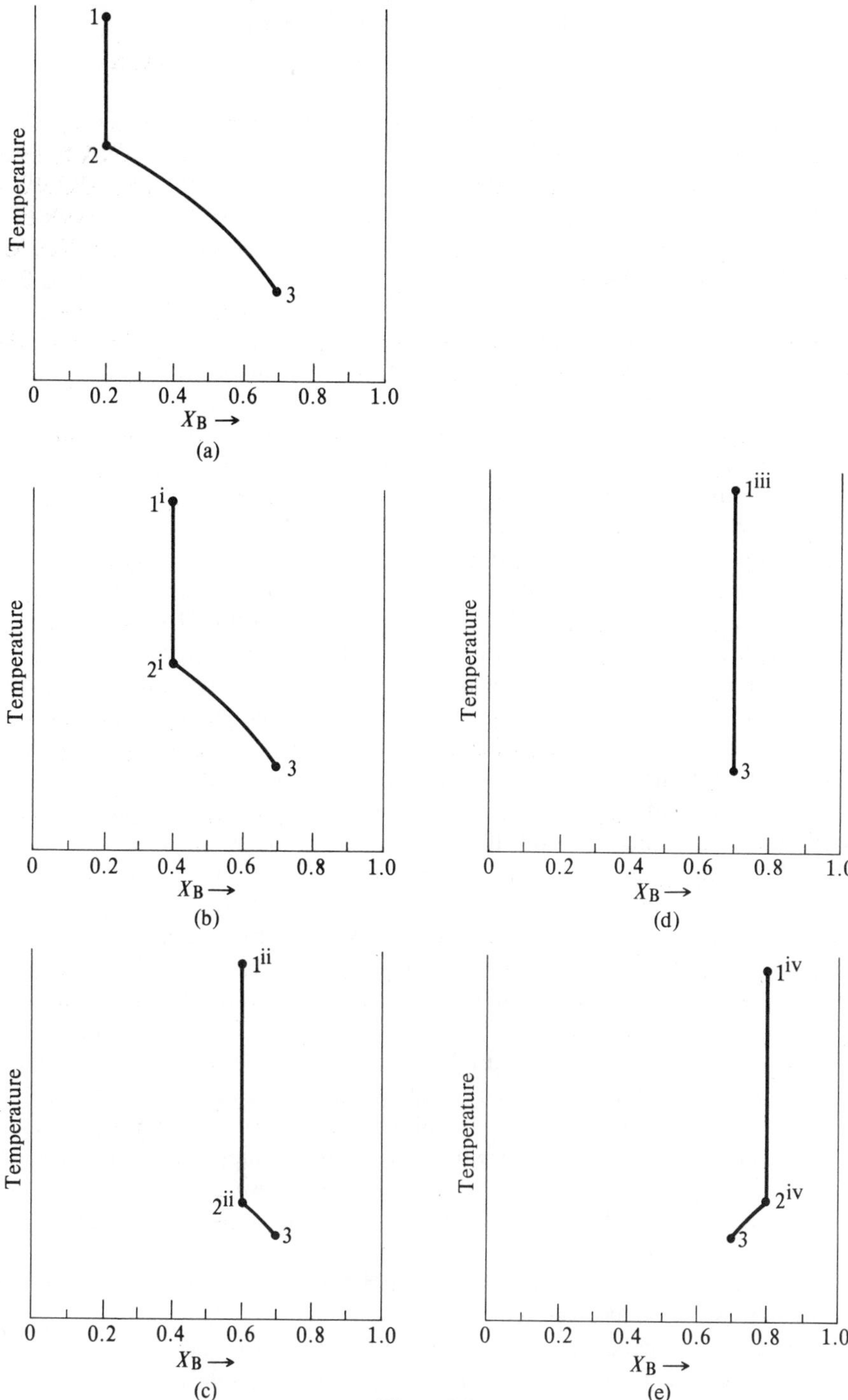

1
2
3
Temperature
0
0.2
0.4
0.6
0.8
1.0
$X_B \rightarrow$
(a)
1^i
2^i
3
Temperature
0
0.2
0.4
0.6
0.8
1.0
$X_B \rightarrow$
(b)
1^{iii}
3
Temperature
0
0.2
0.4
0.6
0.8
1.0
$X_B \rightarrow$
(d)
1^{ii}
2^{ii}
3
Temperature
0
0.2
0.4
0.6
0.8
1.0
$X_B \rightarrow$
(c)
1^{iv}
2^{iv}
3
Temperature
0
0.2
0.4
0.6
0.8
1.0
$X_B \rightarrow$
(e)

Figure 7.2

Since this solution has a smaller excess of A, crystallization of this component from the liquid solution will be longer delayed, $T_{2'} < T_2$. In other words, the solution will reach a lower temperature before becoming saturated. For the same reason more of the eutectic solid can form. Note that the eutectic temperature $T_{3'}$ is the same as that in the first experiment (T_3), and the eutectic concentration is the same in both cases.

Figures 7.1(c) and 7.2(c) trace a third experiment with $X_A = 0.40$. The temperature $T_{2''}$ at which solid A first appears is seen to have shifted still further. Again the eutectic temperature and eutectic composition are the same as observed in the first two experiments.

In a fourth experiment we choose the initial composition of the solution to be the same as the eutectic composition. The behavior observed is recorded in Figs. 7.1(d) and 7.2(d). Now the temperature at which solid first appears coincides with the eutectic temperature.

Figures 7.1(e) and 7.2(e) with $X_A = 0.20$ record our fifth experiment, in which we pass beyond the eutectic composition. Solid first appears at point 2^{iv}, at a temperature above T_3, the eutectic temperature. The solid which appears now is pure B; the solution is saturated with respect to B. This separation of solid B continues, with the temperature falling, until the liquid has adjusted itself to the same eutectic composition and eutectic temperature (point 3) as in the previous experiments. The remaining solution then solidifies without further change in temperature or mole fraction.

Note that in all cases the eutectic has a definite composition and freezes (and conversely melts) sharply like a pure substance, which it is not. This misleading situation can be resolved with a microscope, which shows that the eutectic is a mixture of two kinds of crystals. Also the composition of the eutectic can be changed by changing the pressure of the experiment.

Combining the Fig. 7.2(a–e) curves with information gained from similar experiments, we can obtain a complete temperature-composition diagram for this particular two-component system. In Fig. 7.3, point 3 is again the eutectic; 6 is the normal melting point of pure B ($X_B = 1$); 5 is the normal melting point of pure A ($X_B = 0$). At all points along the curve between points 5 and 3, solid A is in equilibrium with A in solution. Points on this curve give the temperature at which a solution with a particular composition will become saturated and solid A will appear. Conversely this curve gives the saturation concentration—the *solubility*—of A in this solution at particular values of temperature.

Similarly the curve between points 3 and 6 in Fig. 7.3 gives the solubility of B in this solution at given temperatures.

In terms of chemical potentials, at every point on the curve between points 6 and 3, $\mu_B(\text{soln}) = \mu_B(\text{solid})$ and $\mu_A(\text{soln}) < \mu_A(\text{solid})$. At each point on the curve between points 5 and 3, $\mu_A(\text{soln}) = \mu_A(\text{solid})$ and $\mu_B(\text{soln}) < \mu_B(\text{solid})$. At all points above the curves, both $\mu_A(\text{soln}) < \mu_A(\text{solid})$ and

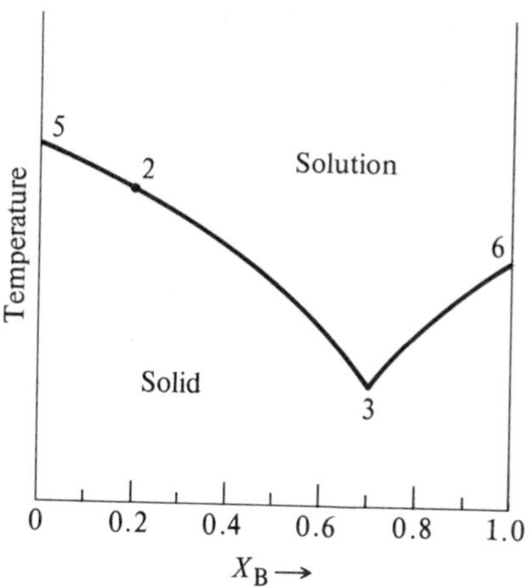

Figure 7.3

$\mu_B(\text{soln}) < \mu_B(\text{solid})$. At point 3, $\mu_A(\text{soln}) = \mu_A(\text{solid})$ and $\mu_B(\text{soln}) = \mu_B(\text{solid})$ simultaneously.

Similar experimental solubility diagrams (see Fig. 7.4) can be determined for all two-component liquid solutions in which the two components show complete mutual solubility in the liquid state but insolubility in the solid state.

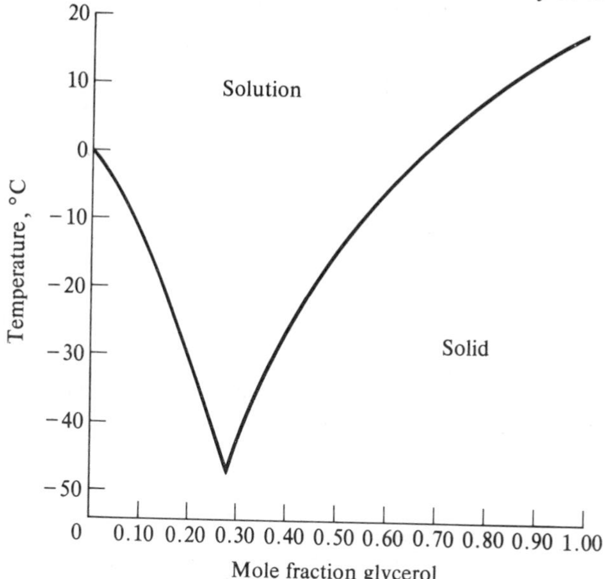

Fig. 7.4. Mutual solubility of the common antifreeze agent glycerol and water (Bosart and Snoddy, *Industrial Engineering Chemistry*, **19**, 506 (1927); Lane, *ibid.*, **17**, 924 (1925).

7.4. Solubility, Thermodynamic Relationships

We can develop equations with the aid of thermodynamics which will allow us to predict the influence of temperature on solubility. When substance B begins to crystallize out, $\mu_{\mathrm{B,\,solid}} = \mu_{\mathrm{B,\,soln}}$ at that temperature and concentration. If we confine our attention to extremely dilute or "ideal" solutions,

$$\mu_{\mathrm{B,\,solid}} = \mu^{\circ}_{\mathrm{B,\,soln}} + RT \ln X_{\mathrm{B}}.$$

But $\mu^{\circ}_{\mathrm{B,\,soln}}$ equals $\overline{G}_{\mathrm{B,\,liquid}}$, the molar free energy of the pure liquid, since $\mu^{\circ}_{\mathrm{B,\,soln}}$ is the chemical potential of B in the solution when $X_{\mathrm{B}} = 1$ (see Section 6.3). Also $\mu_{\mathrm{B,\,solid}}$ equals $\overline{G}_{\mathrm{B,\,solid}}$ since this is a pure phase. Therefore

$$\overline{G}_{\mathrm{B,\,liquid}} - \overline{G}_{\mathrm{B,\,solid}} = -RT \ln X_{\mathrm{B}}$$

$$\ln X_{\mathrm{B}} = -\Delta\overline{G}_{\mathrm{fus,\,B}}/RT. \qquad (117)$$

This is the natural logarithm of the solubility of B in *any* other substance at temperature T. Recall Eq. (49), $\Delta G = \Delta H - T\,\Delta S$; thus

$$\ln X_{\mathrm{B}} = -\Delta\overline{H}_{\mathrm{fus}}/RT + \Delta\overline{S}_{\mathrm{fus}}/R.$$

Differentiating with respect to T,

$$d(\ln X_{\mathrm{B}})/dT = \Delta\overline{H}_{\mathrm{fus}}/RT^2, \qquad (118)$$

if we assume $\Delta\overline{H}_{\mathrm{fus}}$ and $\Delta\overline{S}_{\mathrm{fus}}$ to be independent of temperature. Now if we want to know the solubility of B at a particular temperature T, we integrate from the normal freezing point $T_{f,\mathrm{B}}$ where $X_{\mathrm{B}} = 1$ to the temperature T where the solubility is X_{B}:

$$\int_1^{X_{\mathrm{B}}} d(\ln X_{\mathrm{B}}) = (\Delta\overline{H}_{\mathrm{fus,\,B}}/R) \int_{T_{f,\mathrm{B}}}^{T} (dT/T^2)$$

$$\ln X_{\mathrm{B}} = (\Delta\overline{H}_{\mathrm{fus,\,B}}/R)[-(1/T) + (1/T_{f,\mathrm{B}})]. \qquad (119)$$

Deviations, some rather dramatic, from this equation are commonly observed, especially in regions where the solution is not dilute and for solutions whose components interact strongly. These deviations are not surprising in light of the approximations made in the derivation of this equation.

7.5. Lowering of Freezing Point

The discussion of solubility in the previous section leads us immediately to our second colligative phenomenon, the lowering of the freezing point of the solvent by solute. Referring to Fig. 7.3, if we lower the temperature of pure A

($X_A = 1$, $X_B = 0$) it will solidify (freeze) at its normal freezing point T_5. If, however, we add solute B so that the mole fraction of solvent is decreased to 0.8 ($X_B = 0.2$), then A will begin to solidify (freeze) at a lower temperature T_2. The freezing point of substance A has been lowered from T_5 to T_2 by the addition of solute.

We can use Eq. (119) to predict the temperature at which a liquid will freeze when another substance is added to it. And the equation can be considerably simplified for this purpose.

If $X_{\text{solute}} \ll 1$ (that is, if it is a very dilute solution), $\ln (X_{\text{solvent}}) = \ln (1 - X_{\text{solute}}) \simeq -X_{\text{solute}}$. Also

$$[(1/T) - (1/T_{f,\text{A}})] = -(T_{f,\text{A}} - T)/T_{f,\text{A}}\,T \simeq -\Delta T_f/T_{f,\text{A}}^2$$

if ΔT is small. B is in this case the solute, A the solvent. Then Eq. (119) becomes

$$X_B = (\Delta \bar{H}_{f,\text{A}}/R)\, \Delta T_f/T_{f,\text{A}}^2.$$

As we have previously shown, if B is the solute in a dilute solution, $X_B \simeq m_B M_A/1000$ where m_B is the molality of the solute and M_A the molecular weight of the solvent. Therefore

$$m_B M_A/1000 = (\Delta \bar{H}_{f,\text{A}}/RT_{f,\text{A}}^2)\, \Delta T_f$$

$$\Delta T_f = (M_A RT_{f,\text{A}}^2/1000\, \Delta \bar{H}_{f,\text{A}})m_B. \tag{120}$$

Table 7.1. Values of Freezing Point Depression Constant K_f and Boiling Point Elevation Constant K_b for Selected Solvents (°C/unit molality)

Substance	*Normal melting point, °C*	K_f	*Normal boiling point, °C*	K_b
acetone	—	—	56.2	1.71
benzene	5.45	5.0	80.1	2.53
biphenyl	70.0	8.0	—	—
carbon tetrachloride	—	—	76.8	5.03
chloroform	—	—	61.3	3.63
cyclohexane	6.5	20.0	81.4	2.79
ethanol	—	—	78.4	1.23
ethyl ether	—	—	34.6	2.02
methanol	—	—	64.7	0.86
naphthalene	80.22	6.9	—	—
phenol	40.90	7.0	—	—
water	0	1.86	100	0.51

The quantities within the parentheses of Eq. (120) only depend on the solvent A. This collection of constants is called $K_{f,\,A}$, the freezing-point-lowering constant for substance A, and is characteristic of the particular solvent. The resulting equation is then

$$\Delta T_f = K_{f,\,\text{solvent}}\, m_{\text{solute}}. \tag{121}$$

This equation predicts that, regardless of the nature of the solute, the magnitude of the freezing point depression in a given solvent depends only on the molality of the solute. Remember that this is an approximate equation and can only be expected to give good results if the solution is dilute. Table 7.1 contains values of the constant K_f for a number of solvents.

7.6. Boiling Point Elevation

A similar equation can be derived for the elevation of the boiling point of a solvent upon addition of a nonvolatile solute. Again considering A as the solvent in a dilute solution with solute B, at the boiling point of the solution $\mu_{A,\,\text{soln}} = \mu_{A,\,\text{vapor}}$, and $\overline{G}_{A,\,\text{liquid}} + RT \ln X_{A,\,\text{soln}} = \overline{G}_{A,\,\text{vapor}}$. Consequently

$$\Delta\overline{G}_{\text{vaporization}} = RT \ln X_A. \tag{122}$$

Proceeding as before we obtain the following expression:

$$\ln X_A = (\Delta\overline{H}_{\text{vap}}/R)[(1/T) - (1/T_{b,\,A})]. \tag{123}$$

Note that $T < T_b$ is impossible since $\Delta\overline{H}_{\text{vap}}$ is positive and X_A cannot be greater than 1; consequently $\ln X_A$ must be negative. The boiling point of the solution must then be higher than that of pure A. Adding a solute *elevates* the boiling point, whereas it *lowers* the freezing point, provided the solute in both cases restricts itself to the liquid phase. The boiling point equation is not valid if the vapor being formed contains both components. Similarly the freezing point equation is not valid if the homogeneous crystals being formed contain both components.

If $X_B \ll X_A$ we can continue our derivation exactly as before. We would obtain a similar equation for dilute solutions:

$$\Delta T_b = K_{b,\,\text{solvent}}\, m_{\text{solute}}, \tag{124}$$

where

$$K_{b,\,\text{solvent}} = [MRT_b^2/(\Delta\overline{H}_{\text{vap}}\, 1000)]_{\text{solvent}}. \tag{125}$$

Selected values of K_b are given in Table 7.1.

Example 33 A nonvolatile material (such as a permanent antifreeze) added to the water in an automobile radiator both lowers the freezing point in winter and raises the boiling point in summer. How much would the boiling point be raised if the freezing point is lowered by 20°C? $\Delta\bar{H}_{\text{fusion}} = 1.43$ kcal mole^{-1} and $\Delta\bar{H}_{\text{vap}} =$ 9.71 kcal mole^{-1} for water.

Answer: We could use the value of K_f for water in Table 7.1 (1.86) or we could calculate it from the information given using Eqs. (120) and (121):

$$K_f = \frac{(18 \text{ g mole}^{-1})(1.99 \times 10^{-3} \text{ kcal mole}^{-1} \text{ deg}^{-1})(273°)^2}{1000 \times 1.43 \text{ kcal mole}^{-1}}$$

$$= 1.86,$$

which agrees with the tabular value. Now using Eq. (121)

$$\Delta T_f = 20° = (1.86)m_{\text{solute}}$$

$$m = 10.8 \text{ molal.}$$

Again we could use the tabular value of K_b for water (0.51) or calculate it using Eq. (125):

$$K_b = \frac{(18 \text{ g mole}^{-1})(1.99 \times 10^{-3} \text{ kcal mole}^{-1} \text{ deg}^{-1})(373°)^2}{1000 \times 9.71 \text{ kcal mole}^{-1}}$$

$$= 0.51.$$

Using Eq. (124):

$$\Delta T_b = (0.51)(10.8 \text{ molal}) = 5.5°.$$

A more rapid alternative method for the solution of this problem would be to set up a ratio by dividing Eq. (124) by Eq. (121):

$$\Delta T_b/\Delta T_f = K_b/K_f$$

$$\Delta T_b = (0.51/1.86)20°$$

$$= 5.5°.$$

7.7. Osmotic Pressure

The fourth colligative property, *osmotic pressure* π, is of considerable interest in biological work. A *semipermeable* or *selectively permeable membrane* is one through which certain molecular species can diffuse but others cannot. Such membranes have microscopic pores through which molecules smaller than the pore aperture (5 to 10 Å) may pass. Many membranes of the human body are semipermeable. When such a membrane separates two solution phases, a pressure differential called the *osmotic pressure* develops across the membrane.

Suppose we have only pure water on one side of a cell wall (side I) and an aqueous solution of protein on the other side (side II) and that the cell wall is a semipermeable membrane allowing the passage of water but not of the large protein molecules. At equilibrium all materials which can pass back and forth through the membrane must have equal chemical potentials on both sides of the membrane:

$$\mu_{H_2O}(I) = \mu_{H_2O}(II). \tag{126}$$

Consider first the dependence of the chemical potential on concentration. Since we have pure water on side I, $\mu_{H_2O}(I) = \overline{G}_{H_2O}$. On side II, $\mu_{H_2O}(II) = \mu^{\circ}_{H_2O} + RT \ln X_{H_2O} = \overline{G}_{H_2O} + RT \ln X_{H_2O}$ from Eq. (100). Since X_{H_2O} is necessarily less than unity if some protein is present, $RT \ln X_{H_2O} < 0$ and $\mu_{H_2O}(II) < \overline{G}_{H_2O}$. If the concentration effect were the only one operating, the chemical potential of the water on side II would thus always be less than that on side I. Water would continue to diffuse through the membrane from I to II, moving from a region of high chemical potential to one of lower chemical potential. (Such a flow of solvent is called *osmosis*.) If side II has a definite volume, however, the pressure on the II side increases as water molecules continue to crowd into II. Recall Section 3.8: free energy varies with pressure. As the pressure increases in II the free energy is affected. We should therefore consider the effect of pressure on the chemical potential as well as the effect of concentration. Equation (126) should be written to explicitly express $\overline{G}$ as a function of pressure:

$$\begin{aligned} \mu_{H_2O}(I) &= \mu_{H_2O}(II) \\ \overline{G}_{H_2O}(P_I) &= \overline{G}_{H_2O}(P_{II}) + RT \ln X_{H_2O}. \end{aligned} \tag{127}$$

In other words, water diffuses from I to II until the effects of concentration and of pressure cause the chemical potentials on both sides of the membrane to be equal. Let us now derive an expression which relates the osmotic pressure and the concentration at equilibrium. From Eq. (55) we know that at constant temperature $d\overline{G} = \overline{V}\, dP$. Integrating between P_I and P_{II}:

$$\int_{\overline{G}(P_I)}^{\overline{G}(P_{II})} d\overline{G} = \int_{P_I}^{P_{II}} \overline{V}\, dP$$

$$\overline{G}(P_{II}) - \overline{G}(P_I) = \overline{V}_{H_2O}(P_{II} - P_I). \tag{128}$$

Combining Eqs. (127) and (128) we see that

$$(P_{II} - P_I)\overline{V}_{H_2O} = -RT \ln X_{H_2O}.$$

The pressure difference $(P_{II} - P_I)$ between sides I and II is the pressure differential which must exist across the membrane to prohibit a net flow of solvent from side I to side II; it is the osmotic pressure π. Also, as we have seen, if this is a dilute solution $\ln X_{H_2O} = \ln(1 - X_{\text{solute}}) \simeq -X_{\text{solute}}$ and

$$\pi \simeq RTX_{\text{solute}}/\overline{V}_{\text{solvent}}. \tag{129}$$

Being dilute, $X_{\text{solute}} \simeq n_{\text{solute}}/n_{\text{solvent}}$; substitution gives

$$\pi = \frac{n_{\text{solute}}}{n_{\text{solvent}}} \frac{RT}{\overline{V}_{\text{solvent}}} = \left(\frac{n_{\text{solute}}}{\overline{V}_{\text{solvent}}}\right) RT. \tag{130}$$

The parenthetic term, moles of solute per liter of solvent, expresses concentration but not on one of our standard scales. Since the solution is dilute, however, the volume of the solvent comprises essentially the volume of the solution. Equation (130) then becomes

$$\pi = \frac{n_{\text{solute}}}{V_{\text{solution}}} RT = c_{\text{solute}} RT. \tag{131}$$

By recalling that for a dilute solution $X_{\text{solute}} \simeq m_{\text{solute}} M_{\text{solvent}}/1000$, Eq. (130) can alternatively be written in terms of molality:

$$\pi = [(RTM_{\text{solvent}}/(1000\overline{V}_{\text{solvent}})]m_{\text{solute}} \tag{132}$$

$$\pi = K_{\text{os, solvent}}\, m_{\text{solute}}. \tag{133}$$

Notice that the quantities within the brackets in Eq. (132) collectively called K_{os} in Eq. (133) refer to the solvent only.

Example 34 Water is the universal solvent in biological systems. The osmotic pressure of a body fluid is thus the pressure which must be placed on the fluid to keep water from flowing across a semipermeable membrane separating the fluid from pure water. The osmotic pressure of most human physiological fluids is maintained at a value of 6000 torr. To what molality does this osmotic pressure correspond at 37°C?

Answer:

$$K_{\text{os, }H_2O} = (0.0821)(310)(18)/(1000)(0.018)$$
$$= 25.5 \text{ atm molal}^{-1}.$$

Using Eq. (133), we may now easily calculate the molality.

$$m = \pi/25.5$$
$$= (6000/760)/25.5$$
$$= 0.31 \text{ molal}.$$

Since colligative properties depend only on the number of moles present and not on what the solute is, a given body fluid may contain a number of solutes. However, the total number of moles of all these substances in 1000 g of water must be 0.32 if the solution is to have an osmotic pressure of 6000 torr.

7.8. Colligative Properties, a Summary

The four colligative equations are:

Vapor Pressure Lowering	$\Delta p = K_{vp} m_{\text{solute}}$	(116)
	$= p^{\circ}_{\text{solvent}} X_{\text{solute}}$	(97)
Freezing Point Depression	$\Delta T_f = K_{f,\,\text{solvent}} m_{\text{solute}}$	(121)
Boiling Point Elevation	$\Delta T_b = K_{b,\,\text{solvent}} m_{\text{solute}}$	(124)
Osmotic Pressure	$\pi = K_{\text{os},\,\text{solvent}} m_{\text{solute}}$	(133)
	$= c_{\text{solute}} RT$	(131)

The similarity among these equations is evident. In every case the magnitude of the colligative effect is proportional to the concentration but shows no dependence on the nature of the solute. Note also that the proportionality constants for boiling point and freezing point have fixed values, whereas the other two are temperature-dependent.

7.9. Molecular Weights from Colligative Measurements

Chemists have frequent occasion to determine the molecular weight of a substance whose formula they do not know. If the substance is a gas or can easily be vaporized, they may perform a "vapor density" experiment and solve for the approximate molecular weight M from the ideal gas equation

$$PV = nRT = \frac{W}{M} RT.$$

Such a method is obviously not suitable for solids, but the colligative approach meets this need. The chemist selects an appropriate solvent for the unknown and prepares a solution of known amount of solvent and known weight W of solute. The chemist then measures one or more of the colligative properties, the choice of the property being dependent upon the solvents available, the sensitivity and complexity of the apparatus, and the accuracy required. Each of the four colligative equations contains a concentration term from which the number of moles n of solute present can be determined. Since $n_{\text{solute}} = (W/M)_{\text{solute}}$, the solute's molecular weight M can be calculated readily.

Example 35 An alkaloid of unknown structure isolated from *Vinca rosea* (periwinkle) was found to possess antileukemia activity. In order to determine its molecular weight, 19.0 grams of this alkaloid was added to 100 ml of water.

The change in the boiling point of the solution was found to be 0.060°C while the change in the freezing point was 0.220°C.
a) What is the molecular weight of this alkaloid?
b) What is the osmotic pressure of this solution at 25°C?

Answer: a) Table 7.1 gives for water the values $K_b = 0.51$ and $K_f = 1.86$.

$$\Delta T_b = K_b\, m_{\text{solute}} \quad \text{and} \quad m_{\text{solute}} = \frac{1000\,W_{\text{solute}}}{M_{\text{solute}}\,W_{\text{solvent}}}.$$

Substituting and solving,

$$M = \frac{1000 K_b\, W_{\text{solute}}}{\Delta T_b\, W_{\text{solvent}}} = \frac{1000 \times 0.51 \times 19}{0.060 \times 100} = 1600 \text{ g mole}^{-1}.$$

Similarly for the freezing point experiment,

$$M = \frac{1000 K_f\, W_{\text{solute}}}{\Delta T_f\, W_{\text{solvent}}} = \frac{1000 \times 1.86 \times 19}{0.220 \times 100} = 1600 \text{ g mole}^{-1}.$$

b) $m_{\text{solute}} = \Delta T_f / K_f = 0.220/1.86 = 0.118$ molal

In Eq. (131), $\pi = c_{\text{solute}} RT$, the concentration is measured in moles per liter of solution (molarity), not in molality. Since we are dealing with a solution so dilute that its volume is virtually the volume of the solvent, however, and with water as the solvent, for which 1.0 liter weighs 1.0 kg, it follows that molarity and molality are essentially the same in this case. Accordingly

$$\pi = mRT = 0.118 \times 0.0820 \times 298 = 2.88 \text{ atm}.$$

7.10. Activities from Colligative Property Measurements

As pointed out in Section 6.6 we should use activities (a) in our thermodynamic equations rather than molalities (m) when we are dealing with solutions which are not dilute. The colligative property equations would then be

$$\Delta p = K_{vp}\, a_{\text{solute}} \tag{132}$$

$$\Delta T_b = K_b\, a_{\text{solute}} \tag{133}$$

$$\Delta T_f = K_f\, a_{\text{solute}} \tag{134}$$

$$\pi = K_{\text{os}}\, a_{\text{solute}}. \tag{135}$$

By measuring the colligative properties we can use these equations to determine the activity. In dilute solution the activity will seldom vary significantly from the molality. In more concentrated solutions, however, the activity coefficient $\gamma = a/m$ can be markedly different from unity.

7.11. The van't Hoff Factor

If partial association or dissociation occurs when the solute dissolves, use of the colligative property equations as we have derived them leads to erroneous results. The colligative properties depend upon the number of solute particles present. Upon dissociation the solute molecules may simply produce smaller molecules, or they may form ions. In either case the number of solute particles is increased, and this multiplication of particles will result in all four colligative effects being greater (and equally greater) than would otherwise have been the case. This may be expressed by introducing a factor i, called the *van't Hoff factor*, into the colligative equations such that each effect is increased i-fold:

$$\Delta P = iK_{vp} m_{\text{solute}} \tag{136}$$

$$\Delta T_b = iK_b m_{\text{solute}} \tag{137}$$

$$\Delta T_f = iK_f m_{\text{solute}} \tag{138}$$

$$\pi = iK_{\text{os}} m_{\text{solute}}\,. \tag{139}$$

Thus the solute, of formal concentration m, exhibits colligatively an apparent concentration $m' = im$.

For dissociation, the value of i will lie between the limits 1 (no dissociation) and n, the number of particles resulting from the total dissociation of a single parent molecule. Suppose that when we introduce 5.0 moles of acetic acid into 1000 g of water, complete dissociation occurs:

$$\text{acetic acid} \longrightarrow \text{acetate}^- + H^+.$$

While the formal concentration is 5 molal the apparent concentration m' is 10 molal with $i = n = 2$. If the dissociation is not complete at equilibrium we have acetic acid, acetate, and hydrogen ions present,

$$\text{acetic acid} \rightleftharpoons \text{acetate}^- + H^+;$$

i will be less than 2 and m' will be between 5 and 10 molal.

The fraction or *degree of dissociation* α relates quantitatively the known value of n and the experimental value of i. Before dissociation the concentration of acetic acid is m; afterward, $m - \alpha m$. After dissociation the concentrations of acetic and hydrogen ion are each αm. The concentration of dissociated particles is $2\alpha m$ or, in general, $n\alpha m$. The total concentration of dissociated and undissociated solute particles is $n\alpha m + (m - \alpha m)$, and

$$m' = m(1 + n\alpha - \alpha) \tag{140}$$

$$i = m'/m = (1 + n\alpha - \alpha) \tag{141}$$

$$\alpha = (i - 1)/(n - 1). \tag{142}$$

If association occurs, the number of solute particles is decreased and the colligative properties will consequently all be smaller than would have been expected if no association occurs. The value of i will be less than 1 and will lie between the limits 1 (no association) and $1/p$, where p is the number of parent molecules in each associated complex when association is complete. Suppose we introduce 5.0 moles of benzoic acid into 1000 g of solvent and complete association occurs. Then

2 (C₆H₅—C(=O)—OH) ⟶ C₆H₅—C(=O·HO)(—OH·O=)C—C₆H₅

benzoic acid — benzoic acid dimer

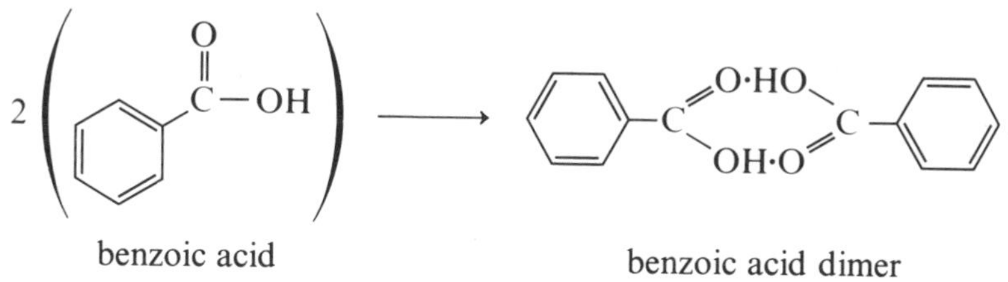

or in general

$$pA \longrightarrow A_p.$$

While the formal concentration is 5 molal, the apparent concentration m' is $5/p$ ($2\frac{1}{2}$ in the case of dimer formation where $p = 2$) with $i = 1/p$ ($\frac{1}{2}$ when $p = 2$). If association is not complete, at equilibrium uncomplexed parent molecules as well as complexes [and in many cases partial complexes of the type $(A)_{p-1}$, $(A)_{p-2}$, etc.] will be present; i will be greater than $1/p$ but less than 1.

The equations derived for dissociation (Eqs. 140–142) also apply for association if we substitute α', the degree of association, for α and $1/p$ for n:

$$m' = m(1 - \alpha' + \alpha'/p) \tag{143}$$

$$i = 1 - \alpha' + \alpha'/p \tag{144}$$

$$\alpha' = (i - 1)/[(1/p) - 1]. \tag{145}$$

If the formula weight is known, the colligative properties can be used to determine the degree of association or dissociation of the solute.

Example 36 $BaCl_2$ dissociates somewhat in aqueous solution:

$$BaCl_2 \rightleftharpoons Ba^{+2} + 2Cl^-.$$

If a weighted 1.50-molal aqueous solution of $BaCl_2$ is observed to freeze at $-6.70°C$, what fraction is apparently dissociated?

Answer: The apparent concentration is

$$m' = \Delta T_f/K_f = 6.70/1.86 = 3.60 \text{ molal},$$

$$i = m'/m = 3.60/1.50 = 2.40, \qquad \text{and}$$

$$\alpha = (i-1)/(n-1) = (2.40-1)/(3-1) = 0.70,$$

or 70% of the $BaCl_2$ molecules have dissociated.

Example 37 Assuming that NaCl completely dissociates in water solution, what molality should a NaCl solution be in order that it be isosmotic with (have the same osmotic pressure as) body fluids?

Answer: In Example 34, we determined that body fluids are 0.31 molal. Since $i = 2$ for the NaCl solution, half as many moles of NaCl will produce the same number of particles as a substance which does not ionize. An isosmotic NaCl solution should therefore be 0.15 molal.

7.12. Ionic Activity

In the previous section the fact that a substance may dissociate into ions when it is added to a solvent was considered, and the effect on the colligative properties caused by the increase in the numbers of particles resulting from such dissociation was discussed. The existence of ions in solution is responsible for a number of solution phenomena including the conduction of electrical current, which will be discussed in the next chapter. In dealing with such phenomena we often encounter cases in which our observations vary significantly from the thermodynamic equations. Such deviations can frequently be minimized by using activities instead of concentrations in our equations (see Section 6.6).

The variation of the activity coefficient from the ideal value in solutions containing ions can be attributed to the interaction of the ions with other ions and with *dipoles*. A molecule possesses a dipole if it is electrically neutral but the centers of positive and negative charges do not coincide. Almost all nonsymmetrical molecules have dipoles (for example, H_2O and CH_3OH), whereas symmetrical molecules do not, CCl_4 and CO_2 for instance. Substances which are good solvents for ionic materials are often *polar* (possess dipoles). Because of the electrostatic interaction with other ions and with the dipoles of the solvent, the "effective concentration" or activity of an ionic species in solution may be quite different from its formal concentration.

We encounter a problem immediately. We have no way of determining the activity of a single ionic species such as Na^+ in a salt solution. Instead we define the *mean ionic activity* $a_\pm$. If the substance AB dissociates in solution to give

$$\text{AB} \longrightarrow \nu_+ \text{C}^+ + \nu_- \text{D}^-,$$

that is, ν_+ "moles" of positive ions and ν_- "moles" of negative ions for each mole of AB, the mean ionic activity is defined as

$$a_\pm = (a_+^{\nu_+} a_-^{\nu_-})^{1/\nu}, \tag{146}$$

where a_+ is the activity of the positive ions and a_- the activity of the negative ions (neither of which can be measured directly), and $\nu = \nu_+ + \nu_-$. Similarly, the *mean ionic activity coefficient* is defined as

$$\gamma_\pm = (\gamma_+^{\nu_+} \gamma_-^{\nu_-})^{1/\nu}, \tag{147}$$

and the *mean ionic molality* as

$$m_\pm = (m_+^{\nu_+} m_-^{\nu_-})^{1/\nu}. \tag{148}$$

These quantities are related by

$$a_\pm = \gamma_\pm m_\pm. \tag{149}$$

The mean ionic activity can be determined from colligative measurements and from electrochemistry. Since the formal concentration is known, m_+ and m_- are easily calculated and Eq. (149) can be used to determine $\gamma_\pm$.

For the dissociation of silver bromide, a salt which finds extensive use in photography,

$$AgBr(s) \rightleftharpoons Ag^+(aq) + Br^-(aq),$$

and the *solubility product* K_{sp} is

$$K_{sp} = (a_{Ag^+})(a_{Br^-})/a_{AgBr}.$$

Since the silver bromide solid is a pure phase, $a_{AgBr} = 1$ and we have

$$K_{sp} = (\gamma_{Ag^+} m_{Ag^+})(\gamma_{Br^-} m_{Br^-}) = (m_{Ag^+} m_{Br^-})(\gamma_{Ag^+} \gamma_{Br^-}).$$

From our definition of $\gamma_\pm$ (Eq. 147), $\gamma_{Ag^+} \gamma_{Br^-} = \gamma_\pm^2$ and

$$K_{sp} = (m_{Ag^+} m_{Br^-})\gamma_\pm^2.$$

7.13. Donnan Membrane Equilibrium

An interesting application of these principles important in biological systems is the *Donnan membrane equilibrium.* Such an equilibrium results when two solutions are separated by a membrane which shows selective permeability toward the ionic species present.

As a simple case, consider a cell membrane separating two solutions of equal volume: on side A there is sodium chloride at concentration C_A, while on side B there is a sodium proteinate at concentration C_B. It is assumed that both the salt NaCl and the proteinate NaPr are completely ionized, and that the pores of the membrane allow free passage to all the ions present except the very large Pr^- ion.

Three considerations determine what will happen. First, there being initially no Cl^- ions in compartment B, an inflow of these ions from compartment A will occur, in a move toward equalization. Second, since electroneutrality of the solution must be maintained throughout, these Cl^- ions must be accompanied by an equal number of Na^+ ions. Third, the flow will continue until the chemical potential of NaCl on each side of the membrane is the same, which requires that the product of the Na^+ concentration and the Cl^- concentration on side A be equal to the product of these two concentrations on side B. The situation may be illustrated as shown in Fig. 7.5.

Since equilibrium requires that

$$(m_{\mathrm{Na}^+})_{\mathrm{A}}(m_{\mathrm{Cl}^-})_{\mathrm{A}} = (m_{\mathrm{Na}^+})_{\mathrm{B}}(m_{\mathrm{Cl}^-})_{\mathrm{B}},$$

then from Fig. 7.5,

$$(C_{\mathrm{A}} - C_x)(C_{\mathrm{A}} - C_x) = (C_{\mathrm{B}} + C_x)(C_x)$$
$$C_{\mathrm{A}}^2 - 2C_{\mathrm{A}}C_x + C_x^2 = C_{\mathrm{B}}C_x + C_x^2$$
$$C_{\mathrm{A}}^2 = C_x(C_{\mathrm{B}} + 2C_{\mathrm{A}})$$
$$C_x/C_{\mathrm{A}} = C_{\mathrm{A}}/(C_{\mathrm{B}} + 2C_{\mathrm{A}}).$$

C_x/C_{A} expresses the fraction of the original NaCl which has migrated from the A solution through the membrane into the B solution.

Donnan theory is important in explaining many phenomena, including ion exchange, diffusion through biological membranes, the swelling of gelatin, and the behavior of macromolecular ions such as proteins.

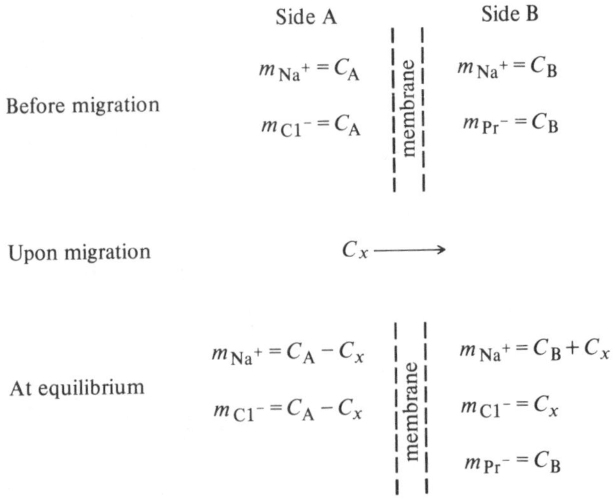

	Side A		Side B
Before migration	$m_{\mathrm{Na}^+} = C_{\mathrm{A}}$ $m_{\mathrm{Cl}^-} = C_{\mathrm{A}}$	membrane	$m_{\mathrm{Na}^+} = C_{\mathrm{B}}$ $m_{\mathrm{Pr}^-} = C_{\mathrm{B}}$
Upon migration	$C_x \longrightarrow$		
At equilibrium	$m_{\mathrm{Na}^+} = C_{\mathrm{A}} - C_x$ $m_{\mathrm{Cl}^-} = C_{\mathrm{A}} - C_x$	membrane	$m_{\mathrm{Na}^+} = C_{\mathrm{B}} + C_x$ $m_{\mathrm{Cl}^-} = C_x$ $m_{\mathrm{Pr}^-} = C_{\mathrm{B}}$

Figure 7.5

Problems

7.1 The extracellular fluid is said to be *isosmotic* when its osmotic pressure is equal to that of the fluid within the cell. Under such conditions there is no net flow of water into or out of the cell. In which direction will water diffuse if the extracellular fluid is *hyperosmotic* (has a higher osmotic pressure than the intracellular fluid) and in which direction when it is *hyposmotic* (has a lower osmotic pressure)?

7.2 (a) What is the boiling point of a typical physiological fluid? See Example 34. (b) At what temperature will such a fluid freeze?

7.3 When a semipermeable membrane separates two solutions of different concentrations (and therefore different osmotic pressures relative to pure solvent), a flow of solvent will occur across the membrane unless additional pressure is applied to one of the solutions. (a) Should additional pressure be applied to the more dilute or to the less dilute solution? The pressure differential, the osmotic pressure, between the solutions is the difference between the osmotic pressures for the two solutions relative to pure solvent. (b) The effective concentration of the extracellular fluid surrounding the capillaries is 0.31 molal while the blood in the capillaries has an effective concentration of 0.35 molal. What is the osmotic pressure of each of these solutions at 37°C relative to pure water? (c) What is the osmotic pressure when these two solutions are separated by a semipermeable membrane (the capillary wall) which is only permeable to water? (d) In which direction will water diffuse? (This tendency is also affected by the hydrostatic pressure produced by the pumping action of the heart.)

7.4 The following are values of ΔT_f for a number of substances in 0.01 molal aqueous solutions:

a) aniline	0.0185
b) citric acid	0.0226
c) picric acid	0.0363
d) D-tartaric acid	0.0234

Assuming the activity coefficient γ is unity, calculate the van't Hoff factor i and the degree of dissociation for each solution.

7.5 The following are experimental values for $\Delta T_f/m$ for 0.001, 0.005, and 0.01 molal aqueous solutions of several substances which are considered to be completely ionized in solution:

Concentration (molal)	0.001	0.005	0.010
KCl	3.66	3.65	3.61
$Pb(NO_3)_2$	5.37	5.09	4.90

In each case give the value of the van't Hoff factor i and calculate the activity coefficient in the 0.01 molal solution.

7.6 Will an egg cook faster if boiled in pure water or in water to which salt has been added? Why?

7.7 Using Fig. 7.4, describe the sequence of events in an automobile radiator containing 0.1 mole fraction glycerol as an *extreme* cold front passes through and the temperature drops from 75°F to −75°F. Also describe the sequence for the same temperature change but with a 0.8 mole fraction glycerol solution.

7.8 Sprinkling salt over ice patches on streets causes the ice to melt since the salt lowers the freezing point of water. The following are values of freezing points of various aqueous $CaCl_2$ solutions.

Weight percent $CaCl_2$	*Freezing point* °C
5.0	27.7
10.0	22.3
15.0	13.5
20.0	−0.4
25.0	−21.0
27.0	−31.2
29.0	−49.4
30.0	−50.8
31.0	−33.2
33.0	−6.9
35.0	14.4
40.0	55.9
100.0	772

(a) Estimate the solubility of $CaCl_2$ in water at (1) 0°C and (2) 25°C. (b) Estimate the temperature and composition of the eutectic. (c) What solid appears when a 25.0 weight percent $CaCl_2$ solution is cooled below −21.0°C? (d) What solid appears when a 33.0 weight percent $CaCl_2$ solution is cooled below −6.9°C?

7.9 A purified protein-type substance, believed to have a molecular weight in the neighborhood of 50,000, is to have this figure more accurately determined. (a) Assuming the instrumentation is available, which of the colligative methods should give best results? To answer this, calculate for a 2% solution of the material in water the boiling point elevation; the freezing point depression; and, in torr at 25°C, the vapor pressure lowering and the osmotic pressure. (b) Suppose osmotic pressure is chosen and measurements on the 2% solution show a pressure differential of 0.01020 atm at 25°C. What is the indicated molecular weight? (c) If the protein material, unknown to the experimenter, actually contains 500 ppm of sodium chloride, what percentage error will this produce in the answer?

7.10 When 25.0 ml of benzene is pipetted onto a sample of 0.370 g of solid propionic acid, the resulting solution is found to freeze at 4.91°C. What does this experiment say about the molecular aggregation of propionic acid in this nonpolar solvent? Benzene has density 0.879, freezing point 5.50°C, and freezing point depression constant 5.12.

7.11 A 10% solution of cane sugar, originally pure, has undergone partial inversion according to the reaction

$$\begin{array}{lcl}\text{sucrose} \quad + \text{ water} & \longrightarrow & \text{glucose} \quad + \text{ fructose} \\ C_{12}H_{22}O_{11} + H_2O & \longrightarrow & C_6H_{12}O_6 + C_6H_{12}O_6 .\end{array}$$

If the boiling point of this solution, after barometric correction, is 100.27°C, (a) what is the average molecular weight of the dissolved material? (b) What fraction of the original sugar has inverted?

7.12 Low temperatures are often handled in industry by circulating brine solutions. Such solutions are usually made of calcium chloride (see Problem 7.8) since, compared to sodium chloride, the calcium salt is more soluble and less corrosive. Compare quantitatively the freezing point lowering capability of these two salts when competing (a) mole for mole, (b) pound for pound, and (c) at saturation. Consult a handbook for eutectic temperatures when solving (c).

8 / Electrochemistry

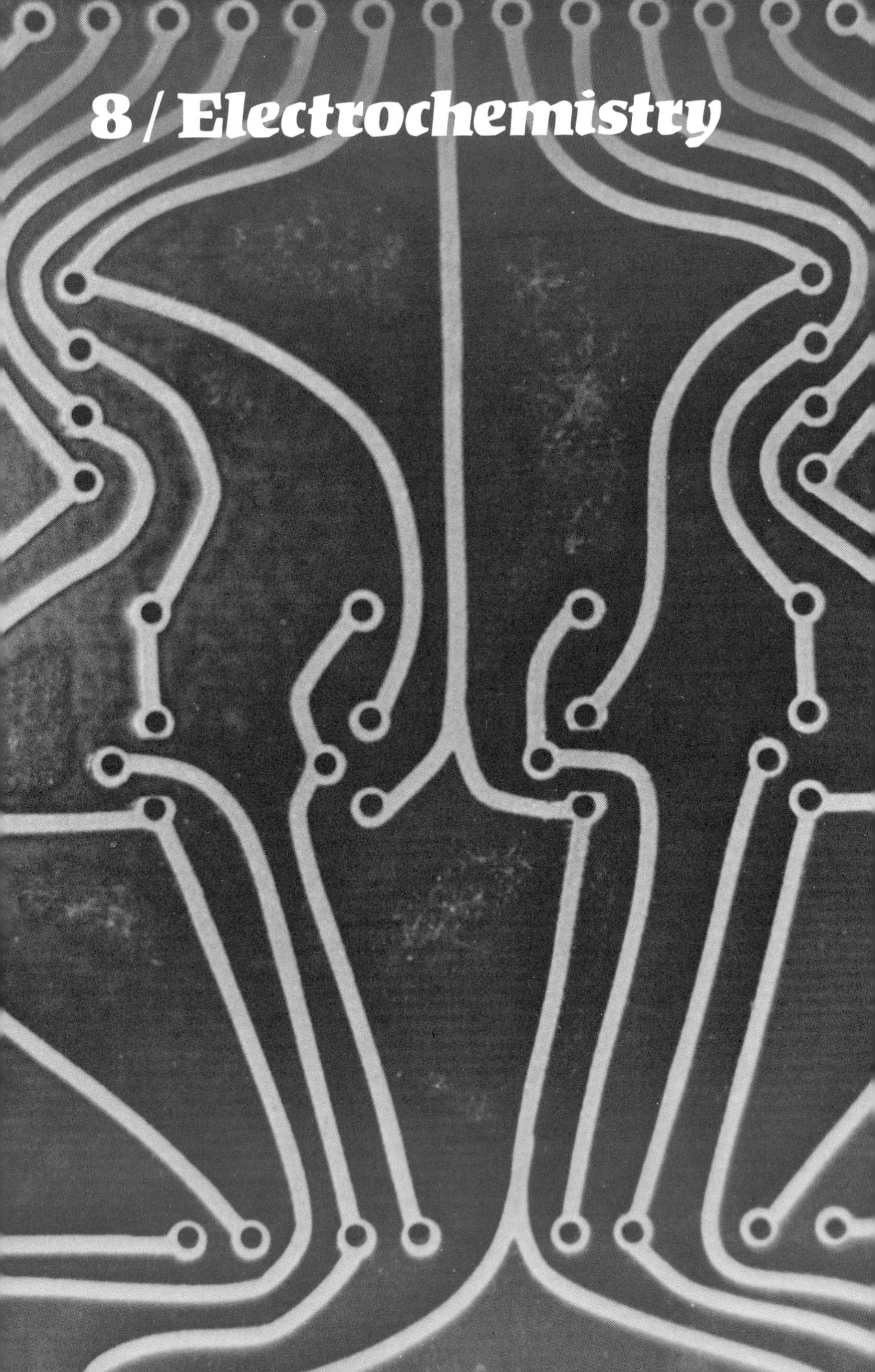

8.1. Conductance

The existence of an electrical potential difference or *voltage* difference V across a material able to conduct electricity causes a *current* flow through the material. The amount of current I is given by *Ohm's law*:

$$I = V/R, \tag{150}$$

where R is the *resistance* of the material to the flow of current. The *conductance* C is defined as the reciprocal of the resistance:

$$C = 1/R, \tag{151}$$

so that Ohm's law may also be written

$$I = VC. \tag{152}$$

In the case of a metallic substance such as a copper wire, the electrical current is carried through the material by the movement of electrons. If the conducting medium is an ionic solution, the current is composed of positive and negative ions migrating in opposite directions through the solution. In such a conducting solution there may be an additional factor: chemical reactions occur if the ions give up or receive electrons. The study of such reactions is of great interest to chemists and to scientists in related fields. We shall see that this study, known as *electrochemistry*, makes important application of thermodynamic principles as well as being a major source of thermodynamic information.

8.2. Oxidation and Reduction

A chemical reaction which involves the transfer of electrons from one atom, molecule, or ion to another is called an *oxidation-reduction* (or *redox*) reaction. For example,

$$Pb^{+4} + Hg \longrightarrow Pb^{+2} + Hg^{+2}.$$

Here two electrons move from each uncharged mercury atom, leaving Hg^{+2}, and attach themselves to the Pb^{+4} ion, resulting in Pb^{+2}. There is a change in the *oxidation state* of both lead and mercury. The Hg atom is *oxidized*: it *loses* two electrons and goes to the higher oxidation state Hg^{+2}. Conversely Pb^{+4} is *reduced*: it *gains* two electrons, dropping to the lower oxidation state Pb^{+2}. In such a process the electrons lost by the material being oxidized must equal in sum total those gained by the material being reduced; that is, the "electron accounting" must always be complete. Thus oxidation and reduction must be simultaneous and equivalent.

Example 38 Assign oxidation numbers to the key atoms in the following reactions and indicate the oxidation-reduction process occurring in each:

a) $H_2(g) + S(s) \rightarrow H_2S(g)$

b) $H_2O_2(l) \rightarrow H_2O(l) + \frac{1}{2}O_2(g)$

c) $2K_2Cr_2O_7(s) + 3S(s) \rightarrow 3SO_2(g) + 4KCrO_2(s)$

Answer: a) Hydrogen is oxidized from a zero oxidation state in H_2 to $+1$ in H_2S, while sulfur is reduced from zero in S(s) to -2 in H_2S.

b) One oxygen atom in hydrogen peroxide is reduced from -1 to -2 in water while the other oxygen atom is oxidized from -1 to zero in O_2. Hydrogen remains $+1$ throughout.

c) Chromium changes from a $+6$ state in $K_2Cr_2O_7$ to a $+3$ state in $KCrO_2$. Sulfur changes from a zero state in S(s) to a $+4$ state in SO_2. The oxidation states of potassium and oxygen do not change.

8.3. Electrochemical Cells

Any spontaneous redox reaction can theoretically be made into a current-producing electrochemical cell, with the reduction part of the reaction physically separated from the oxidation part and the electron transfer taking place via an external conductor such as a copper wire. A schematic drawing of such a cell is shown in Fig. 8.1. The barrier separating the two compartments or *half-cells* permits electrical contact but prevents mixing of the two solutions.

The *half-cell reaction* occurring on the right (A) in the cell pictured involves the gain of one or more electrons by a chemical species. For example,

$$Pb^{+4} + 2e^- \longrightarrow Pb^{+2}.$$

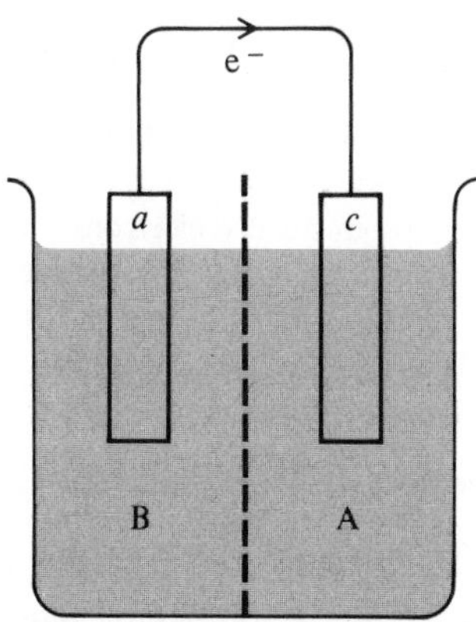

Figure 8.1

The half-cell reaction on the left (B) involves electron loss. For example:

$$Hg \longrightarrow Hg^{+2} + 2e^-,$$

where e^- is an electron. The total cell reaction is obtained by adding the two half-cell reactions with a cancellation of electrons:

$$Pb^{+4} + Hg \longrightarrow Pb^{+2} + Hg^{+2}.$$

It is accomplished with a flow of electrons through the wire from left to right. Electrons are given up to the external conductor at an *electrode a*, the *anode*; thus oxidation occurs at the anode. Reduction occurs on the right as electrons are picked up from the *electrode c*, the *cathode*. Each electron carries a charge of -1.6×10^{-19} coul. The flow of Avogadro's number N of electrons (1.0 "mole" of electrons) is equivalent to $-96{,}487$ coul, the charge on Avogadro's number of electrons. The magnitude of this value is called the *faraday F*. The amount of material which reacts to produce or absorb one faraday (N electrons) is called an *equivalent*.

Example 39 In the presence of an enzyme catalyst, acetaldehyde CH_3CHO can be reduced to ethanol CH_3CH_2OH. A similar reaction can be made to occur in an electrochemical cell:

$$CH_3CHO + H_2 \longrightarrow CH_3CH_2OH.$$

If 1.0 mole of acetaldehyde is converted to 1.0 mole of ethanol how many electrons will be transmitted through the external circuit? The *gram equivalent weight* of a substance is the weight in grams of 1.0 equivalent of that substance. What are the equivalent weights of hydrogen, acetaldehyde, and ethanol in this cell?

Answer: The oxidation state of hydrogen changes from zero to $+1$ in this reaction. When one molecule of CH_3CHO reacts, 2 atoms of hydrogen react; thus 2 electrons are lost by hydrogen and 2 are picked up by CH_3CHO. For 1.0 mole of reaction, $2N$ electrons will be transferred.

Since 1.0 mole of H_2 produces $2N$ electrons, one equivalent weight of H_2 would be half a mole of hydrogen. The gram equivalent weight of H_2 is then the weight in grams of $\frac{1}{2}$ mole H_2 or $M_{H_2}/2 = 1$ g. Similarly the equivalent weights of CH_3CHO and CH_3CH_2OH are equal to their molecular weights divided by 2.

The spontaneous electrical cell outlined above is called a *galvanic cell*. Batteries are galvanic cells. It is also possible to bring about a non-spontaneous reaction by doing electrical work on the cell, using an external source such as a motor-driven generator. This type of electrochemical cell is called an *electrolytic cell*.

8.4. Electrical Work, Free Energy

The flow of electrons through the external conductor during a spontaneous chemical reaction can be used to perform electrical work W_{el}. The amount of such work which a galvanic cell can perform is given by

$$W_{el} = qV, \tag{153}$$

where q is the amount of charge moving through an electrical potential or voltage difference V. This charge will be equal to n, the number of equivalents which react, multiplied by $-F$, the charge on 1.0 equivalent:

$$q = -nF = -n(96487 \text{ coul equiv}^{-1}). \tag{154}$$

A voltage difference between two electrodes causes electrons to flow spontaneously through the external conductor. This difference exists because one half-cell reaction has a greater affinity for electrons than the other. In our thermodynamic language we say that there is a free-energy difference between the two electrodes. We could measure the voltage difference between the two with a voltmeter and calculate the W_{el} via Eqs. (153) and (154). This would be the electrical work performed when n equivalents react—when nN electrons flow through the external conductor impelled by that particular voltage difference.

Of greater interest is the maximum amount of work such a redox reaction can supply. Recall that under isothermal conditions the greatest amount of work a system can do in changing from one given state to another is the magnitude of the reversible work W_{rev}. If no pressure-volume work is done during the process (for reactions in solution there would be very little if any change in volume and therefore $W_{PV} \simeq 0$), then $W_{rev} = W_{net,\,rev} = \Delta G$. In our case

$$\Delta G = W_{net,\,rev} = W_{el,\,rev}. \tag{155}$$

Only when the voltage difference between the two electrodes is measured in a reversible manner will the W_{el} calculated via Eqs. (153) and (154) be $W_{el,\,rev}$ and thus ΔG. Ordinary voltmeters require a current flow to actuate the needle; this drain reduces the voltage being measured, and thus the measurement is not made reversibly. We need to measure the electrical potential difference with a device which does not require a flow of current. This can be accomplished with a potentiometer (Fig. 8.2). When connected as shown, the cell whose reversible voltage ε is being measured is seen to be in opposition to a controllable fraction of the greater voltage V from the external battery. In operation, the sliding contact c is adjusted until the galvanometer G shows zero deflection. Since the slide wire is uniform, the resistance over any given

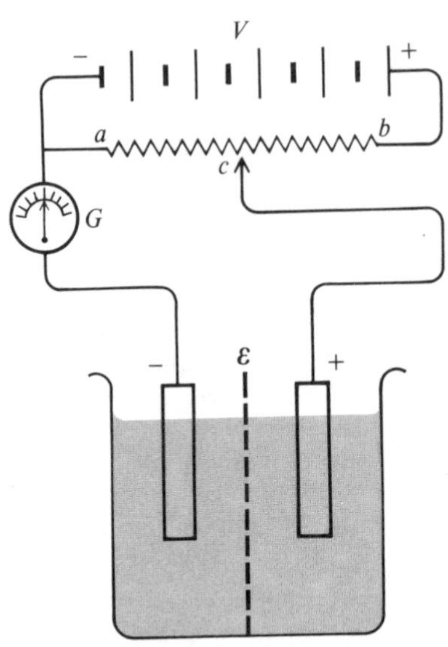

Fig. 8.2. Here V represents a battery of known voltage greater than that being measured. Unit $\overline{ab}$ is a high-resistance wire with a sliding contact c. Instrument G is a galvanometer, which sensitively detects a flow of current in either direction.

section is proportional to the length of that section. The potentiometer equation then becomes

$$V/\varepsilon = \overline{ab}/\overline{ac}.$$

Knowing the voltage V and the length $\overline{ab}$, we can measure the length ac and quickly calculate the unknown potential:

$$\varepsilon = (V/\overline{ab})\overline{ac}. \tag{156}$$

By inserting a variable resistance at point b, it is further possible to make $V/\overline{ab} = 1$. The potentiometer is then said to be "direct reading."

The reversible voltage thus measured is called the *electromotive force* or *emf* and will be represented by ε. If we use ε in Eqs. (153) and (154) we obtain the reversible electrical work which is ΔG:

$$\Delta G = W_{\text{el, rev}} = -nF\varepsilon. \tag{157}$$

From a measurement of ε for a redox reaction we can therefore directly obtain the value of ΔG.

8.5. Temperature Dependence of emf; Determination of ΔS and ΔH

Equation (58) gives the variation of ΔG with T:

$$(\partial\, \Delta G/\partial T)_p = -\Delta S. \tag{58}$$

By incorporating $\Delta G = -nF\varepsilon$, we obtain an equation for the temperature dependence of the electromotive force:

$$\left(\frac{\partial \varepsilon}{\partial T}\right)_p = \Delta S/nF. \tag{158}$$

We can conversely obtain ΔS via Eq. (158) by measuring ε at various temperatures. We can then find ΔH by using Eq. (49):

$$\Delta H = \Delta G + T\, \Delta S \tag{49}$$

Example 40 The temperature dependence of the emf of an electrochemical cell can often be written in an equation of the form

$$\varepsilon = (a + bT + cT^2 + dT^3) \text{ volt}, \tag{159}$$

where a, b, c, and d are constants. A certain commercially available battery was found to have $a = 1.19237$, $b = -1.537 \times 10^{-4}$, $c = 2.73 \times 10^{-8}$, and $d = 1.78 \times 10^{-11}$. Determine ΔG, ΔH, and ΔS for this cell at 27°C if $n = 3$.

Answer: Since 96,487 coul equiv^{-1} is equal to 23.06 kcal volt^{-1} equiv^{-1}, and recalling that $\Delta G = -nF\varepsilon$,

$$\begin{aligned}\Delta G &= -3(23.06)[1.19237 - (1.537 \times 10^{-4})(300)\\ &\quad + (2.73 \times 10^{-8})(300)^2 + (1.78 \times 10^{-11})(300)^3]\\ &= -79.52 \text{ kcal}.\end{aligned}$$

Using Eq. (158) in the form $\Delta S = nF(\partial\varepsilon/\partial T)$ and substituting the derivative of Eq. (159),

$$\begin{aligned}\Delta S &= nF(0 + b + 2cT + 3\,dT^2)\\ \Delta S &= 3(23.06)[-1.537 \times 10^{-4} + 2(2.73 \times 10^{-8})T + 3(1.78 \times 10^{-11})T^2]\\ &= -9.17 \text{ cal deg}^{-1}.\end{aligned}$$

Using Eq. (49),

$$\Delta H = \Delta G + T\,\Delta S = -79.52 + 300(-9.17 \times 10^{-3}) = -82.27 \text{ kcal}.$$

8.6. Spontaneity and emf

Since $\Delta G < 0$ for a spontaneous isothermal constant-pressure process when no work other than pressure-volume work is involved ($W_{\text{net}} = 0$), it follows that $\varepsilon > 0$ for a process which is spontaneous under these same conditions.

In other words, the reaction for which we have determined ε using an electrochemical cell will proceed spontaneously from left to right as written at constant temperature and pressure when the reactants are placed together without an external circuit and electrodes if $\varepsilon > 0$.

We must, of course, be certain of our measuring procedure. Our convention here is to (1) take the redox reaction as written, (2) separate it into two half-cell reactions, (3) construct an electrochemical cell with these half-cells, (4) connect the positive terminal of the potentiometer to the half-cell in which reduction would occur if the reaction as written is spontaneous and connect the negative pole to the half-cell in which oxidation would occur (as in Fig. 8.2), and (5) obtain the potentiometer readings when the galvanometer is balanced at zero current flow. If the reaction as written is spontaneous, the electromotive force of the cell can be obtained from the readings of the balanced potentiometer. The sign of this emf will be positive. If the reaction as written is not spontaneous, however, oxidation instead of reduction is the process occurring in the compartment connected to the positive terminal, and reduction occurs in the compartment connected to the negative terminal. Consequently no adjustment of the sliding contact ($\overline{ac}$) will lead to balance. Only when the connections are reversed can a reading be obtained and an emf determined. The emf for the reaction *as written* is then negative.

Proceeding by these numbered steps, consider the reaction of silver with zinc chloride in solution (1):

$$2Ag + Zn^{+2} + 2Cl^{-} \longrightarrow Zn + 2AgCl.$$

The half-cell reactions are (2):

$$2Ag + 2Cl^{-} \longrightarrow 2AgCl + 2e^{-}$$

$$Zn^{+2} + 2e^{-} \longrightarrow Zn.$$

We now construct the cell (3) and connect (4) the zinc half-cell to the potentiometer terminal marked +, since this half-cell involves reduction *if the reaction is spontaneous as written.* Conversely, the silver chloride half-cell connects to the terminal marked −. We then attempt to balance the potentiometer (5) but discover that the galvanometer cannot be made to read zero current flow. When we switch the two wires leading to the electrodes, however, a zero current reading is obtained, leading to an emf of 0.985 volts. A minus sign is assigned to this emf. Thus the reaction *as written* has an electromotive force of −0.985 volts and is not spontaneous. The reverse reaction,

$$Zn + 2AgCl \longrightarrow 2Ag + 2Cl^{-} + Zn^{+2},$$

has an emf of +0.985 and is spontaneous.

Example 41 In a study of physiological electron transport, an electrochemical cell is constructed from a cytochrome b half-cell and a cytochrome c half-cell. The variable resistance can be adjusted to give a zero galvanometer reading when the positive pole of the battery of the potentiometer is connected to the electrode of the cytochrome b half-cell. In the total spontaneous cell reaction is cytochrome c reduced or is it oxidized?

Answer: The potentiometer can be balanced only when its positive pole is connected to the electrode of the half-cell in which reduction is spontaneously occurring. In this example, reduction must therefore be occurring in the cytochrome b half-cell and oxidation in the cytochrome c half-cell. The spontaneous total-cell reaction is cytochrome b (ox) + cytochrome c (red) → cytochrome b (red) + cytochrome c (ox).

8.7. The Nernst Equation

Just as ΔG depends on concentration, the electromotive force depends on the relative amounts of reactants and products present:

$$\Delta G = \Delta G^\circ + RT \ln M$$

$$-nF\varepsilon = -nF\varepsilon^\circ + RT \ln M$$

$$\varepsilon = \varepsilon^\circ - (RT/nF) \ln M. \tag{160}$$

Here ε° is the emf measured with the reactants and products all in their standard states. When we are dealing with ions and molecules in solution, the standard state is unit molality or unit activity. Equation (160) is known as the *Nernst equation.* At 298°K

$$\varepsilon = \varepsilon^\circ - (0.05915/n) \log M.$$

Example 42 Dumped into streams as a part of industrial waste, liquid mercury has proved to be a troublesome pollutant. It is transformed by natural processes into various chemical compounds harmful to biological systems. One such process which has been proposed is the spontaneous oxidation of mercury by dissolved oxygen in water:

$$2Hg(l) + O_2(aq) + 2H_2O(l) \longrightarrow 2Hg^{+2}(aq) + 4OH^-(aq). \tag{i}$$

Measured in an electrochemical cell, the standard potential for the following reaction is found to be -0.453 volt:

$$2Hg(l) + O_2(g) + 2H_2O(l) \longrightarrow 2Hg^{+2}(aq) + 4OH^-(aq). \tag{ii}$$

The Henry's law constant for oxygen is 4.0×10^4 atm. Assume the pH ($-\log m_{H^+}$) is 7 and p_{O_2} is 0.20 atm.

a) What is the emf when Hg^{+2} is at a concentration of 10^{-5} molality?
b) What is the concentration of Hg^{+2} at equilibrium?

Answer: a) The Nernst equation for Reaction (i) is

$$\varepsilon_i = \varepsilon_i^\circ - (0.05915/4)\log[(m_{Hg^{+2}})^2(m_{OH^-})^4/m_{O_2}]. \quad \text{(A)}$$

Our first task is the determination of the standard potential ε_i°. The standard potential is given for Reaction (ii), but this reaction differs in that it contains gaseous oxygen at 1 atm pressure, whereas Reaction (i) under standard conditions involves dissolved oxygen at unit molality. From a free-energy point of view:

$$2Hg(l) + O_2(aq, 1\ m) + 2H_2O(l) \xrightarrow{\Delta G_i^\circ} 2Hg^{+2}(aq, 1\ m) + 4OH^-(aq, 1\ m)$$

$$\downarrow \Delta G^\circ \qquad \nearrow \Delta G_{ii}^\circ$$

$$2Hg(l) + O_2(g, 1\ atm) + 2H_2O(l)$$

Using Hess's law reasoning, $\Delta G_i^\circ = \Delta G_{ii}^\circ + \Delta G^\circ$, and

$$\Delta G_{ii}^\circ = -nF\varepsilon_{ii}^\circ = -4(23.06)(-0.453) = 41.78\ \text{kcal}.$$

Here ΔG° is simply the free-energy change under standard conditions accompanying the transfer of 1.0 mole of oxygen from an aqueous solution at 1.0 molal concentration to the gas phase with a partial pressure of 1.0 atm. Recall that

$$\Delta G^\circ = -RT\ln K_{eq} = -RT\ln(p_{O_2}/m_{O_2}). \quad \text{(B)}$$

Henry's law relates the partial pressure and the mole fraction of a solute at equilibrium, $p_{O_2} = K_{O_2, H_2O}X_{O_2}$. Since this is a dilute solution,

$$X_{O_2} \simeq m_{O_2}M_{H_2O}/1000.$$

Therefore, at equilibrium,

$$K_{eq} = \frac{p_{O_2}}{m_{O_2}} = \frac{K_{O_2, H_2O}M_{H_2O}}{1000} = \frac{(4.0\times10^4)(18)}{1000} = 720,$$

and Eq. (B) becomes

$$\Delta G^\circ = -2.303RT\log 720 = -3.95\ \text{kcal}.$$

Now $\Delta G_i^\circ = 41.78 - 3.95 = 37.83$ kcal, and using $\Delta G^\circ = -nF\varepsilon^\circ$, $\varepsilon_i^\circ = -0.410$ volt.

In order to use Eq. (A) to calculate the emf we must also determine m_{O_2} from p_{O_2}. This is done through a straightforward application of Henry's law, employing the relationship between molality and mole fraction for a dilute solute:

$$X_{O_2} = p_{O_2}/K_{O_2, H_2O}$$
$$m_{O_2} = p_{O_2}1000/K_{O_2, H_2O}M_{H_2O}$$
$$m_{O_2} = (0.20)(1000)/(4.0\times10^4)(18) = 2.78\times10^{-4}.$$

The final value we need for use in the Nernst equation (A) is the hydroxide ion concentration. If $\text{pH} = -\log m_{H^+} = 7$, then $m_{H^+} = 10^{-7}$, and since $(m_{H^+})(m_{OH^-}) = 10^{-14}$, $m_{OH^-} = 10^{-7}$.

Now substituting into the Nernst equation we have

$$\varepsilon_i = -0.410 - (0.05915/4) \log [(10^{-5})^2(10^{-7})^4/(2.78 \times 10^{-4})]$$
$$\varepsilon_i = -0.410 + 0.496 = 0.086 \text{ volt.}$$

With $\varepsilon > 0$, Hg^{+2} could be spontaneously produced under these conditions.

b) $\Delta G^\circ = -RT \ln K$

$$\varepsilon^\circ = -0.410 = (RT/nF) \ln K = (0.05915/4) \log K = 0.0148 \log K.$$
$$K = 2.0 \times 10^{-28} = (m_{OH^-})^4(m_{Hg^{+2}})^2/m_{O_2} = (10^{-28})(m_{Hg^{+2}})^2/(2.78 \times 10^{-4})$$
$$(m_{Hg^{+2}})^2 = 5.7 \times 10^{-4}$$
$$m_{Hg^{+2}} = 2.4 \times 10^{-2} \text{ molal.}$$

There would thus be a significant concentration of Hg^{+2} ion at equilibrium.

8.8. Standard Half-cell Potentials

It is possible to tabulate certain values of ε which will allow us to calculate ε (and hence ΔG) for a reaction without having to actually set up an electrochemical cell and measure the reversible electrical potential.

Consider the three cells illustrated in Fig. 8.3. Reaction 3 is Reaction 1 minus Reaction 2. As a result of reasoning similar to that of Hess's Law, $\Delta G_3 = \Delta G_1 - \Delta G_2$ and, since $\Delta G = -nF\varepsilon$, we may write $\varepsilon_3 = \varepsilon_1 - \varepsilon_2$. Hess's Law may thus be directly applied to cell potentials.

Note that Fig. 8.3 follows the convention that oxidation occurs in the half-cell on the left and reduction in the half-cell on the right.

Recall that we have tabulated values of ΔG_f° for substances to facilitate the calculation of ΔG for reactions of interest. Similarly we shall tabulate *standard half-cell potentials* to enable us to determine ε for redox reactions. We cannot measure the potential of a half-cell; a reduction half-cell never exists without an oxidation half-cell. The electromotive force we measure is the emf of the combination of the two half-cells; it is the difference in the potential between the two half-cells. In order that we may tabulate values which we shall call half-cell potentials, we arbitrarily choose one half-cell to which we assign a potential of zero. Suppose B in Fig. 8.3 is the half-cell chosen to be assigned a zero potential. When half-cell B is coupled with half-cell A the total-cell potential difference is ε_1; that is, the difference in potential between A and B is ε_1. Since we have assigned B to be zero volts we can then record ε_1 volts as the half-cell potential of A. We obtain an emf ε_2 when the reference half-cell B with assigned potential zero is coupled with C; we then

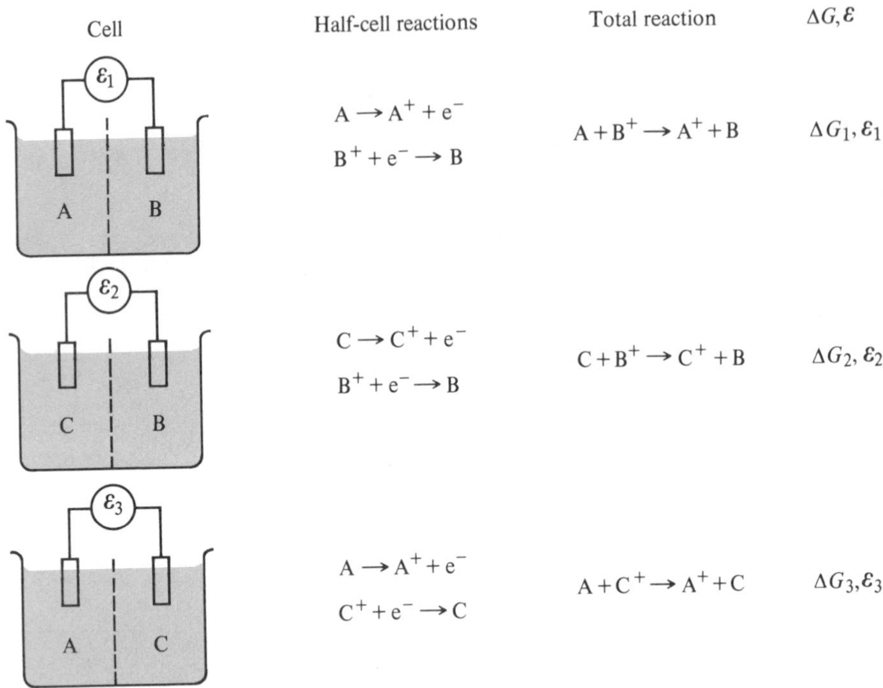

Figure 8.3

similarly assign ε_2 as the potential of the half-cell reaction of C. If we need to know the emf of a cell comprised of the A and C half-cells we use Hess's Law and subtract their half-cell potentials, ε_1 and ε_2.

The *standard hydrogen electrode* SHE (see Fig. 8.4) is chosen as the half-cell to which a potential of zero is assigned. We measure the value of the cell emf when another half-cell is linked with the SHE under standard conditions. We then assign the observed emf to that half-cell as its standard half-cell potential.[1] Tabulations of these standard values may be used to calculate the standard potentials of cells made from various combinations of these half-cells.

Tables of *standard oxidation potentials*, E°

$$A \longrightarrow A^{+n} + ne^-$$
$$B^{-n} \longrightarrow B + ne^-$$

[1] In actual measurements, the inconvenience of working with a gas electrode leads to the practice of using a reference electrode such as the saturated calomel electrode for a working half-cell, then correcting the results to the SHE basis. Referred to the SHE, the saturated calomel has an ε_{298}° of $+0.2458$ volt.

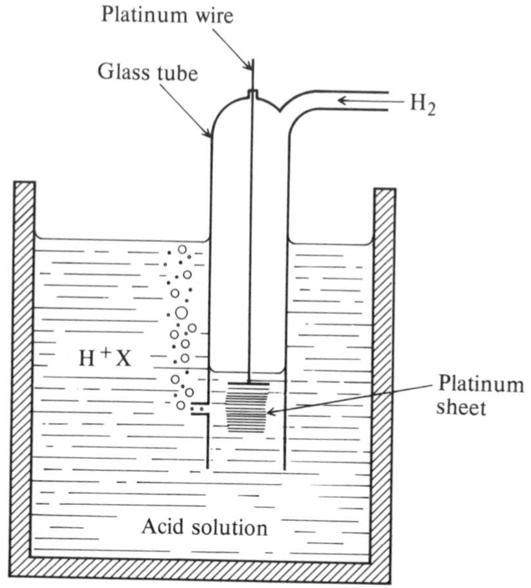

Fig. 8.4. Standard hydrogen electrode. Adapted from G. W. Castellan, *Physical Chemistry*, 2nd ed. (1971), Addison-Wesley, Reading, Mass.

as well as of *standard reduction potentials*, V°

$$A^{+n} + ne^- \longrightarrow A$$
$$B + ne^- \longrightarrow B^{-n}$$

are available. For a given half-cell, $E^\circ = -V^\circ$. Standard reduction potentials for a number of half-cells are given in Table 8.1(a). In these half-cells all concentrations are standard (1.0 molal for solutes, 1.0 atm for gases). In Table 8.1(b), however, the values given are the *biological standard half-cell reduction potentials* V'. The concentrations of reactants and products in these half-cells are all the usual standard values except for H^+, which has a value of 10^{-7} molal (pH 7) unless otherwise designated. The emf values obtained using these half-cell potentials are the biological standard values ε'. These may be converted to ε° values by using the Nernst equation:

$$\varepsilon^\circ = \varepsilon' + \frac{RT}{nF} \ln M,$$

in which all the concentration terms in M are 1.0 except for H^+, which is usually 10^{-7}.

Table 8.1(a). Standard Half-Cell Reduction Potentials (25°C, aq)

Half-cell reaction	*V°, volt*
$Ca(OH)_2 + 2e^- \rightarrow 2OH^- + Ca$	−3.02
$Ca^{+2} + 2e^- \rightarrow Ca$	−2.866
$Na^+ + e^- \rightarrow Na$	−2.714
$2H_2O + 2e^- \rightarrow H_2 + 2OH^-$	−0.82806
$Zn^{+2} + 2e^- \rightarrow Zn$	−0.7628
$Cr^{+3} + 3e^- \rightarrow Cr$	−0.744
$Fe^{+2} + 2e^- \rightarrow Fe$	−0.4402
$PbSO_4 + 2e^- \rightarrow Pb + SO_4^{-2}$	−0.358
$PbCl_2 + 2e^- \rightarrow 2Cl^- + Pb$	−0.268
$Ni^{+2} + 2e^- \rightarrow Ni$	−0.250
$CO_2(g) + 2H^+ + 2e^- \rightarrow HCOOH(aq)$	−0.199
$Pb^{+2} + 2e^- \rightarrow Pb$	−0.126
$Fe^{+3} + 3e^- \rightarrow Fe$	−0.036
$2H^+ + 2e^- \rightarrow H_2(g)$	0.000
$Cu^{+2} + e^- \rightarrow Cu^+$	0.153
$AgCl + e^- \rightarrow Cl^- + Ag$	0.2222
$Cu^{+2} + 2e^- \rightarrow Cu$	0.337
$O_2 + 2H_2O + 4e^- \rightarrow 4OH^-$	0.401
$Cu^+ + e^- \rightarrow Cu$	0.521
$Fe^{+3} + e^- \rightarrow Fe^{+2}$	0.771
$Ag^+ + e^- \rightarrow Ag$	0.800
$Hg^{+2} + 2e^- \rightarrow Hg$	0.854
$HNO_2 + H^+ + e^- \rightarrow NO + H_2O$	0.99
$Br_2(l) + 2e^- \rightarrow 2Br^-$	1.0652
$O_2 + 4H^+ + 4e^- \rightarrow 2H_2O$	1.229
$Cl_2(g) + 2e^- \rightarrow 2Cl^-$	1.3595
$Pb^{+4} + 2e^- \rightarrow Pb^{+2}$	1.69

In the table we find the following:

$$Zn^{+2} + 2e^- \longrightarrow Zn \qquad V^\circ = -0.763 \text{ volt}$$

$$Pb^{+2} + 2e^- \longrightarrow Pb \qquad V^\circ = -0.126 \text{ volt}$$

If these two half-cells are combined to form an electrochemical cell or if the materials are mixed together in a beaker, both processes could not occur as written. Either (case 1) lead will be reduced and zinc oxidized or (case 2) lead will be oxidized and zinc reduced.

Table 8.1(b). Biological Standard Half-cell Reduction Potentials (oxidized form/reduced form, pH 7)

Half-cell	*n*	*V′, volt*
acetic acid + CO_2 + $2H^+$/pyruvic acid + H_2O	2	−0.70
acetic acid + $2H^+$/acetaldehyde + H_2O	2	−0.60
acetyl CoA + $2H^+$/acetaldehyde + CoA	2	−0.41
pyruvic acid + CO_2 + $2H^+$/malic acid	2	−0.33
NAD^+ + H^+/NADH	2	−0.32
acetaldehyde + $2H^+$/ethanol	2	−0.20
pyruvic acid + $2H^+$/lactic acid	2	−0.19
oxaloacetic acid + $2H^+$/malic acid	2	−0.17
α-ketoglutaric acid + NH_4^+ + H^+/glutamic acid + H_2O	2	−0.14
pyruvic acid + NH_4^+ + H^+/alanine + H_2O	2	−0.13
fumaric acid + $2H^+$/succinic acid	2	0.03
cytochrome b (Fe^{+3}/Fe^{+2})	1	0.07 (pH 7.4)
dehydroascorbic acid + $2H^+$/ascorbic acid	2	0.08 (pH 6.4)
cytochrome b_2 (Fe^{+3}/Fe^{+2})	1	0.12 (pH 7.4)
methemoglobin/hemoglobin (Fe^{+3}/Fe^{+2})	1	0.17
cytochrome c (Fe^{+3}/Fe^{+2})	1	0.22
cytochrome a (Fe^{+3}/Fe^{+2})	1	0.29
$\frac{1}{2}O_2$ + $2H^+$/H_2O	2	0.82

W. M. Latimer, *Oxidation Potentials*, 2nd ed., Englewood Cliffs, N.J.: Prentice Hall.
A. J. de Bethane, T. S. Light, and N. Swedeman, *Journal of the Electrochemical Society*, **106**, 616 (1959).
A. White, P. Handler, and E. L. Smith, *Principles of Biochemistry*, 3rd ed., New York: McGraw-Hill (1959).

case 1:

$$\begin{array}{lr} Zn \rightarrow Zn^{+2} + 2e^- & -V^\circ = 0.763 \text{ volt} \\ Pb^{+2} + 2e^- \rightarrow Pb & V^\circ = -0.126 \text{ volt} \\ \hline Pb^{+2} + Zn \rightarrow Pb + Zn^{+2} & \varepsilon^\circ = 0.637 \text{ volt} \end{array}$$

case 2:

$$\begin{array}{lr} Zn^{+2} + 2e^- \rightarrow Zn & V^\circ = -0.763 \text{ volt} \\ Pb \rightarrow Pb^{+2} + 2e^- & -V^\circ = 0.126 \text{ volt} \\ \hline Pb + Zn^{+2} \rightarrow Pb^{+2} + Zn & \varepsilon^\circ = -0.637 \text{ volt} \end{array}$$

Since $\varepsilon > 0$ for a spontaneous process we can say that case 1 is the spontaneous process: Pb^{+2} is capable of spontaneously oxidizing Zn under standard conditions. When the concentrations of the materials are not standard we must use the Nernst equation to determine ε. If we wish to determine ε at a temperature other than standard we *must* employ Eq. (158):

$$\left(\frac{\partial \varepsilon}{\partial T}\right)_p = \frac{\Delta S}{nF} \tag{158}$$

The Nernst equation *cannot* be used alone to find the influence of temperature.

Example 43 Determine the standard cell potentials for the following reactions using the half-cell potentials of Table 8.1.
a) $Na(s) + \frac{1}{2}Cl_2(g) \rightarrow Na^+(aq) + Cl^-(aq)$
b) $Hg(l) + \frac{1}{2}O_2(g) + H_2O(l) \rightarrow Hg^{+2}(aq) + 2OH^-(aq)$ (See Example 34.)

Answer: a) We need the value of the half-cell potential for

$$\tfrac{1}{2}Cl_2 + e^- \longrightarrow Cl^- \tag{i}$$

so that the stoichiometry of the total reaction will be correct and so that the number of electrons lost in the oxidation half-cell will be the same as the number gained in the reduction half-cell. The half-cell listed in Table 8.1 is

$$Cl_2 + 2e^- \longrightarrow 2Cl^- \tag{ii}$$

with $V^\circ = 1.3595$. What is the potential for half-cell (i)?

The free-energy change for the production of 2.0 moles of product is twice that for the production of 1.0 mole because free energy is an extensive property. Therefore $\Delta G_{ii}^\circ = 2\,\Delta G_i^\circ$ for the above half-cell reactions, and $n_{ii}\varepsilon_{ii}^\circ F = 2n_i\varepsilon_i^\circ F$. Since $n_i = 1$ and $n_{ii} = 2$, $\varepsilon_{ii}^\circ = \varepsilon_i^\circ$ or $V_i^\circ = V_{ii}^\circ$.

$Na \rightarrow Na^+ + e^-$	$-V^\circ = 2.714$
$\frac{1}{2}Cl_2 + e^- \rightarrow Cl^-$	$V^\circ = 1.360$
$Na + \frac{1}{2}Cl_2 \rightarrow Na^+ + Cl^-$	$\varepsilon^\circ = 4.074$ volt

b)

$Hg \rightarrow Hg^{+2} + 2e^-$	$-V^\circ = -0.854$
$\frac{1}{2}O_2 + H_2O + 2e^- \rightarrow 2OH^-$	$V^\circ = 0.401$
$Hg + \frac{1}{2}O_2 + H_2O \rightarrow Hg^{+2} + 2OH^-$	$\varepsilon^\circ = -0.453$ volt

Problems

8.1 Prove that 96,487 coul equiv^{-1} is equal to 23.06 kcal volt^{-1} equiv^{-1}.

8.2 Physiological poisoning by cyanide results from the formation of complexes between cytochrome oxidase and cyanide ions. Methemoglobin also forms such complexes with cyanide, but these do not cause the body to suffer severe pathological consequences. Therefore one method of treating cyanide poisoning is to increase the amount of methemoglobin, which takes up the cyanide and reduces the amount of harmful cyanide-oxidase. Methemoglobin may be produced by administering a nitrite. Assuming the nitrite is hydrolyzed to nitrous acid HNO_2, calculate ε and ΔG for the oxidation of hemoglobin to methemoglobin by nitrous acid when all solutes are at a concentration of 0.01 molal and the pH is 7.

8.3 Table 8.1 contains standard half-cell reduction potentials when the solvent is water. These values are somewhat different in other solvents. For the half-cell

$$Cu^{+2} + 2e^- \longrightarrow Cu(s),$$

for instance, the half-cell potential in ethanol is 0.21 volts. Calculate the free-energy difference between Cu^{+2} (1 m, aq) and Cu^{+2} (1 m, ethanol). In which of these solutions is the copper ion more stable? Can you explain why?

8.4 One of the steps in the citric acid cycle is

$$\text{succinic acid} + \text{FAD} \longrightarrow \text{fumaric acid} + \text{FADH}_2,$$

where FAD is a flavoprotein coenzyme. The two half-cell reactions are

$$\text{oxidation: succinic acid} \longrightarrow \text{fumaric acid} + 2H^+ + 2e^-.$$

$$\text{reduction: FAD} + 2H^+ + 2e^- \longrightarrow \text{FADH}_2.$$

What is the minimum value which the standard reduction potential of the FAD half-cell may have for the total reaction to be spontaneous under standard conditions?

8.5 In glycolysis, 2.0 moles of adenosine triphosphate and 2.0 moles of pyruvate are produced for each mole of glucose metabolized. The fate of pyruvate in animals depends on the conditions. (a) Under *aerobic* conditions (oxygen available), pyruvic acid is converted to acetyl coenzyme A by the following reaction:

$$\text{pyruvic acid} + \text{CoA} + \text{NAD}^+ \longrightarrow \text{acetyl CoA} + CO_2 + \text{NADH} + H^+.$$

Acetyl coenzyme A then enters the citric acid cycle. Using the values of V' in Table 8.1(b), calculate ε° and ΔG° for this process. Also determine ΔG for this reaction under more physiological conditions: 37°C, $p_{CO_2} = 0.05$ atm, pH 7, and with all other concentrations 0.01 molal ($n = 2$). (b) Under anaerobic conditions (no oxygen available), pyruvic acid is converted to lactic acid:

$$\text{pyruvic acid} + \text{NADH} + H^+ \longrightarrow \text{lactic acid} + \text{NAD}^+.$$

Anaerobic conditions result when the body is engaged in exercise. The rapid production of energy by glycolysis in the muscles results in a large amount of pyruvic acid being formed. This pyruvic acid in turn is converted into lactic acid. (1) Calculate ε° and ΔG° for this reaction. (2) Determine ΔG for the reaction at 37°C, pH 7 with all other concentrations 0.01 molal ($n = 2$).

The lactic acid may be removed by the blood stream. If lactic acid is produced more rapidly than it is removed from the muscle, the excess lactate causes cramps. When aerobic conditions are reestablished, the above reaction may be reversed, pyruvic acid being produced from lactic acid. (3) What is the minimum concentration of lactic acid necessary to make this reverse process spontaneous at pH 7 when all other species have a concentration of 0.01 molal?

8.6 In some anaerobic microorganisms such as yeast, glycolysis proceeds in much the same manner as in animals to the point where pyruvic acid is formed. In animals (see Problem 8.5) pyruvic acid is converted into acetyl coenzyme A or into lactic acid. In yeast it is converted into ethanol (recall our sociologically important reaction of previous chapters) in the following sequence of reactions:

$$\text{pyruvic acid} \longrightarrow \text{acetaldehyde} + CO_2$$
$$\text{acetaldehyde} + \text{NADH} + H^+ \longrightarrow \text{ethanol} + \text{NAD}^+.$$

Using Table 8.1(b), calculate ΔG° for the conversion of pyruvic acid into ethanol. (Note that with a change in enzymes our exercises could make us drunk instead of giving us cramps!)

8.7 One of the steps of the citric acid cycle is

$$\text{malic acid} + \text{NAD}^+ \rightleftharpoons \text{oxaloacetic acid} + \text{NADH} + H^+.$$

(a) Calculate the equilibrium constant K for this reaction. (b) Is this reaction spontaneous under standard conditions? In the citric acid cycle the concentration of malic acid can be maintained at a high enough level by preceding reactions and that of oxaloacetic acid at a low enough level by subsequent reactions for this reaction to be spontaneous. (c) If the concentrations of oxaloacetic acid, NADH, and NAD^+ are 0.01 molal and the pH is 7.0, what is the minimum concentration of malic acid which would make the reaction spontaneous? (d) If the concentrations of malic acid, NADH, and NAD^+ are 0.01 molal and the pH is 7.00, what is the maximum concentration of oxaloacetic acid for which this reaction would be spontaneous?

8.8 The electrodes of an electrochemical cell are made of solid iron. One dips into a solution which is 0.1 molal in Fe^{+2}; the other into a solution 0.01 molal in Fe^{+3}. The two half-cell reactions are

$$Fe^{+2} + 2e^- \longrightarrow Fe$$
$$Fe^{+3} + 3e^- \longrightarrow Fe.$$

(a) Write the balanced total reaction. (b) Which electrode is the anode and which the cathode? (c) Determine ΔG for the spontaneous reaction. (d) Calculate the equilibrium constant K.

8.9 From the data of Table 8.1(a) determine the free-energy change for the following dilution:

$$\text{HCl (0.1 m)} \longrightarrow \text{HCl (0.01 m)}.$$

Describe an electrochemical cell in which the emf of this reaction can be measured. Such a cell is called a *concentration cell.*

8.10 (a) Calculate the solubility product K_{sp} for AgCl using Table 8.1(a). $K_{sp} = m_{Ag^+} m_{Cl^-}$; Ag and AgCl are solids. (b) Propose electrochemical experiments for the determination of ΔH° and ΔS° for the dissociation of AgCl:

$$\text{AgCl(s)} \longrightarrow Ag^+\text{(aq)} + Cl^-\text{(aq)}.$$

8.11 It has been suggested that the mercurous ion should be written Hg_2^{++} rather than Hg^+. Describe precisely how a concentration cell experiment could be set up to prove which is true.

8.12 The *calomel electrode* is easily constructed and is often used as a standard or reference electrode. The half-cell reaction is

$$Hg_2Cl_2(s) + 2e^- \longrightarrow 2Hg(s) + 2Cl^-(aq).$$

The standard half-cell potential is 0.285 volts. The saturated calomel electrode is constructed so that there is excess solid KCl present. The solution is therefore saturated with KCl. This cell has a half-cell potential of 0.246 volts. Calculate the concentration of Cl^- ions in this saturated solution and determine the apparent solubility product constant K_{sp} for KCl in this solution.

8.13 Ammonia is readily "cracked" at elevated temperature into nitrogen and hydrogen; the nitrogen is inert, and any remaining ammonia may be removed by bubbling the gas mixture through sulfuric acid. Suppose that, working at 25°C and 760 torr, this mixture of nitrogen and hydrogen gases is used to operate a hydrogen electrode, the other half-cell being a saturated calomel reference electrode (see Problem 8.12). If a certain solution, tested with this set-up, gives an apparent pH of 5.0 when the gas is falsely assumed to be made up of hydrogen only, what is the true pH?

PART TWO

Chemical Kinetics

9 / Reaction Rates

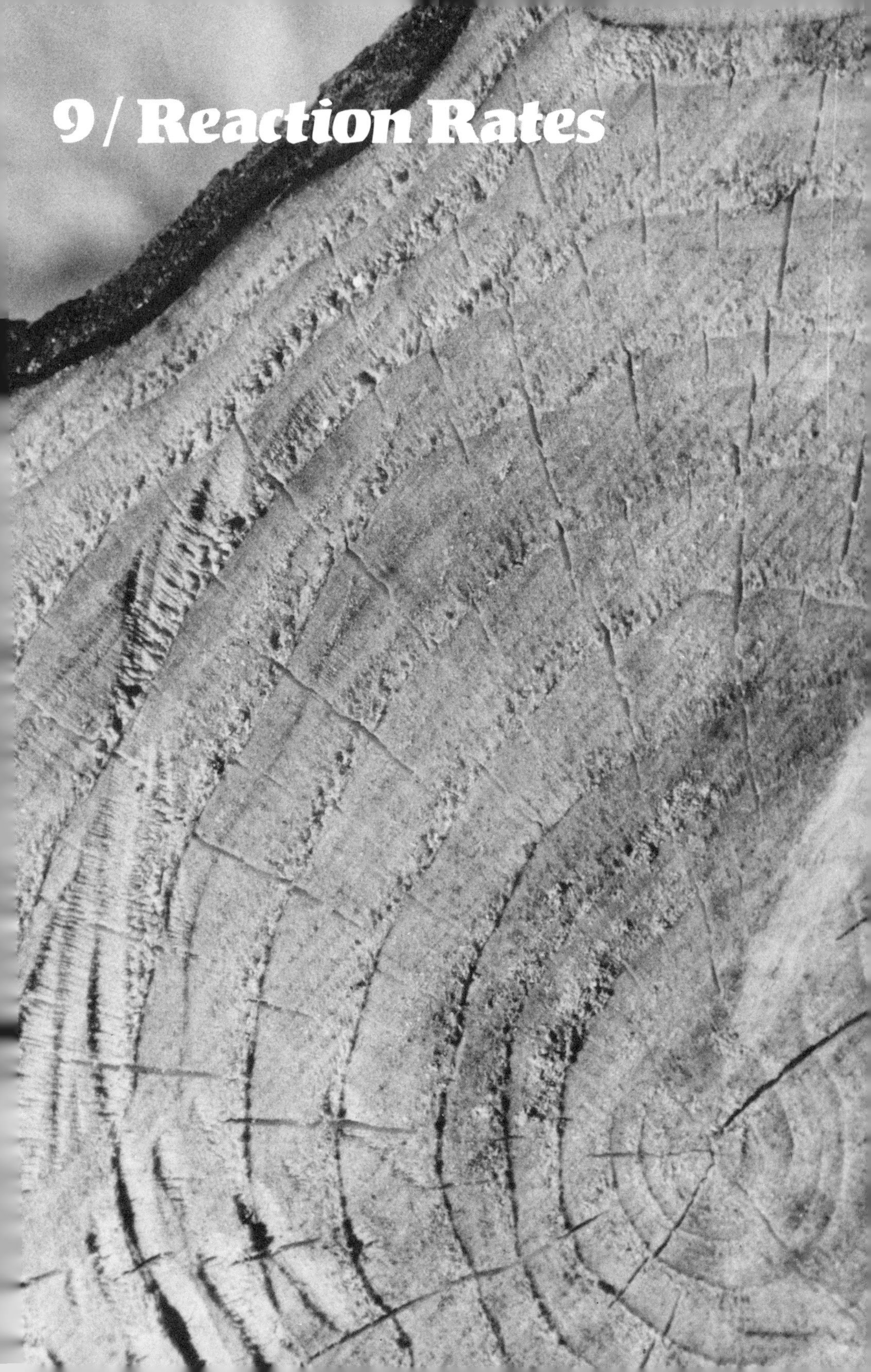

9.1. Introduction

Our study of thermodynamics has provided answers to some very important chemical questions. On the basis of free-energy considerations, we are able now to predict whether a given reaction is possible, and also to predict the relative amounts of products and reactants at equilibrium. In addition, we can predict whether changes in the experimental conditions will increase or decrease the amount of product at equilibrium. Thermodynamics provides us with methods which are of great power and utility in dealing with chemical processes.

But thermodynamics cannot provide all the needed information. Thermodynamics, regardless of the inclusion of "dynamics" in its name, deals primarily with unchanging equilibrium situations. It considers chemical changes by concerning itself with initial and final macroscopic states, ignoring the dynamic microscopic processes occurring between these states. Thermodynamics can tell us that it is possible for a reaction to occur, but not how fast it occurs, nor how it occurs, nor even whether it occurs. As a macroscopic science, thermodynamics deals with neither the mechanism nor the time element.

For a complete understanding of chemical reactivity, we need more than the knowledge gained from thermodynamic considerations; we need methods which will allow us to examine the details of dynamic chemical processes. We need information about how fast the reaction proceeds, and how and why a variation of experimental conditions or the concentration of the various participants in the reaction, will affect its rate. The field of study which deals with such questions is *chemical kinetics.*

There are two major concerns in most kinetic studies of chemical processes. The first is experimental determination of the *rate* of the reaction and the dependence of this rate upon the concentrations of the reacting substances. The second is the use of this data in an attempt to explain the molecular pathway or *mechanism* by which the chemical change takes place.

9.2. Reaction Rate

The *rate* of a reaction is simply how fast it occurs. More precisely, the reaction rate is the speed with which a reactant is disappearing or a product is appearing; it is the change in concentration of reactant or product which occurs in a unit time interval. Reaction rate thus has the units of moles per liter per second (or moles liter^{-1} min^{-1}, moles liter^{-1} hr^{-1}, etc.).

For the stoichiometrically simple process

$$A \longrightarrow B,$$

the reaction rate is $[A]_2$, the concentration of A at time t_2 at the end of a given interval of reaction, minus $[A]_1$, its concentration at the beginning of that interval (at t_1), divided by the time interval $(t_2 - t_1)$:

$$\frac{[A]_2 - [A]_1}{t_2 - t_1} = \frac{\Delta[A]}{\Delta t}.$$

Since $[A]_2$ is less than $[A]_1$ we conventionally introduce a minus sign so that the rate will be a positive quantity. The *average rate* in the given time interval is then $-\Delta[A]/\Delta t$.

Alternatively for this reaction we could define the rate with respect to the appearance of product. Following a similar line of reasoning the average rate R_{avg} would be

$$R_{avg} = \frac{\Delta[B]}{\Delta t} = -\frac{\Delta[A]}{\Delta t}.$$

Of much more practical application and importance is the *instantaneous rate* R_{inst}, defined as the rate at a particular instant. Mathematically

$$R_{inst} = \lim_{\Delta t \to 0} \frac{\Delta[B]}{\Delta t} = \frac{d[B]}{dt},$$

evaluated at time t. Henceforth we shall refer to this simply as the rate R:

$$R = -\frac{d[A]}{dt} = \frac{d[B]}{dt}.$$

For example, the rate of the citric acid cycle reaction

$$\underset{\text{oxaloacetic acid}}{HO_2CCH_2COCO_2H} \longrightarrow CO_2 + \underset{\text{pyruvic acid}}{CH_3COCO_2H}$$

is $R = -d[\text{oxaloacetic acid}]/dt = d[CO_2]/dt = d[\text{pyruvic acid}]/dt$.

If the stoichiometry of the reaction is more complicated, however, we face a problem in defining a unique rate. For instance if the reaction under consideration is of the form

$$A \longrightarrow 2B,$$

the quantity $-d[A]/dt$ will be half as large as $d[B]/dt$ since the product B is appearing twice as fast as the reactant A is disappearing; for every mole of A which disappears, 2 moles of B appear. We cannot then simply use

$$R = -\frac{d[A]}{dt} = \frac{d[B]}{dt}$$

for the rate but must weight the terms stoichiometrically. The rate in terms of the disappearance of reactant is

$$R_1 = -\frac{d[\mathrm{A}]}{dt} = \frac{1}{2}\frac{d[\mathrm{B}]}{dt},$$

while the rate in terms of appearance of product is

$$R_2 = \frac{d[\mathrm{B}]}{dt} = -2\frac{d[\mathrm{A}]}{dt}.$$

Obviously $R_1 \neq R_2$, but rather $R_1 = \frac{1}{2}R_2$. We shall employ a convention so that when we speak of *the* rate of a reaction our language will be unambiguous. We shall adopt the practice of dividing each differential rate term by its stoichiometric coefficient in the reaction equation *as written*. For the above reaction *the* rate would be

$$R = -\frac{d[\mathrm{A}]}{dt} = \frac{1}{2}\frac{d[\mathrm{B}]}{dt},$$

which corresponds to R_1 above.

For the general reaction

$$a\mathrm{A} + b\mathrm{B} \longrightarrow g\mathrm{G} + h\mathrm{H},$$

when x amount of rightward reaction occurs, ax moles of A and bx moles of B are used up, while gx moles of G and hx moles of H appear. The rate in terms of appearance or disappearance of the various products and reactants will therefore be different. Employing our convention, the rate is

$$R = -\frac{1}{a}\frac{d[\mathrm{A}]}{dt} = -\frac{1}{b}\frac{d[\mathrm{B}]}{dt} = \frac{1}{g}\frac{d[\mathrm{G}]}{dt} = \frac{1}{h}\frac{d[\mathrm{H}]}{dt}. \tag{161}$$

The situation may be pictured as follows:

Reaction:	$a\mathrm{A}$	+	$b\mathrm{B}$	$\longrightarrow$	$g\mathrm{G}$	+	$h\mathrm{H}$
Initial concentration	$[\mathrm{A}]_0$		$[\mathrm{B}]_0$		$[\mathrm{G}]_0$		$[\mathrm{H}]_0$
Concentration at time t	$([\mathrm{A}]_0 - ax)$		$([\mathrm{B}]_0 - bx)$		$([\mathrm{G}]_0 + gx)$		$([\mathrm{H}]_0 + hx)$

The rate at time t is

$$R = \frac{1}{g}\frac{d([\mathrm{G}]_0 + gx)}{dt} = \frac{1}{h}\frac{d([\mathrm{H}]_0 + hx)}{dt} = -\frac{1}{a}\frac{d([\mathrm{A}]_0 - ax)}{dt}$$

$$= -\frac{1}{b}\frac{d([\mathrm{B}]_0 - bx)}{dt}.$$

Since all initial concentrations are mathematical constants, it follows that $R = dx/dt$.

Given this definition of x and the convention of sign, the rate is seen to be unambiguously expressed as $R = dx/dt$ regardless of the molecular species whose concentration is measured in the experiment. Again, for this general case, we may write

$$R = \frac{dx}{dt} = -\frac{1}{a}\frac{d[\mathrm{A}]}{dt} = -\frac{1}{b}\frac{d[\mathrm{B}]}{dt} = \frac{1}{f}\frac{d[\mathrm{F}]}{dt} = \frac{1}{g}\frac{d[\mathrm{G}]}{dt}. \tag{161}$$

Example 44 Write the rate in terms of disappearance of reactants and appearance of products for each of the following reactions:
a) $H_2(g) + \frac{1}{2}O_2(g) \rightarrow H_2O(g)$
b) $2H_2(g) + O_2(g) \rightarrow 2H_2O(g)$
c) glucose(aq) $\rightarrow$ 2 ethanol(aq) + $2CO_2(g)$
d) $CH_3COOH(aq) + 2O_2(g) \rightarrow 2CO_2(g) + 2H_2O(l)$

Answer:
a) $R = -d[H_2]/dt = -2d[O_2]/dt = d[H_2O]/dt$
b) $R = -\frac{1}{2}d[H_2]/dt = -d[O_2]/dt = \frac{1}{2}d[H_2O]/dt$
c) $R = -d[\text{glucose}]/dt = \frac{1}{2}d[\text{ethanol}]/dt = \frac{1}{2}d[CO_2]/dt$
d) $R = -d[CH_3COOH]/dt = -\frac{1}{2}d[O_2]/dt = \frac{1}{2}d[CO_2]/dt = \frac{1}{2}d[H_2O]/dt.$

9.3. Factors Affecting Reaction Rate

The speed of a chemical reaction is determined by four principal factors: (1) the nature of the reaction, (2) the concentrations of the reacting species, (3) the temperature, and (4) the action of catalysts.

The nature of the reaction is of course primary. Different processes differ enormously in their inherent rate, other conditions being equal. We know that most ionic reactions are essentially instantaneous, whereas it takes 4.9 billion years for only half of a sample of radioactive uranium 238 to decay. For convenience, most laboratory studies have been made on processes which have reaction times between a few seconds and a few days.

9.4. Dependence of Rate on Concentration

An early generalization on the influence of concentration was the *Law of Mass Action* (C. M. Guldberg and P. Waage, 1863): the rate of a chemical reaction is proportional to the concentrations of the reacting substances, each raised to some power. A great many reactions are found to obey this principle.

For instance, the rate of the decomposition reaction

$$N_2O_5 \longrightarrow 2NO_2 + \tfrac{1}{2}O_2$$

is found experimentally to be proportional to the concentration of N_2O_5:

$$R = -\frac{d[N_2O_5]}{dt} \propto [N_2O_5].$$

We write this proportionality as an equality by introducing a constant k called the *rate constant*:

$$R = k[N_2O_5].$$

Such an expression which relates the reaction rate to the concentration of reaction participants is called a *rate equation* or a *rate law*. The relation which actually obtains must be found by experiment for each reaction.

As another example, the rate equation for the rearrangement of benzidine is found by experiment to be $R = k[\text{benzidine}][H^+]^2$: the rate is proportional to the concentration of benzidine and to the second power of the concentration of hydrogen ion.

In general a reaction of the type $a\text{A} + b\text{B} \rightarrow$ Products which follows the Law of Mass Action will have a rate expression

$$R = -\frac{1}{a}\frac{d[\text{A}]}{dt} = -\frac{1}{b}\frac{d[\text{B}]}{dt} = k[\text{A}]^m[\text{B}]^n, \tag{162}$$

where m and n are not necessarily equal to a and b respectively, nor are m and n necessarily integers.

The *order* of the above reaction *with respect to* A is m; with respect to B it is n. The *overall order* is $m + n$. This latter quantity is often simply called the *reaction order*. If $m + n = 1$, the reaction is said to be first-order; if $m + n = 2$, the reaction is second-order, etc.

Example 45 What is the order with respect to the individual reactants and the overall order for the two following reactions:

a) $N_2O_5 \rightarrow 2NO_2 + \frac{1}{2}O_2 \qquad R = k[N_2O_5]$

b)

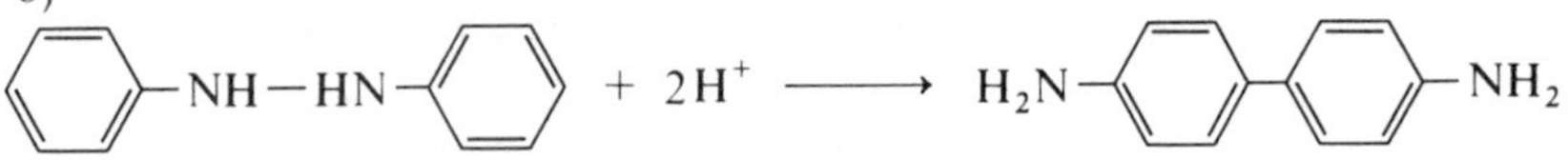

$$\text{R} = [\text{benzidine}][H^+]^2$$

Answer: a) This reaction is first-order with respect to dinitrogen pentoxide and is first-order overall.

b) This reaction is first-order with respect to benzidine and second-order with respect to hydrogen ion. It is third-order overall.

It should be emphasized that the rate equation with its rate constant and order is an experimental finding; it is *not* predictable from the stoichiometry of the balanced reaction equation.

For the special case of an *elementary reaction* the rate equation will always be of the form of Eq. (162), and m and n will be equal to a and b respectively. An elementary reaction is one whose mechanism is exactly what the stoichiometry of the reaction equation would suggest. *If* the following reactions are elementary, then the rate expressions obtained will be those given:

1. 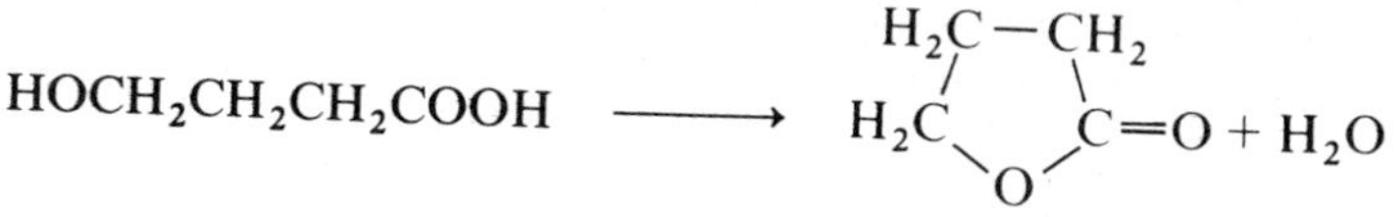 $HOCH_2CH_2CH_2COOH \longrightarrow$ (lactone ring: H_2C-CH_2 / H_2C and $C{=}O$ joined through O) $+ H_2O$

$$R = k[HOCH_2CH_2CH_2COOH]$$

2. 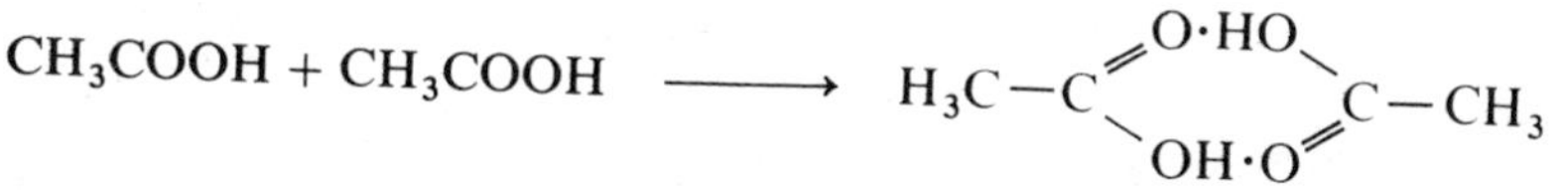 $CH_3COOH + CH_3COOH \longrightarrow H_3C-C(=O\cdot HO)(-OH\cdot O=)C-CH_3$ (hydrogen-bonded dimer)

$$R = k[CH_3COOH]^2$$

3. $CH_2{=}CH_2 + H_2 \rightarrow CH_3{-}CH_3$ $\quad R = k[CH_2{=}CH_2][H_2]$
4. $2NO + O_2 \rightarrow 2NO_2$ $\quad R = k[NO]^2[O_2]$
5. $aA + bB \rightarrow$ Products $\quad R = k[A]^a[B]^b$

Such elementary reactions occur by repetition of a single step whose mechanism is simply the coming together of the requisite number of reactant molecules in a single encounter, resulting in the formation of products. For example the first reaction above, if elementary, occurs as the ends of single molecules of hydroxybutyric acid interact to form a lactone ring. The third reaction above would require the coming together of two molecules, one of ethylene and one of hydrogen, the continued progress of this reaction consisting in the repetition of its one-plus-one basic step.

For elementary reactions the sum $m + n = a + b$ is called the *molecularity* of the reaction. This is the number of reactant molecules which come together in the reaction's single step. If this unit act involves the participation of only one molecule, the reaction is said to be *unimolecular*; if the unit act involves the conjunction of just two molecules (of the same or different species) the process is *bimolecular*; if three molecules, *termolecular*. Being predicated on three-body collisions (a relatively infrequent event), termolecular reactions are rare; higher molecularities are unknown.

Example 46 Give the molecularity of each of elementary reactions 1–4.

Answer:

1. unimolecular
2. bimolecular
3. bimolecular
4. termolecular

In citing the molecularity of a reaction, one is claiming a knowledge of its mechanism, a claim which is difficult to prove with certainty. (We shall discuss the problems associated with the assignment of mechanisms in the next chapter.) The quantity which can be found by experiment is the order of the reaction.

Although a great many reactions obey the Law of Mass Action prediction (Eq. 162), there are numerous others whose rate expressions are not of such simple form. For instance the rate equation for the reaction $H_2(g) + Br_2(g) \rightarrow 2HBr(g)$ is

$$R = \frac{1}{2}\frac{d[HBr]}{dt} = \frac{k[H_2][Br_2]^{1/2}}{1 + k'[HBr]/[Br_2]}.$$

This reaction is first-order with respect to hydrogen gas, but it would be impossible to assign the order with respect to bromine and to hydrogen bromide. Likewise the overall order is meaningless in this case. The reaction mechanism of this as well as other complex reactions is composed of more than one single, elementary process. The complex nature of the rate equation results from the combination of the rates of all the individual elementary reactions which comprise the total mechanism. This situation will be explored in Chapter 10.

9.5. Rate Laws

Many reactions have simple one-step mechanisms and therefore have rate equations of the form of Eq. (162). Also the rate equations of other reactions with more complex mechanisms sometimes reduce to this form. Moreover the elaborate rate expressions of complex reactions are the end result of a set of equations similar to Eq. (162), since these reactions result from a series of simple elementary encounters. Consequently it will be advantageous to examine this equation further for some special cases.

Zero-Order

Consider the reaction $aA + bB \rightarrow$ Products. Suppose that this reaction is experimentally determined to proceed at a constant rate

$$R = -\frac{1}{a}\frac{d[A]}{dt} = k_0[A]^0[B]^0 = k_0, \tag{163a}$$

which does not depend on the concentration of any of the reaction participants. Comparison of this equation with Eq. (162) indicates that m and n are zero; the order of the reaction is thus zero.

Equation (163a) is called the zero-order *differential rate law*. This equation may be placed in another form by integration. The initial ($t = 0$) concentration of A is $[A]_0$; at the end of a time interval during which an amount x of reaction occurs (ax moles of A are used up) the concentration is $[A]_t = [A]_0 - ax$.

Then

$$-\frac{1}{a}\,d[A] = k_0\,dt \tag{163b}$$

$$-\int_{[A]_0}^{[A]_t} d[A] = ak_0 \int_0^t dt$$

$$[A]_0 - [A]_t = ak_0\,t \tag{164a}$$

$$x = k_0\,t. \tag{164b}$$

This last equation is the zero-order *integrated rate law*. The units of k_0 are moles liter^{-1} sec^{-1}.

Although reactions which have an overall order of zero are rare, it is not unusual to find the order with respect to a particular reactant to be zero (that is, the rate of the reaction does not depend on the concentration of that reactant). For example, we shall see in Chapter 12 that under certain conditions the initial rate of the enzyme catalyzed reaction

$$\text{substrate} \xrightarrow{\text{enzyme}} \text{product}$$

is equal to k[enzyme]. The reaction is therefore first-order with respect to enzyme concentration but zero-order with respect to substrate concentration.

The decomposition of various gases on the surface of metallic catalysts often exhibits zero-order kinetics.

First-order

In a first-order reaction, $m + n = 1$. The most common case of this is that in which the reaction rate is found by experiment to depend on the first power of the concentration of only one reactant; for the reaction $aA + bB \rightarrow$ Products, the differential rate law is then

$$R = -\frac{1}{a}\frac{d[A]}{dt} = k_1[A]^1[B]^0 = k_1[A]. \tag{165a}$$

The integrated rate law may be readily obtained:

$$-\frac{1}{a}\frac{d[A]}{[A]} = k_1\,dt \tag{165b}$$

$$-\int_{[A]_0}^{[A]_t} \frac{d[A]}{[A]} = ak_1 \int_0^t dt$$

$$\ln\left(\frac{[A]_0}{[A]_t}\right) = \ln\left(\frac{[A]_0}{[A]_0 - ax}\right) = ak_1 t \tag{166}$$

Since any logarithm is a pure number and thus unitless, the units of k_1 are sec^{-1}. Upon rearranging Eq. (165):

$$ak_1 = -\left(\frac{d[A]}{[A]}\right)/dt$$

We see that, for an infinitesimal amount of reaction, ak_1 expresses the fraction reacting per unit time. If for example a first-order rate constant is 0.002 sec^{-1} and a is 1, this means that $\frac{2}{10}$ of 1% of the material present will react in the ensuing second. This would be true regardless of the size of the sample and regardless of the concentration units.

Example 47 At a certain temperature the rate constant for the unimolecular decomposition of dinitrogen pentoxide mentioned above is 1.00×10^{-5} sec^{-1}.

a) What is the rate if the concentration of N_2O_5 is (1) 0.25 mole $liter^{-1}$? (2) 0.50 mole $liter^{-1}$?

b) If the initial concentration of N_2O_5 is 0.50 mole $liter^{-1}$ what will its concentration be after 100 hr?

Answer: a) The rate R is equal to $k[N_2O_5]$. Thus

1) $R = (1.00 \times 10^{-5}\ sec^{-1})[0.25\ mole\ liter^{-1}]$
$= 2.5 \times 10^{-6}\ mole\ liter^{-1}\ sec^{-1}$

2) $R = (1.00 \times 10^{-5}\ sec^{-1})[0.50\ mole\ liter^{-1}]$
$= 5.0 \times 10^{-6}\ mole\ liter^{-1}\ sec^{-1}$.

b) The reaction as previously written in the text is

$$N_2O_5 \longrightarrow 2NO_2 + \tfrac{1}{2}O_2.$$

The rate is thus

$$R = -d[N_2O_5]/dt = k[N_2O_5].$$

Equation (166) becomes

$$\ln\left(\frac{[N_2O_5]_0}{[N_2O_5]_t}\right) = kt$$

$$\ln\left(\frac{0.50}{[N_2O_5]_t}\right) = (1.00 \times 10^{-5}\ sec^{-1})(360{,}000\ sec)$$

$$\ln [N_2O_5] = \ln (0.50) - 3.60$$

$$\log [N_2O_5] = \log (0.50) - 3.60/2.3$$

$$= -0.30 - 1.57 = -1.87$$

$$[N_2O_5] = 1.3 \times 10^{-2}\ mole\ liter^{-1}.$$

Second-order

There are a number of ways in which the sum of m and n can be equal to two. The two most common are

$$-\frac{1}{a}\frac{d[A]}{dt} = k_2[A]^2 \qquad (167)$$

and

$$-\frac{1}{a}\frac{d[A]}{dt} = k_2'[A][B]. \tag{168}$$

Examples of these two types of second-order reaction are

$$2NO_2 \longrightarrow 2NO + O_2$$

and

$$NO_2 + CO \longrightarrow CO_2 + NO.$$

Experimentally the respective rate equations are found to be

$$-\tfrac{1}{2}d[NO_2]/dt = k[NO_2]^2$$

and

$$d[CO_2]/dt = k[NO_2][CO].$$

Again the integrated rate laws can be obtained. In the first case (Eq. 167),

$$-\int_{[A]_0}^{[A]_t} \frac{d[A]}{[A]^2} = ak_2 \int_0^t dt$$

$$1/[A]_t - 1/[A]_0 = ak_2 t \tag{169a}$$

$$x/\{[A]_0([A]_0 - ax)\} = k_2 t. \tag{169b}$$

The second case (Eq. 168) is somewhat more complicated to integrate. For the simplest case in which the stoichiometry of the reaction is A + B → Product, at any instant $[A] = ([A]_0 - x)$ and $[B] = ([B]_0 - x)$. Since $-d[A]/dt = dx/dt$,

$$\int_0^x \frac{dx}{([A]_0 - x)([B]_0 - x)} = k_2' \int_0^t dt. \tag{170}$$

By the method of partial fractions

$$\frac{1}{[B]_0 - [A]_0}\left(\ln\frac{[B]_0 - x}{[A]_0 - x} - \ln\frac{[B]_0}{[A]_0}\right) = k_2' t$$

$$\frac{1}{[B]_0 - [A]_0} \ln\frac{[A]_0([B]_0 - x)}{[B]_0([A]_0 - x)} = k_2' t \tag{171}$$

For the special case in which the initial concentrations of A and B are identical ($[A]_0 = [B]_0$), Eq. (170) becomes

$$\int_0^x \frac{dx}{([A]_0 - x)^2} = k_2' \int_0^t dt.$$

Integration leads to Eq. (169).

For the general second-order reaction of type aA + bB → Products with differential rate law

$$-\frac{1}{a}\frac{d[A]}{dt} = k_2'[A][B],$$

the integrated rate law will be

$$k_2' t = \frac{1}{a[\mathrm{B}]_0 - b[\mathrm{A}]_0} \ln \frac{[\mathrm{A}]_0([\mathrm{B}]_0 - bx)}{[\mathrm{B}]_0([\mathrm{A}]_0 - ax)}. \tag{172}$$

The units of k_2 and k_2' are liters mole^{-1} sec^{-1}.

Example 48 The rate constant at 25°C for the second-order reaction between the two gaseous pollutants nitrous oxide and ozone

$$\mathrm{NO} + \mathrm{O_3} \longrightarrow \mathrm{NO_2} + \mathrm{O_2}$$

is 1.20×10^{10} cm^3 mole^{-1} sec^{-1}.

a) If the initial concentrations of NO and of O_3 are both 0.10 mole liter^{-1}, what is the concentration of O_3 after 1.0 sec in a constant-volume reaction vessel?
b) If the initial concentrations of nitrous oxide and ozone are both 1.0 g liter^{-1}, find the concentration of O_2 after 10^{-5} sec.

Answer: a) Since the initial concentrations are equal, Eq. (169) may be used.

$$\frac{1}{[\mathrm{O_3}]_t} - \frac{1}{0.10 \text{ mole liter}^{-1}} = \frac{1.20 \times 10^{10} \text{ cm}^3 \text{ mole}^{-1} \text{ sec}^{-1}}{1000 \text{ cm}^3 \text{ liter}^{-1}} t$$

$$\frac{1}{[\mathrm{O_3}]} = (1.20 \times 10^7 \text{ liters mole}^{-1} \text{ sec}^{-1})(1 \text{ sec})$$

$$[\mathrm{O_3}] = 8.30 \times 10^{-8} \text{ mole liter}^{-1}.$$

b) The initial concentrations of nitrous oxide and ozone are:

$$[\mathrm{NO}]_0 = (1.0 \text{ g liter}^{-1})/(30 \text{ g mole}^{-1}) = 0.033 \text{ mole liter}^{-1}$$

$$[\mathrm{O_3}]_0 = (1.0 \text{ g liter}^{-1})/(48 \text{ g mole}^{-1}) = 0.021 \text{ mole liter}^{-1}.$$

We cannot use Eq. (169) because the initial concentrations are not equal; however, since $a = b = 1$, Eq. (171) may be used:

$$\left[\frac{1}{(0.021 - 0.033) \text{ mole liter}^{-1}}\right] \ln \left[\frac{0.033(0.021 - x)}{0.021(0.033 - x)}\right]$$

$$= (1.2 \times 10^7 \text{ liters mole}^{-1} \text{ sec}^{-1})(10^{-5} \text{ sec})$$

$$2.3 \log \left(\frac{6.3 \times 10^{-4} - 0.033x}{6.3 \times 10^{-4} - 0.021x}\right) = (1.2 \times 10^2 \text{ liters mole}^{-1})(-0.012 \text{ mole liter}^{-1})$$

$$\log \left(\frac{6.3 \times 10^{-4} - 0.033x}{6.3 \times 10^{-4} - 0.021x}\right) = -.63$$

$$\frac{6.3 \times 10^{-4} - 0.033x}{6.3 \times 10^{-4} - 0.021x} = 0.23$$

$$x = 1.7 \times 10^{-2}.$$

The concentration of O_2 is equal to x and is therefore 1.7×10^{-2} mole liter^{-1} after 10^{-5} sec.

Higher Orders

We could continue to write differential rate laws and to obtain by integration the integrated rate laws for reactions of higher orders. These integrations become more complex as the order increases.

A general equation is available for the simplest case. When the rate depends on the concentration of only one reactant, that is $-d[\mathrm{A}]/dt = k_n[\mathrm{A}]^n$ $(n \neq 1)$, with the stoichiometric factor $a = 1$, the differential form can be written

$$dx/dt = k_n([\mathrm{A}]_0 - x)^n, \tag{173}$$

and this can be integrated to give

$$\frac{1}{n-1}\left\{\frac{1}{([\mathrm{A}]_0 - x)^{n-1}} - \frac{1}{[\mathrm{A}]_0{}^{n-1}}\right\} = k_n t. \tag{174}$$

The units of k_n are moles^{1-n} liter^{n-1} sec^{-1}.

Example 49 Give the integrated rate law and the units of the rate constant for the third-order $(n = 3)$ reaction whose differential rate expression is $-d[\mathrm{A}]/dt = k_3[\mathrm{A}]^3$.

Answer: From Eq. (174),

$$\frac{1}{2}\left\{\frac{1}{([\mathrm{A}]_0 - x)^2} - \frac{1}{[\mathrm{A}]_0^2}\right\} = k_3 t \tag{175}$$

The units of k_3 are liter^2 mole^{-2} sec^{-1}.

Noninteger Rate Laws

Even though in practice the value of the reaction order obtained is rarely exactly integral due to experimental error, the value is usually reasonably close to an integer. In some cases, however, the value is found to be consistently a noninteger. Since molecularity must be integral, a noninterger order indicates that the reaction mechanism must be complex.

For half-integer reaction orders (order $= p + \frac{1}{2}$ where p is an integer) with the differential rate equation $d[\mathrm{A}]/dt = k_{p+1/2}[\mathrm{A}]^{p+1/2}$ the integrated rate law can be obtained from Eq. (174), in which $n = p + \frac{1}{2}$:

$$\frac{1}{p-\frac{1}{2}}\left\{\frac{1}{([\mathrm{A}]_0 - x)^{p-1/2}} - \frac{1}{[\mathrm{A}]_0^{p-1/2}}\right\} = k_{p+1/2}\, t. \tag{176}$$

When $p = 0$, $n = \frac{1}{2}$ (half-order):

$$2\{[\mathrm{A}]_0{}^{1/2} - ([\mathrm{A}]_0 - x)^{1/2}\} = k_{1/2}\, t. \tag{177}$$

The units of $k_{1/2}$ are mole$^{1/2}$ liter$^{-1/2}$ sec^{-1}.

For example, for the reaction

$$CH_3CHO \longrightarrow CH_4 + CO,$$

the rate expression is found to be

$$-d[CH_3CHO]/dt = k[CH_3CHO]^{3/2}$$

Half-integer orders may not occur in such a straightforward manner. For instance the rate of decomposition of phosgene in the reaction

$$COCl_2 \longrightarrow CO + Cl_2$$

is given by

$$-d[COCl_2]/dt = k[COCl_2][Cl_2]^{1/2}$$

Also note that the integer orders discussed above can arise from noninteger orders with respect to the individual reactants:

$$-d[A]/dt = k_1[A]^{1/2}[B]^{1/2} \qquad \text{(first-order)}$$

An example of this will be found below. Although rate expressions with noninteger powers are relatively rare, they are encountered and must be dealt with.

Negative Orders

Rate expressions also occur with orders less than zero with respect to individual reactants. For the transformation of ozone into oxygen,

$$2O_3 \longrightarrow 3O_2,$$

the experimental rate expression is

$$\frac{1}{3}\frac{d[O_2]}{dt} = -\frac{1}{2}\frac{d[O_3]}{dt} = k[O_3]^2/[O_2]$$

$$= k[O_3]^2[O_2]^{-1}.$$

The reaction is second-order with respect to O_3 and negative first-order with respect to O_2. Overall, the reaction is first-order.

In the presence of alcohol the aminolysis of some esters, for example

$$CH_3\overset{\overset{\displaystyle O}{\|}}{C}OCH_3 + CH_3CH_2NH_2 \xrightarrow{CH_3CH_2OH} CH_3\overset{\overset{\displaystyle O}{\|}}{C}NHCH_2CH_3 + CH_3OH$$

has the rate expression

$$-d[CH_3CH_2NH_2]/dt = \frac{k[CH_3CH_2NH_2]^{3/2}[CH_3\overset{\overset{\displaystyle O}{\|}}{C}OCH_3]}{[CH_3CH_2OH]^{1/2}}.$$

The rate is $\frac{3}{2}$-order with respect to amine, first-order with respect to ester, and reciprocal $\frac{1}{2}$-order with respect to ethanol. The overall order is second.

Zero, negative, and noninteger orders are not generally applicable to elementary, single-step reactions. We shall see how they result from more complex mechanisms.

9.6. Half-Life

Before discussing the methods available for determining the rate expressions of reactions, let us look at a particularly useful quantity which can be obtained from the integrated rate laws.

The *half-life* τ of a reaction is the time required for the concentration of reactant to decrease by half. When $t = \tau$, $[A] = \frac{1}{2}[A]_0$.

For a first-order reaction we can obtain the half-life relationship from the integrated rate law (Eq. 166):

$$\ln\left\{\frac{[A]_0}{\frac{1}{2}[A]_0}\right\} = ak_1\tau_1$$

$$\tau_1 = \ln 2/ak_1 = 0.693/ak_1. \tag{178}$$

For a second-order reaction with rate equal to $k_2[A]^2$ or to $k_2'[A][B]$ with $[A]_0 = [B]_0$ and $a = b$ (Eq. 169a):

$$\tfrac{1}{2}[A]_0/\{[A]_0(\tfrac{1}{2}[A]_0)\} = ak_2\tau_2$$

$$\tau_2 = 1/ak_2[A]_0. \tag{179}$$

The half-life of a first-order reaction depends only upon the rate constant; for a second-order reaction it is also inversely proportional to the initial concentration. A determination of half-life is often a straightforward way of establishing the order of a reaction and of calculating its rate constant. For more complicated cases or when reactants are mixed in other than their stoichiometric proportions, the half-life concept is not particularly helpful.

Example 50 From the half-life data obtain the correct experimental reaction order and the value of the rate constant for the following reaction:

$$NO + O_3 \longrightarrow NO_2 + O_2$$

$[NO]_0 = [O_3]_0$, moles liter^{-1}:	1.0	0.10	0.010
τ, sec:	8.3×10^{-8}	8.4×10^{-7}	8.3×10^{-6}

Answer: Obviously τ depends on the initial concentration in this case. Therefore the reaction is not first-order (Eq. 178). We can test for second-order by calculating values of k_2 using each initial concentration (Eq. 179):

$[NO]_0$:	1.0	0.10	0.010
k_2:	1.20×10^7	1.19×10^7	1.20×10^7

The reaction is thus second-order with a rate constant of 1.2×10^7 liters mole^{-1} sec^{-1}. (See Example 48.)

9.7. Obtaining Experimental Rate Data

The dependence of reaction rate on concentration is of great usefulness in proposing mechanisms for reactions. We must, of course, first determine this dependence by experiment.

The data for determining the rate of a reaction consist of the concentration of a reactant or product at the end of successive time intervals during the course of the reaction. (A plot of data from a typical reaction appears in Fig. 9.5.)

Determining concentration while the reaction is proceeding often presents difficulties. Much effort and ingenuity have been expended developing techniques for accomplishing this. Let us consider a few of them.

The concentration may be determined by extracting a small amount of the reacting mixture and the amount of a particular species determined by titration or other analytical technique. Unless the reaction is a slow one and the analysis is performed swiftly, the concentration in the extracted portion continues to change as the reaction proceeds; such an analysis is subject to large error. As we shall see, reaction rates can be slowed considerably by cooling. If the portion extracted is immediately plunged into an ice bath and the analysis performed rapidly, accuracy may be improved. Such methods do suffer from inaccuracies, however. Methods which allow instantaneous determination of the concentration without disturbing the reaction mixture are preferable when they can be accomplished.

Any measurable property of the mixture which is a consequence of a reactant or product and whose change is proportional to the change in concentration of a reactant or product may be used. For instance, in the reaction considered earlier,

$$\underset{\text{glucose}}{C_6H_{12}O_6(s)} \longrightarrow \underset{\text{ethanol}}{2C_2H_5OH(l)} + 2CO_2(g), \tag{180}$$

a gas is produced. The increase in the pressure in a constant-volume reaction vessel can be measured and related via an equation of state to the moles of gas produced. Consequently the concentration of gas present at various times during the reaction may be easily determined. Alternatively the change in volume of a constant-pressure reaction vessel may be related to the amount of gas produced (or consumed) in a reaction. With this technique the rate of the process may be followed by observing the rate of volume change.

If ions are consumed or produced in a reaction, the rate of the reaction can be followed by observing the rate of change in the *conductance* of the solution. A typical cell which may be used for such measurements is shown schematically in Fig. 9.1. In order to relate conductance directly to concentration a graph or table of conductance versus concentration must be constructed using solutions with known concentrations of the ion (or ions) of interest. Other electroanalytical methods are used as well. These include potentiometry and polarography.

One particularly common method for determining concentration is spectrophotometry. We shall discuss in detail the absorption of electromagnetic radiation (including visible light) by chemical species in a later chapter. At the moment, it is sufficient for our purposes to know that chemical species absorb electromagnetic radiation and that each material has its own characteristic *absorption spectrum.* The absorption spectrum is a plot of the

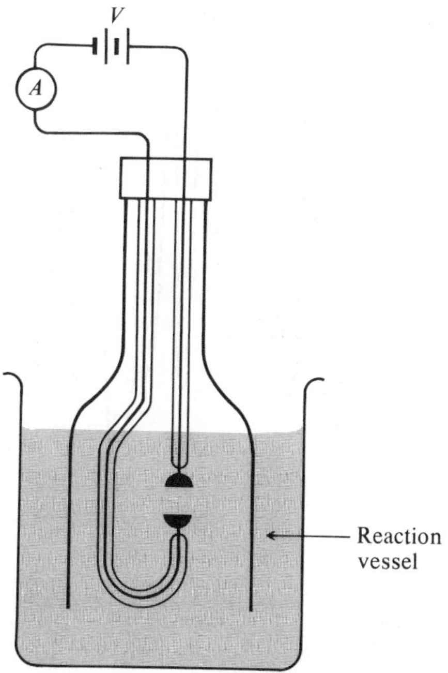

Fig. 9.1. A typical cell for conductometric measurements. Device A is an ammeter for measuring current flow.

amount of radiation absorbed versus the wavelength of the radiation. Figure 9.2 is a typical absorption spectrum.

Beer's law relates the amount of radiation of a given wavelength absorbed to the concentration of material:

$$\log (P_0/P) = \varepsilon bc = A. \tag{181}$$

P is the *power* of the radiation beam after passing through the sample material, P_0 the power of the beam when no sample is present. The power (the energy of the radiation striking a unit area in unit time) can be measured using a photo cell. If the material absorbs radiation at a particular wavelength, P will be less than P_0 and $\log (P_0/P)$ will be larger than zero. This log term is called the *absorbance*. The absorbance is proportional to b, the length of the path traveled by the radiation through the sample, and to c, the concentration in moles liter^{-1} of the species absorbing the radiation. The proportionality constant ε is called the *molar absorptivity* or *molar extinction coefficient*; ε is a function of the wavelength and of the absorbing material. A

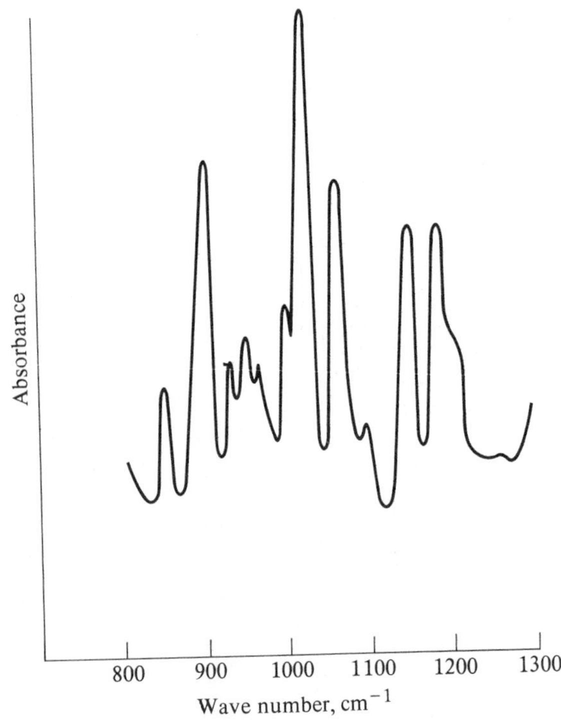

Fig. 9.2. A portion of the infrared absorption spectrum of polystyrene.

plot of ε versus wavelength for a particular material is essentially the absorption spectrum of that material (Fig. 9.2).

If we choose a wavelength for which ε is large for the chemical species of interest, the amount of radiation absorbed (the absorbance) will be directly proportional to the concentration of that material. If this substance is either a reactant or product in a reaction we are studying, we can determine the rate of the reaction by observing the rate of appearance or disappearance of this substance as given by the rate of increase or decrease of the absorbance at the chosen wavelength.

9.8. Use of Rate Data in Determining Rate Equations: Simple Cases

How may the rate data obtained by experiment be used in determining the rate expression? Let us first examine a rather straightforward case.

In a previous example we considered one possible problem which complicates the use of ethers for anesthesia in the tropics. Also of some concern is their thermal decomposition. Consider for instance the thermal decomposition of dimethylether:

$$CH_3OCH_3 \longrightarrow CH_4 + H_2 + CO.$$

C. N. Hinshelwood and P. J. Askey studied the kinetics of this reaction at 504°C (*Proceedings of the Royal Society*, **A115**, 215, 1927.) At this temperature all reactants and products are gases. We could follow the rate of the reaction by observing the rate of pressure change in a constant-volume reaction container. When x mole of ether decomposes, $3x$ mole of products is formed, a net increase of $2x$ mole. For each x mole of reaction the pressure change (assuming ideal gas behavior) will be:

$$\Delta P = \Delta nRT/V = 2xRT/V.$$

With concentration in moles liter^{-1}, the change in concentration of ether with a given change in total pressure is

$$[\text{ether}]_t - [\text{ether}]_0 = -x/V = -\Delta P/2RT. \qquad \text{(A)}$$

The concentration of ether at time t is thus

$$[\text{ether}]_t = [\text{ether}]_0 - \Delta P/2RT.$$

A sample set of data is given in Table 9.1.

If this were a zero-order reaction, the integrated rate law for zero-order (Eq. 164a) would require that $[\text{ether}]_0 - [\text{ether}]_t = x = k_0 t$. In other words. if this reaction is zero-order, a plot of the concentration of reactant versus time would be a straight line with slope $([\text{ether}]_t - [\text{ether}]_0)/t$ equal to $-k_0$.

Table 9.1. Rate Data for the Reaction $CH_3OCH_3 \rightarrow CH_4 + H_2 + CO$

Time t, sec	*Total pressure* P_t, atm	$\Delta P_t = P_t - P_0$	$(P_0 - \frac{1}{2}\Delta P)_t$	$\ln (P_0 - \frac{1}{2}\Delta P)_t$	$(P_0 - \frac{1}{2}\Delta P)_t^{-1}$
0	0.50	0.00	0.50	−0.69	2.0
200	0.58	0.08	0.46	−0.78	2.2
500	0.69	0.19	0.41	−0.89	2.4
800	0.79	0.29	0.36	−1.02	2.8
1000	0.84	0.34	0.33	−1.11	3.0
1200	0.90	0.40	0.30	−1.20	3.3
1500	0.97	0.47	0.26	−1.35	3.8
1800	1.03	0.53	0.23	−1.47	4.3
2000	1.07	0.57	0.21	−1.56	4.8

(Conversely a plot of the concentration of any one of the products versus time would be a straight line with slope k_0.) Since $[\text{ether}]_t - [\text{ether}]_0$ is given by Eq. (A) above, a plot of P_t versus time should be a straight line if the reaction is zero-order. The slope would be equal to $+2RTk_0$. Figure 9.3(a) is such a plot using the data of Table 9.1.

Similarly, if this reaction is first-order, the integrated rate law (Eq. 166) requires that the plot of ln [ether] versus t be a straight line with slope $-k_1$:

$$\text{slope} = \frac{\ln [\text{ether}]_t - \ln [\text{ether}]_0}{t} = \ln \left(\frac{[\text{ether}]_t}{[\text{ether}]_0}\right)/t = -k_1.$$

For this particular reaction a plot of $\ln (P_0 - \frac{1}{2}\Delta P)$ versus time should yield a straight line with slope equal to $-k_1$. This is true because if only ether is present initially, then $P_0 = RT[\text{ether}]_0$ and thus $(P_0 - \frac{1}{2}\Delta P) = RT[\text{ether}]_t$. Figure 9.3(b) is such a plot.

If this reaction is second-order, the integrated rate law (Eq. 169a) requires that a plot of the reciprocal of the ether concentration versus time must be a straight line with slope k_2. A plot of $(P_0 - \frac{1}{2}\Delta P)^{-1}$ versus time should be a straight line if this reaction is second-order. The slope of such a graph would be k_2/RT. Figure 9.3(c) is such a plot.

From Fig. 9.3 we can see that this reaction is first-order with respect to ether. The differential rate expression is

$$-d[\text{ether}]/dt = k[\text{ether}].$$

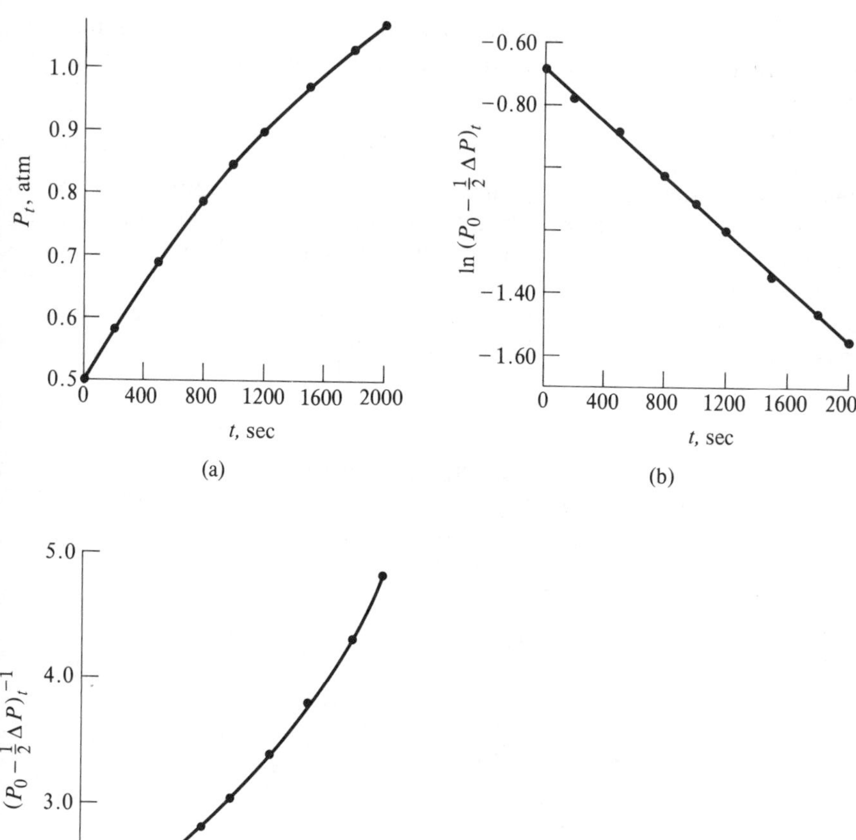

(a)

(b)

(c)

Figure 9.3

We can calculate the rate constant for this reaction from the slope of the line in Fig. 9.3(b). As stated above, the slope is equal to $-k$. The slope is -4.2×10^{-4} sec^{-1}.

The complete rate equation for this reaction is

$$R = (4.2 \times 10^{-4}\ \text{sec}^{-1})[\text{ether}].$$

The use of the integrated rate equations is the most straightforward method of determining order and the rate constant for reactions with relatively simple reaction equations.

Another approach for these reactions is the use of the half-life as mentioned above (Section 9.6) and illustrated in Example 50. When we determine the time required for half of the dimethyl ether to decompose in three separate experiments with initial pressures of 0.25 atm, 0.5 atm, and 1.0 atm, the half-lives observed are all essentially equal to 1650 sec. This confirms the order since the half-life of a first-order reaction does not depend on the initial concentration (see Eq. 178). The rate constant may also be calculated from the observed half-life:

$$k = 0.693/\tau = 4.2 \times 10^{-4}\ \text{sec}^{-1}.$$

Example 51 Natural rubber is a polymer of isoprene and of the isoprene dimer dipentene. The dimerization and the polymerization of isoprene are spontaneous processes. From the following rate data, obtain the correct experimental differential rate expression and the value of the rate constant at 527°C for the gas phase dimerization of isoprene:

2 H_3C–C(=CH_2)–CH=CH_2 ⟶ dipentene

CH_3
C
H_2C CH
H_2C CH_2
CH
C
H_3C CH_2

isoprene dipentene

time, sec:	0	0.1	0.2	0.3	0.4	0.5
[isoprene], moles liter^{-1}:	1.00	0.57	0.39	0.30	0.24	0.20

Answer: Figures 9.4(a), 9.4(b) and 9.4(c) are the plots for zero-order, first-order, and second-order kinetics respectively. Interpretation of these plots based on the preceding discussion gives the following rate expression for this reaction:

$$-\frac{d[\text{isoprene}]}{dt} = k[\text{isoprene}]^2.$$

The rate constant may be calculated from the slope of the curve of Fig. 9.4(c) or directly from the data using the integrated rate law for second-order. Its value is

$$\begin{aligned} k &= \{(1/[\text{isoprene}]_t) - (1/[\text{isoprene}]_0)\}/t \\ &= (1/0.39) - (1/1.00)/0.2 = 7.9\ \text{liters mole}^{-1}\ \text{sec}^{-1}. \end{aligned}$$

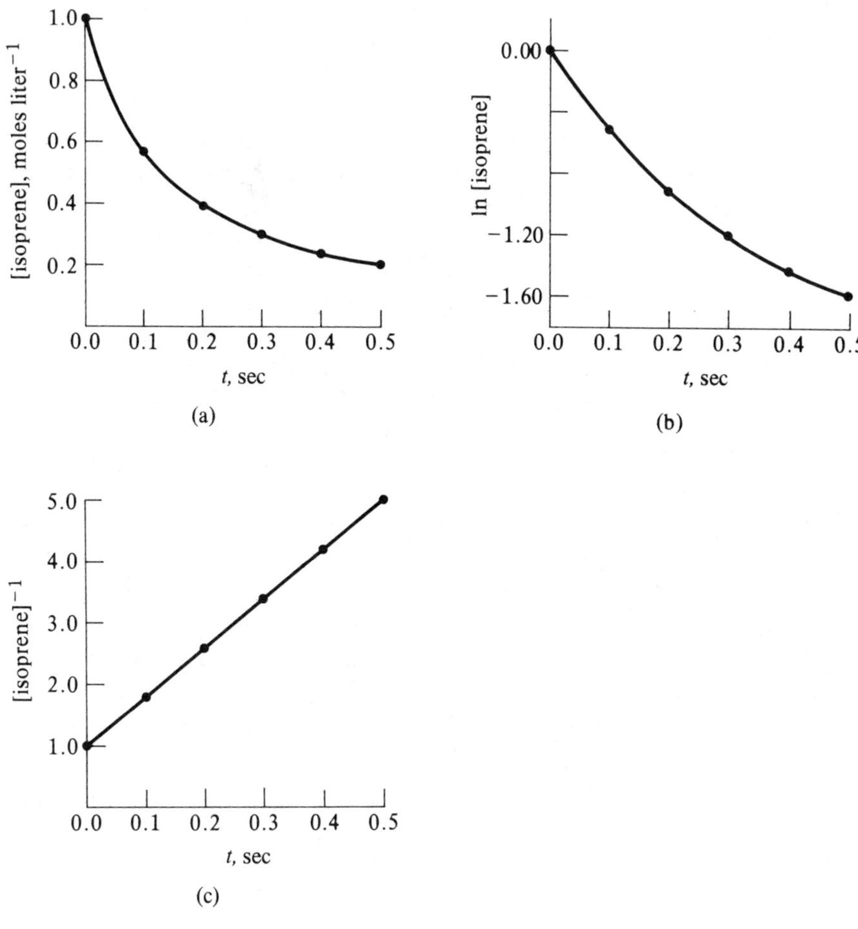

Figure 9.4

9.9. Determining Rate Equations: More Complex Cases

Due to the complex nature of the rate equations, in many cases the approaches outlined in the previous section are not adequate.

The hypohalites such as NaOCl, NaOBr, and NaOI are rather interesting ionic molecules in which the halides are in +1 oxidation states. They are thus easily reduced and are used widely as oxidizing agents in industrial processes. Certain processes require the use of a particular hypohalite (for example the iodoform reaction). The kinetics of the conversion of hypochlorite to hypoiodite has been studied by Y. T. Chia and R. E. Connick (*Journal of Physical Chemistry*, **63**, 1518, 1959). The reaction in basic aqueous solution is

$$I^- + OCl^- \longrightarrow OI^- + Cl^-.$$

The reaction may be followed by choosing a radiation wavelength which only the OI^- ion absorbs. The necessary rate information may then be obtained by observing the rate of increase of absorbance at this wavelength.

Let us suppose that in several preliminary experiments we discover that the rate of this reaction depends on the concentration of iodide, of hypochlorite, and of hydroxide. To determine exactly the dependence of the rate on the concentrations of these species we first measure the *initial rate* at various initial concentrations of the reactants and of OH^-. We do this for a particular set of concentrations by first recording the increase in absorbance as a function of time as the reaction proceeds. Figure 9.5 is a plot of the concentration of OI^- versus time obtained from such an absorbance study by using Beer's law (Eq. 181). In this particular case the initial concentrations $[I^-]_0 = [OCl^-]_0 = [OH^-]_0 = 0.1$ mole liter^{-1}. Recall that the average rate (Eq. 160) is given by $\Delta[OI^-]/\Delta t$. The average rate for the first 0.2 sec is the slope of the straight line from the point on the curve at 0 sec to the point at 0.2 sec (line *a* of Fig. 9.5). The instantaneous rate at t is $\lim_{\Delta t \to 0} \frac{\Delta[OI^-]}{\Delta t} = d[OI^-]/dt$ evaluated at t. This corresponds mathematically to the slope of the tangent line drawn to the curve at t. The initial rate is

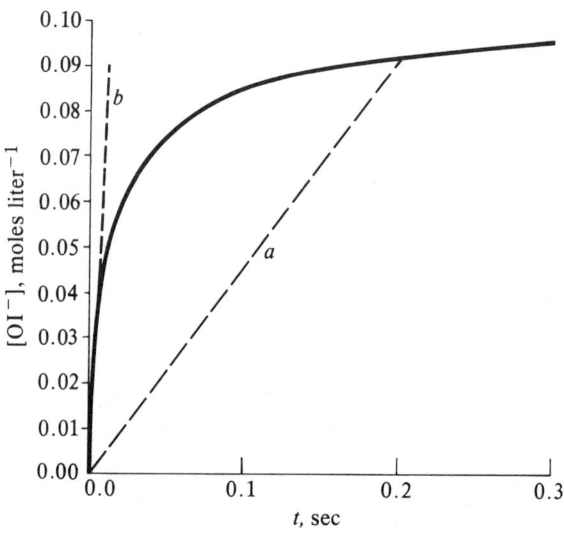

Figure 9.5

then the slope of this line at $t = 0$ (line b of Fig. 9.5). In this particular case the initial rate is 6.0 moles liter^{-1} sec^{-1}.

This process is repeated systematically to obtain the data in Table 9.2. Note that we first obtain initial rates for a series of mixtures in all of which the initial concentrations of OCl^- and OH^- are 0.1 molar but for which the concentration of I^- is different in each mixture. We then obtain the initial rates for a series in which the initial concentrations of I^- and OH^- are 0.1 molar but the concentration of OCl^- is varied. In the third series $[OH^-]_0$ is varied while $[I^-]_0$ and $[OCl^-]_0$ are held constant.

Table 9.2. Rate Data for the Iodide-Hypochlorite Reaction

a) $[OCl^-]_0 = [OH^-]_0 = 0.1$ mole liter^{-1}					
$[I^-]_0$, mole liter^{-1}:	0.01	0.03	0.05	0.07	0.10
Initial rate, moles liter^{-1} sec^{-1}:	0.6	1.8	3.0	4.2	6.0
b) $[I^-]_0 = [OH^-]_0 = 0.1$ mole liter^{-1}					
$[OCl^-]_0$, mole liter^{-1}:	0.01	0.03	0.05	0.07	0.10
Initial rate, moles liter^{-1} sec^{-1}:	0.6	1.8	3.0	4.2	6.0
c) $[I^-]_0 = [OCl^-]_0 = 0.1$ mole liter^{-1}					
$[OH^-]_0$, mole liter^{-1}:	0.01	0.03	0.05	0.07	0.10
Initial rate, moles liter^{-1} sec^{-1}:	60.0	20.0	12.0	8.6	6.0

Figure 9.6 contains the plots of initial rate versus initial concentration for the four sets of data in Table 9.2. The appearance of these graphs can help determine the order with respect to each of the three substances which participate in the reaction.

Note that the initial rate in all three cases depends on the concentration. Whereas the rate increases with increasing $[I^-]_0$ and with increasing $[OCl^-]_0$, it decreases with increasing $[OH^-]_0$. The rate must be directly related to $[I^-]$ and to $[OCl^-]$ but reciprocally related to $[OH^-]$. The rate law is therefore of the form

$$R = k\,\frac{[I^-]^m[OCl^-]^n}{[OH^-]^p}. \tag{182}$$

We must now determine m, n, p, and the rate constant k.

Note that the data for $[I^-]$ and $[OCl^-]$ in Table 9.2(a) and 9.2(b) are identical, as are the plots of the two sets of data (Figs. 9.6a and 9.6b). Everything we have to say about one will apply to the other as well.

The initial rate R_0 is always given by

$$R_0 = k\,\frac{[I^-]_0^m[OCl^-]_0^n}{[OH^-]_0^p}. \tag{183}$$

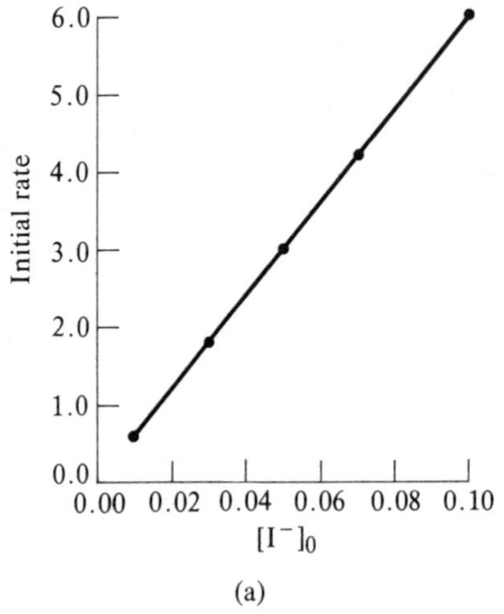

(a)

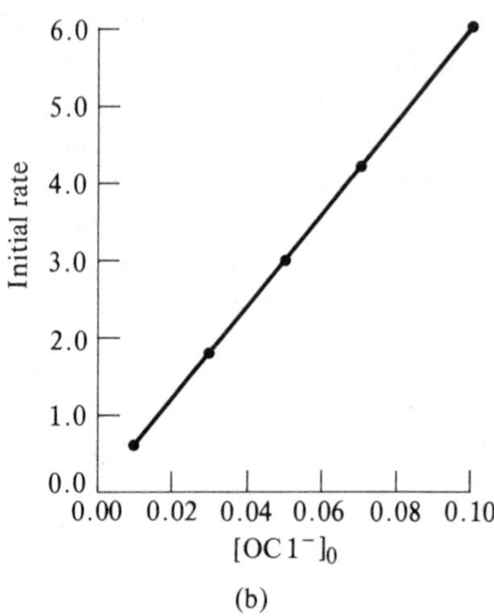

(b)

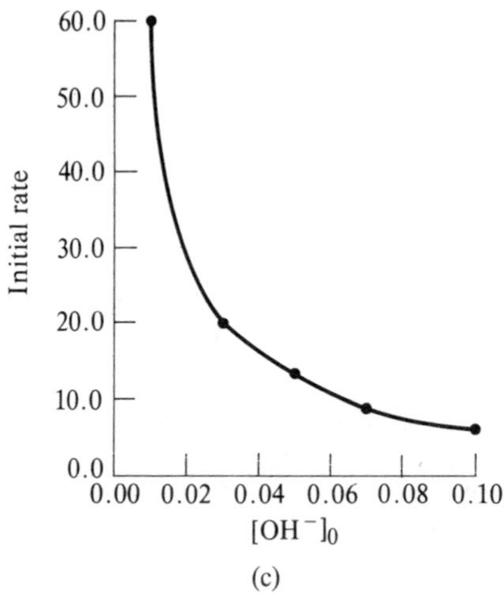

(c)

Figure 9.6

For the set of data in which $[OCl^-]_0 = [OH^-]_0 = 0.1$ mole liter^{-1}, Eq. (183) reduces to

$$R_0 = k(0.1)^{n-p}[I^-]_0^m = k'[I^-]_0^m. \tag{184}$$

Therefore the initial rate data of Table 9.2(a) give us direct information regarding the order with respect to $[I^-]$.

The curve in Fig. 9.6(a) is a straight line with a slope of 60.0. The differential rate law for first-order reactions (Eq. 165) requires that a plot of rate versus concentration for a first-order reaction be a straight line with slope equal to k. In our case then the reaction is first-order with respect to $[I^-]$ ($m = 1$), and $k' = 60.0\ \text{sec}^{-1}$.

Since the rates with respect to $[I^-]$ and to $[OCl^-]$ are the same, the reaction is also first-order with respect to $[OCl^-]$ ($n = 1$).

Now consider the data of Table 9.2(c), plotted in Fig. 9.6(c). Since $[I^-]_0 = [OCl^-]_0 = 0.1$ for this set of data, Eq. (183) becomes

$$R_0 = k(0.1)^2/[OH^-]_0^p = k''[OH^-]_0^{-p} \tag{185}$$

for this case. First test for $p = 1$ by plotting the initial rate versus $[OH^-]_0^{-1}$ (Fig. 9.7). If this is the correct order, Eq. (185) (with $p = 1$) requires that such a plot be a straight line with a slope of k''. Indeed the curve of Fig. 9.7 is a straight line with slope 0.60. This indicates that p is equal to 1 and k'' is 0.60 mole^2 liter^{-2} sec^{-1}.

The differential rate law for this reaction is therefore

$$\frac{d[OI^-]}{dt} = k[I^-][OCl^-]/[OH^-]. \tag{186}$$

We may calculate the rate constant from Eq. (184) or Eq. (185):

Equation (184): $k(0.1\ \text{molar})^{n-p} = k'$
$k = 60\ \text{sec}^{-1}/(0.1\ \text{molar})^{1-1} = 60\ \text{sec}^{-1}$.

Equation (185): $k(0.1\ \text{molar})^2 = k''$
$k = 0.60\ \text{mole}^2\ \text{liter}^{-2}\ \text{sec}^{-1}/0.01\ \text{mole}^2\ \text{liter}^{-2} = 60\ \text{sec}^{-1}$.

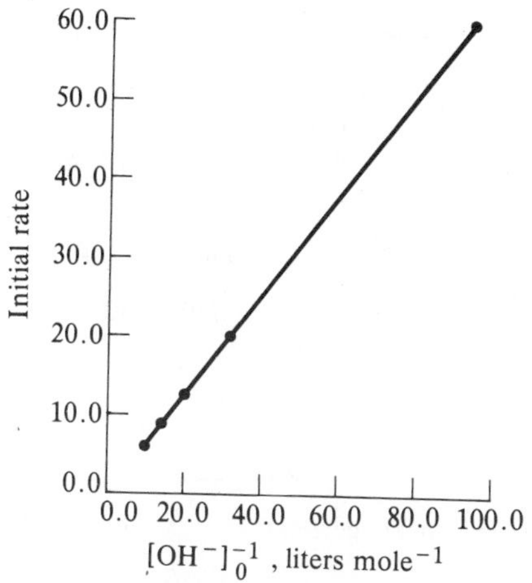

Figure 9.7

The total rate equation for this reaction is thus

$$R = 60 \text{ sec}^{-1} \frac{[\text{I}^-][\text{OCl}^-]}{[\text{OH}^-]}. \tag{187}$$

Example 52 Using the following rate data, obtain the correct experimental differential rate expression and the value of the rate constant for the following reaction in which the poison gas phosgene is synthesized at 350°C:

$$CO + Cl_2 \longrightarrow COCl_2.$$

Experiment number:	*1*	*2*	*3*	*4*
$[CO]_0$, moles liter^{-1}:	0.10	0.10	0.050	0.050
$[Cl_2]_0$, moles liter^{-1}:	0.10	0.050	0.10	0.050
R_0, moles liter^{-1} sec^{-1}:	1.2×10^{-2}	4.26×10^{-3}	6.0×10^{-3}	2.13×10^{-3}

Answer: Let us assume the rate expression has the form

$$R = k[CO]^m[Cl_2]^n.$$

Now examine the data from the first and the third experiments. The concentration of B is 0.1 in both of these; [A] is different.

$$1: \quad 1.2 \times 10^{-2} = k(0.10)^m(0.10)^n = \{k(0.10)^n\}(0.10)^m$$

$$3: \quad 6.0 \times 10^{-3} = k(0.050)^m(0.10)^n = \{k(0.10)^n\}(0.050)^m$$

Therefore

$$\frac{1.2 \times 10^{-2}}{6.0 \times 10^{-3}} = \frac{(0.10)^m}{(0.050)^m} = \left(\frac{0.10}{0.050}\right)^m = 2^m$$

$$2 = 2^m.$$

The value of m must be 1. This conclusion may be confirmed by examining the data of the second and the fourth experiments.

Proceeding similarly, we solve for n using the data from the first and the second experiments:

$$1: \quad 1.2 \times 10^{-2} = \{k(0.10)\}(0.10)^n$$

$$2: \quad 4.26 \times 10^{-3} = \{k(0.10)\}(0.050)^n$$

$$\frac{1.2 \times 10^{-2}}{4.26 \times 10^{-3}} = \left(\frac{0.10}{0.050}\right)^n$$

$$2.82 = 2^n.$$

Obviously n cannot be an integer; rather it is equal to $\frac{3}{2}$. The differential rate expression is thus

$$R = d[COCl_2]/dt = k[CO][Cl_2]^{3/2}. \tag{188}$$

The value of the rate constant can be calculated using any one of the data points:

$$4:\quad 2.13 \times 10^{-3} \text{ mole liter}^{-1} \text{ sec}^{-1} = k(0.05 \text{ mole liter}^{-1})(0.05 \text{ mole}^{-1})^{3/2}$$

$$k = 3.8 \text{ liter}^{3/2} \text{ mole}^{-3/2} \text{ sec}^{-1}.$$

The reaction is thus first-order with respect to carbon monoxide, 3/2-order with respect to chlorine and 5/2-order overall.

Problems

9.1 Derive the half-life relationship for zero-order reactions.

9.2 In Problem 8.2 we discussed the use of isopropyl nitrite in the treatment of cyanide poisoning. The "shelf-life" of pharmacological agents is of some concern in hospitals. From the following data determine (a) the reaction order of the spontaneous decomposition of isopropyl nitrite and (b) the time required for 10% of the original nitrite to decompose.

time, hr:	0	10^4	2×10^4	5×10^4	10^5
$[\text{nitrite}]_t$, moles liter^{-1}:	1.000	0.964	0.931	0.836	0.697

9.3 We have mentioned the enzyme-catalyzed isomerization of maleic acid to fumaric acid. In the systematic study of this and related reactions, a kinetic study was made of the gas phase isomerization of methylmaleic acid to methylfumaric acid. Using the following data, determine the order and the rate constant.

$[\text{methylmaleic acid}]_0$, moles liter^{-1}:	2.0	1.5	1.0	0.5
half-life τ, hr:	1.070	1.069	1.071	1.069

9.4 (a) Dog fanciers hold that an animal known to have been pure bred for seven generations is thereby a purebred regardless. Such blue-ribbon canines could still have what percentage of mutt in their inheritance? (b) Workers in radiochemistry use a rule of thumb that the activity from any radioisotope sample will be relatively negligible after ten half-lives. What fraction will actually remain? Caution: Here "relatively" means compared to the initial activity; it does *not* mean that after ten half-lives the residual radioactivity will be so slight as to be harmless.

9.5 Since the body concentrates iodine into the thyroid gland, it is logical to combat thyroid cancer by administering radioactive iodine. For this purpose $Na^{131}I$ is often used, in which the ^{131}I has a half-life of 7.80 days. Using this treatment and barring losses by excretion, what fraction of the original radioactivity will have decayed after 3 weeks? How long will it be until only 2.0% of the original activity remains?

9.6 The hydrolysis of phosphate esters is of great importance in biological systems. The more detailed *in vitro* chemical information about the reaction of phosphates with water which can be obtained, the greater our ability to interpret biological reactions. With this in mind, studies of the hydrolysis of a number of relatively simple phosphate esters have been undertaken. For instance, the hydrolysis of dimethyl phosphate is found to be first-order with respect to dimethyl phosphate with a rate constant of 4.2×10^{-6}

sec^{-1}. See C. A. Bunton, M. M. Mhala, K. G. Oldham, and C. A. Vernon, *Journal of the Chemical Society of London*, 3293 (1960). (a) Calculate the initial rate of the hydrolysis when the concentration of ester is 0.01, 0.05, and 0.10. (b) Calculate the time required for 25%, 50%, and 75% of the ester to be hydrolyzed at each of these three initial concentrations.

9.7 In chapter 12 we shall discuss the general mechanism of enzyme-catalyzed reactions. In general it involves initially the formation of an enzyme-substrate complex ES by the coming together of an enzyme molecule E and a substrate molecule S

$$E + S \longrightarrow ES.$$

This initial step is extremely fast but can be followed using a special technique called relaxation spectrometry. The following data may be obtained for the formation of such a complex by the enzyme ribonuclease and the substrate cytidine-2′-phosphate (See Gordon G. Hammes, *Accounts of Chemical Research*, **1**, 326 (1968).)

$[E]_0 = [S]_0$, moles liter^{-1}:	0.001	0.005	0.010	0.015
half-life, sec:	10^{-4}	2×10^{-5}	10^{-5}	6.7×10^{-6}

Determine the *overall* order and the rate constant.

9.8 (a) Do the results of Problem 9.7 allow us to say anything about the order with respect to enzyme and the order with respect to substrate? (b) Three additional experiments are performed for this reaction and the following data are obtained:

Experiment number:	*1*	*2*	*3*
$[S]_0$, moles liter^{-1}:	0.010	0.010	0.005
$[E]_0$, moles liter^{-1}:	0.010	0.005	0.010
R_0, moles liter^{-1} sec^{-1}:	10^3	5×10^2	5×10^2

Calculate the order with respect to enzyme and with respect to substrate and thus obtain the differential rate law.

9.9 In the reaction of Problems 9.7 and 9.8 how long is required for half of the enzyme initially present to react when the initial concentrations are $[S]_0 = 0.010$ mole liter^{-1} and $[E]_0 = 0.005$?

9.10 When a material not generated by the body is introduced into it, the living organism moves to break down and excrete the added substance. The rate of excretion is usually proportional to the concentration present; thus the elimination process is essentially first-order, and a "biological half-life" can be assigned. Suppose in such an experiment a subject took a massive dose of vitamin C, then underwent blood and urine analyses at successive intervals, with the following results:

Time of analysis, hours o'clock	Mon. 1300	Mon. 1700	Mon. 2300	Tues. 0600	Tues. 1100	Tues. 2300	Wed. 0800	Wed. 2300
Instrument reading	4.84	3.84	2.72	1.81	1.35	0.68	0.40	0.17

What is the biological half-life of vitamin C? (The instrument reading is proportional to the concentration of vitamin C in the blood.)

9.11 All living matter contains both ^{14}C and ^{12}C in its organic molecules. In life processes the $^{14}C/^{12}C$ ratio is kept constant through interchange with the fixed level of $^{14}CO_2$ in the atmosphere. When the plant or animal dies this replenishment ceases, and the ^{14}C radioactivity dwindles with a half-life of 5760 years. Using careful techniques (because of the low levels of activity to be measured) it is therefore possible to establish how long it has been since a given tissue sample was alive. Suppose this "^{14}C dating" is applied to a snip of the flaxen wrapping of a mummy from an early civilization. It shows in 1974 a $^{14}C/^{12}C$ ratio equal to 67.0% of the normal value. What was the approximate date of the mummy's burial?

9.12 A set of experiments were performed in an *in vitro* study of the enzyme-catalyzed dehydrogenation of succinic acid to fumaric acid with the concentration of the substrate (succinic acid) much smaller than that of the enzyme. From the data obtained determine the reaction order with respect to substrate. The enzyme concentration is the same in all the experiments.

$[\text{Substrate}]_0$, moles liter^{-1}:	0.06	0.08	0.10
Half-life, sec:	6×10^3	8×10^3	10^4

9.13 In Example 52 we determined the rate expression for the synthesis of phosgene. Using the following data, determine the rate expression and the rate constant for the decomposition of phosgene at 451°C.

$$COCl_2 \longrightarrow CO + Cl_2$$

Experiment number:	*1*	*2*	*3*	*4*
$[COCl_2]_0$, moles liter^{-1}:	0.16	0.16	0.04	0.04
$[Cl_2]_0$, moles liter^{-1}:	0.16	0.04	0.16	0.04
R_0, moles liter^{-1} sec^{-1}:	1.9×10^{-2}	9.6×10^{-3}	4.8×10^{-3}	2.4×10^{-3}

9.14 Combining the principles involved in Problems 9.5 and 9.10 leads to the concept of "effective half-life" in radiobiological work. This denotes the time required for an administered radioactivity to drop to half its initial level due to the combined effects of radioactive decay plus biological elimination. For example iodine, with a biological half-life of 26 days and a radioactive half-life (131-isotope) of 7.80 days, has an effective half-life of 6.0 days. (a) Show mathematically how this latter figure is correct. (b) Workers with radio-iodine know that an animal used in an experiment with I-131 should not be re-used for at least two months in another experiment employing that isotope. Correlate this rule with other information within this group of problems.

9.15 Suppose an experimental animal is injected with a sodium iodide solution containing both ^{131}I and ^{128}I in the mole proportions of 3 : 1. An initial radioactivity indication of 1.21×10^4 is measured. (a) Taking into account the biological half-life of iodine (26 days), the radioactive half-life of ^{131}I (7.80 days), and the radioactive half-life of ^{128}I (25.0 min), calculate the expected activity reading every 20 minutes for the first 3 hr, then every hour for the next 3 hr, then every 24 hr for the remainder of the week, followed by a final reading 10 days after the original maximum. Approximations may be used if an explicit justification is given. The ratio of radioactivity of two materials is directly proportional to the ratio of their molar concentrations and inversely proportional to the ratio of their half-lives. (b) Plot the log of the activity value on the ordinate against time on the abscissa. Does a straight line result? Why? Explain why this result does, or does not, reconcile with the observations made in Problem 9.14(a).

10 / Mechanisms of Chemical Reactions

10.1. Introduction

The previous chapter dealt primarily with the rate and the order of chemical reactions as experimental observables. Just as with the thermodynamic state functions, however, we are interested in the molecular behavior which underlies these observables. We would like to explain the macroscopic experimental facts in terms of a microscopic model.

We shall again operate with atoms and molecules, and we shall postulate various types and sequences of interactions between them to explain the observed rate behavior. As we shall see, these postulates lead to predictions which are in good agreement with experimental results. This does not mean that they are therefore correct, but rather that they are reasonable and consistent. In some cases alternative mechanisms for reactions may be postulated which fit the experimental data equally well. For the experimental results to be found compatible with a proposed mechanism is a necessary but not sufficient condition for the proposed mechanism to be correct.

10.2. Elementary Mechanisms

In the previous chapter we briefly discussed the concept of the elementary reaction. Such a process is pictured as taking place through the repeated occurrence of a single molecular event. In a unimolecular reaction, this event is the spontaneous decomposition of a single molecule. In a bimolecular process, two molecules come into intimate contact (they collide) resulting in the breaking of some molecular bonds and the forming of others, with the effect that reactant molecules are rearranged into product. A termolecular reaction requires the simultaneous collision of three molecules.

This model agrees well with the rate laws. Recall that for elementary reactions the molecularity and the order are the same. If we double the concentration of the reactant in a unimolecular reaction (such as the decomposition of dimethyl ether considered in Chapter 9), the first-order rate equation (Eq. 165) indicates that a doubling of rate should be expected, since the rate is proportional to the first power of the concentration. On the molecular level this is explained by the fact that since the reaction depends on the rate of decomposition of single molecules, if there are twice as many molecules present in a given volume (resulting from a doubling of concentration) twice as many molecules will react in that volume in a given time interval.

Suppose the chemical nature of a particular molecular species is such that a single molecule has a 50% chance of decomposing within one second. If the concentration is such that there are two molecules in a given volume, then one of these molecules may be expected to decompose in the next second (Fig. 10.1a).

We now double the concentration so that there are four molecules in this volume. With a 50% chance for each molecule to decompose in one second, two molecules in this volume may be expected to react within the next second and the rate (proportional to the number of molecules which react in one second within a unit volume) will double (Fig. 10.1b).

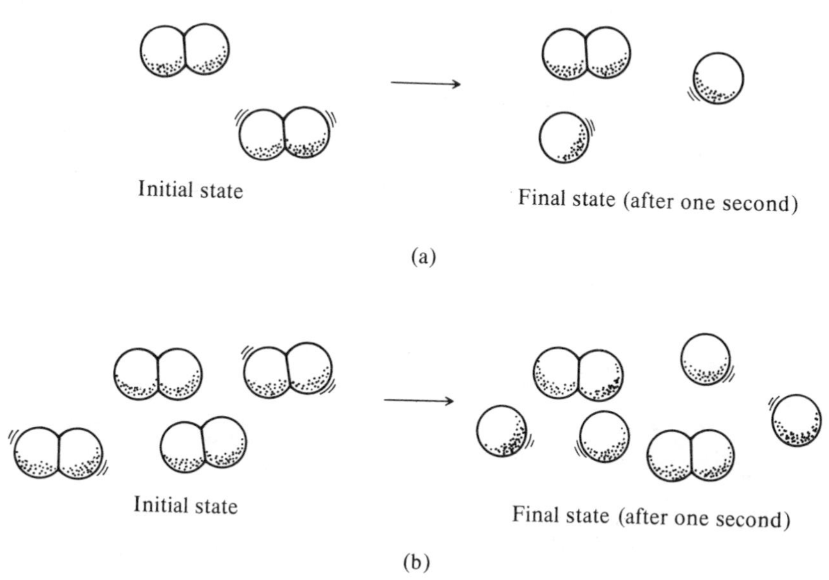

Fig. 10.1. (a) Unimolecular reaction with two molecules in a unit volume. (b) Unimolecular reaction with four molecules in a unit volume.

For the bimolecular reaction between the two gaseous pollutants nitric oxide and ozone,

$$NO + O_3 \longrightarrow NO_2 + O_2,$$

the second-order rate expression applies:

$$R = k_2[NO][O_3].$$

Doubling the concentration of NO or of O_3 results in doubling the reaction rate. Doubling the concentration of both NO and O_3 quadruples the rate.

Figure 10.2 is a schematic drawing of a molecular model for this reaction. The concentration dependence may be explained by realizing that if the reaction requires the collision of one molecule of NO with one molecule of O_3, then the rate will be proportional to the number of collisions between NO and O_3 molecules in a unit volume in unit time.

Doubling the concentration of O_3 doubles the number of molecules of O_3 with which a given NO molecule can collide and react. This doubles the frequency of such reactive collisions in a given volume—and thus doubles the rate. A similar consequence results from doubling the concentration of NO molecules.

Doubling the concentration of both nitric oxide and ozone molecules produces a doubling and a redoubling of the collision frequency. The net result would be a rate four times that originally observed.

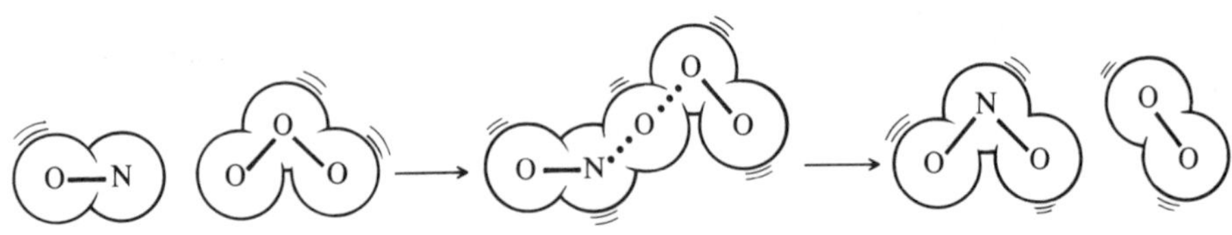

Fig. 10.2. The second-order, bimolecular reaction of NO and O_3 may be pictured as the collision of a molecule of NO and a molecule of O_3. This intimate contact results in the rearranging of bonds, after which the product molecules NO_2 and O_2 pull apart.

Whenever the experimental rate expression is found to have the form of Eq. (162), with the exponents (*m*, *n*, etc.) equal to the stoichiometric coefficients (*a*, *b*, etc.) of the reactants, it is tempting to assume the reaction is elementary and has a straightforward one-step mechanism. Recall, however, that while it is necessary that our mechanistic predictions agree with experiment, the fact that they do so does not insure that the mechanism is correct. We can never say with absolute certainty that a reaction actually occurs by a particular mechanism. We *can* often design further experiments to prove which of several kinetically acceptable mechanisms could *not* be correct (see Section 10.9). From the rate information all we can say, with certainty, is that if the reaction's stoichiometric coefficients are not the same as its rate exponents, the process *cannot* have an elementary mechanism. If they do agree, the most we can say, even so, is that the process *may* be elementary. Our subsequent discussion of complex mechanisms will help explain how, under certain conditions, rate expressions of the simple form of Eq. (162) may arise for reactions which do not have a simple one-step mechanism.

A more detailed discussion of collision theory will be undertaken in Chapter 11. It is sufficient at this point to be able to visualize how reactions proceed in terms of the behavior of individual molecules. For elementary reactions, this visualization is quite straightforward; for processes with complex rate expressions, more ingenuity is required.

10.3. Complex Mechanisms

Most reactions are not simple, one-step processes. Their complex nature and the resulting complicated rate expressions cannot be explained as the repetition of a single one-step molecular collision. To interpret complex reactions, we postulate that their mechanisms involve a set of related elementary, single-step processes.

There are a number of reasonable ways of combining elementary reactions to arrive at a possible mechanism for a chemical reaction. In subsequent sections, we shall consider the most important types: opposed or reversible reactions, side or concurrent reactions, consecutive or sequential reactions, and chain reactions. As we shall see, the experimental rate expression will often enable us to choose among several feasible mechanisms.

10.4. Opposed or Reversible Reactions

Up to this point, we have concerned ourselves mainly with one-way reactions and have thought along the general lines of A + B → Products. The nature of the products has been of little concern. With the process thus visualized as irreversible, it has been sufficient to consider only the rate at which A and B disappear. This viewpoint can be justified, however, only for a reaction whose equilibrium constant is extremely large and which therefore goes essentially to completion, or for the very early stages of a reacting system when very little product has been formed. Almost all reactions are in fact reversible; as soon as "products" are formed, they begin to revert to "reactants"; the forward reaction is coupled with an opposed, reverse reaction. For instance, in the reaction of ozone with nitric oxide:

$$\text{Forward:}\quad O_3 + NO \xrightarrow{k_f} O_2 + NO_2$$

$$\text{Reverse:}\quad O_2 + NO_2 \xrightarrow{k_r} O_3 + NO$$

This relationship is familiarly written as

$$O_3 + NO \underset{k_r}{\overset{k_f}{\rightleftharpoons}} O_2 + NO_2,$$

which explicitly indicates the dual direction. If both forward and reverse reactions are elementary one-step processes, the forward reaction proceeds at a rate R_f equal to $k_f[O_3][NO]$, while the reverse reaction has rate $R_r = k_r[O_2][NO_2]$. The rate of formation of products R will be equal to the rate of the elementary forward step minus the rate of the elementary reverse step, the rate, that is, at which products are created minus the rate at which they are destroyed:

$$R = \frac{d[O_2]}{dt} = -\frac{d[O_3]}{dt} = k_f[O_3][NO] - k_r[O_2][NO_2]. \qquad (189)$$

If $k_f \gg k_r$ or if our observations are made so early in the reaction that $[O_2]$ and $[NO_2]$ are quite small, Eq. (189) becomes

$$R \approx k_f[O_3][NO].$$

As reaction proceeds, the concentration of products ($[O_2]$ and $[NO_2]$) grows and the concentration of reactants ($[O_3]$ and $[NO]$) dwindles. Consequently $k_f[O_3][NO]$ diminishes in magnitude, while $k_r[O_2][NO_2]$ increases. The effective rate therefore tapers off, until at some point the forward rate equals the reverse rate and there is no further net progress. This situation is familiarly known as the equilibrium state. Thus equilibrium is really dynamic. Instead of being a state where reaction has ceased, equilibrium is a state in which the forward and reverse rates are equal. All reaction participants are being formed as fast as they are being destroyed. Thus there is no further change in the various concentrations. At equilibrium we may write

$$R_f = R_r$$

$$k_f[O_3][NO] = k_r[O_2][NO_2]$$

$$\frac{k_f}{k_r} = \frac{[O_2][NO_2]}{[O_3][NO]}.$$

The expression on the right is the familar formulation of the equilibrium constant K_{eq}. It follows that the equilibrium constant is in fact the ratio of the forward and reverse rate constants:

$$K_{eq} = k_f/k_r. \qquad (190)$$

The usual method of attack in the kinetic study of a reaction whose mechanism is thought to involve opposed forward and reverse steps is to first determine k_f by measurements made in the early stages of reaction, when the reverse reaction may be ignored. Then k_r may be found by dividing k_f by the equilibrium constant, this latter value measured in a separate experiment or calculated from tabulated thermodynamic data.

Example 53 The reaction of ozone and nitric oxide is thought to proceed by a one-step mechanism:

$$O_3 + NO \longrightarrow NO_2 + O_2.$$

The observed rate is $R = k_1[O_3][NO]$ with k_1 equal to 1.2×10^{10} liters mole^{-1} sec^{-1}. This reaction is potentially reversible. Using the data of Table 3.1, explain why the observed rate does not appear to be influenced by the accumulated nitrogen dioxide and oxygen.

Answer: From Table 3.1 we can calculate the value of the equilibrium constant for this reaction:

$$\begin{aligned} \Delta G^\circ &= \Delta G_f^\circ(NO_2) + \Delta G_f^\circ(O_2) - \Delta G_f^\circ(O_3) - \Delta G_f^\circ(NO) \\ &= (12.39 + 0.0 - 39.06 - 20.72)\ \text{kcal mole}^{-1} \\ &= -47390\ \text{cal mole}^{-1} \\ \Delta G^\circ &= -RT \ln K_{eq} \\ \log K_{eq} &= 47390\ \text{cal mole}^{-1}/(2.303)(1.987\ \text{cal deg}^{-1}\ \text{mole}^{-1})(298^\circ) \\ &= 34.75 \\ K_{eq} &= 5.6 \times 10^{34}. \end{aligned}$$

Since we know $k_1(= k_f)$, k_r may now be calculated using Eq. (190):

$$\begin{aligned} k_r &= k_f/K_{eq} \\ &= 1.2 \times 10^{10}\ \text{liter mole}^{-1}\ \text{sec}^{-1}/5.6 \times 10^{34} \\ &= 2.1 \times 10^{-25}\ \text{liter mole}^{-1}\ \text{sec}^{-1}. \end{aligned}$$

The overall rate is then

$$R = 1.2 \times 10^{10}[O_3][NO] - 3.75 \times 10^{-25}[O_2][NO_2].$$

Obviously the second term will only become important when the concentrations of O_2 and NO_2 are extremely large relative to those of O_3 and NO. For reactions with very large equilibrium constants (such as this one), the reverse rate may therefore be ignored.

The above discussion and the relationships derived (including Eq. 190) also apply to reversible reactions whose forward and reverse steps are not single-step processes.

Example 54 The reaction which produces phosgene from carbon monoxide and chlorine is reversible:

$$CO + Cl_2 \underset{k_r}{\overset{k_f}{\rightleftharpoons}} COCl_2 .$$

The forward and reverse rates are found by experiment to be respectively

$$R_f = d[COCl_2]/dt = k_f[CO][Cl_2]^{3/2}$$
$$R_r = -d[COCl_2]/dt = k_r[COCl_2][Cl_2]^{1/2}.$$

(a) Explain why neither process can be a single-step, elementary reaction. (b) Show that Eq. (190) nevertheless holds for this reaction. (c) Calculate the equilibrium constant if at 25°C

$$k_f = 77.8 \text{ liter}^{3/2} \text{ mole}^{-3/2} \text{ sec}^{-1} \text{ and } k_r = 0.30 \text{ liter}^{1/2} \text{ mole}^{-1/2} \text{ sec}^{-1}.$$

Answer: a) In neither case are the stoichiometric coefficients equal to the kinetic exponents. In fact, in the reverse reaction the rate depends on the concentration of one of its products. If the reactions were one-step processes, the rates would be

$$R_f = k_f[CO][Cl_2]$$
$$R_r = k_r[COCl_2].$$

b) The net rate of production of phosgene is

$$R = d[COCl_2]/dt = R_f - R_r = k_f[CO][Cl_2]^{3/2} - k_r[COCl_2][Cl_2]^{1/2}.$$

At equilibrium $R = 0$ and $R_f = R_r$.

$$k_f[CO][Cl_2]^{3/2} = k_r[COCl_2][Cl_2]^{1/2}$$

$$\frac{k_f}{k_r} = \frac{[COCl_2][Cl_2]^{1/2}}{[CO][Cl_2]^{3/2}} = \frac{[COCl_2]}{[CO][Cl_2]} = K_{eq} .$$

c)

$$K_{eq} = k_f/k_r$$
$$= 77.8/0.30$$
$$= 2.6 \times 10^2.$$

10.5. Side or Concurrent Reactions

Rate expressions for some complex reactions can be explained by postulating that the products observed were formed by the initial reactants simultaneously undergoing several different reactions. For example, in solution the aldehydic form of glucose reacts to both α-glucose and β-glucose:

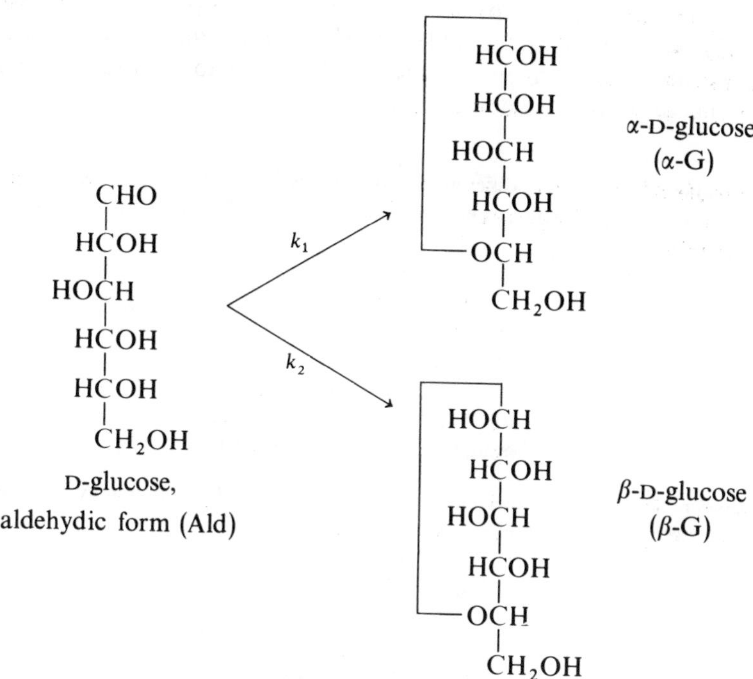

The rate expressions[1] for these processes respectively are found to be

$$R_1 = d[\alpha\text{-G}]/dt = k_1[\text{Ald}]. \tag{191a}$$

$$R_2 = d[\beta\text{-G}]/dt = k_2[\text{Ald}]. \tag{191b}$$

The overall rate, as measured by the rate of disappearance of aldehyde, is the sum of the individual rates:

$$\begin{aligned} -d[\text{Ald}]/dt &= k_1[\text{Ald}] + k_2[\text{Ald}] \\ &= (k_1 + k_2)[\text{Ald}] \\ &= k'[\text{Ald}]. \end{aligned} \tag{192}$$

The first-order rate constant k' is the sum of the rate constants k_1 and k_2 of the two simultaneous reactions. The experimental determination of any two of the three constants k', k_1, or k_2 allows the calculation of the third, since $k' = k_1 + k_2$.

[1] The subscripts on the rate constants are used here and in subsequent sections as an identifying device, not to indicate the reaction's kinetic order as in Chapter 9.

This is an example of how a reaction actually involving more than one step may appear to be uncomplicated, with a rate expression of the form of Eq. (162). Note also that the additivity of the rates means additivity of the rate constants only if the molecularity with respect to each reactant has the same value in the two elementary side reactions.

Example 55 Write the differential rate expressions for the appearance of the products and the disappearance of reactant in the following concurrent reaction mechanism:

2A acetic acid monomer	$\xrightarrow{k_1}$	AA acetic acid dimer
3A acetic acid monomer	$\xrightarrow{k_2}$	AAA acetic acid trimer

Answer: The rate expressions for the appearance of product may be written immediately if these reactions are elementary:

$$R_1 = \frac{d[\mathrm{AA}]}{dt} = k_1[\mathrm{A}]^2$$

$$R_2 = \frac{d[\mathrm{AAA}]}{dt} = k_2[\mathrm{A}]^3.$$

Writing the rate of disappearance of reactant requires more care. In the first reaction:

$$R_1 = -\frac{1}{2}\frac{d[\mathrm{A}]}{dt} = k_1[\mathrm{A}]^2$$

$$\left(-\frac{d[\mathrm{A}]}{dt}\right)_1 = 2k_1[\mathrm{A}]^2.$$

For the second reaction:

$$R_2 = -\frac{1}{3}\frac{d[\mathrm{A}]}{dt} = k_2[\mathrm{A}]^3$$

$$\left(-\frac{d[\mathrm{A}]}{dt}\right)_2 = 3k_2[\mathrm{A}]^3.$$

The total rate of disappearance of acetic acid is

$$-\frac{d[\mathrm{A}]}{dt} = \left(-\frac{d[\mathrm{A}]}{dt}\right)_1 + \left(\frac{-d[\mathrm{A}]}{dt}\right)_2 = 2k_1[\mathrm{A}]^2 + 3k_2[\mathrm{A}]^3.$$

10.6. Consecutive or Sequential Reactions

One of the most productive ways of combining elementary single-step reactions into defensible mechanisms is to postulate a consecutive reaction sequence. Such sequential reactions can lead to rather complicated rate expressions. Simplifying procedures are sometimes available, however. Consider the following sequence mechanism proposed for the reaction of nitric oxide and oxygen:

$$\begin{array}{lrcl} \text{Reaction:} & 2NO + O_2 & \longrightarrow & 2NO_2 \\ \text{Mechanism:} & NO + O_2 & \longrightarrow & NO_3 \\ & NO_3 + NO & \longrightarrow & 2NO_2 \end{array}$$

A transitory species such as NO_3, present neither at the beginning nor at the end of the reaction, is called an *intermediate*.

If the second step of this mechanism is many times slower than the first, it is called the *rate-determining step*. Like a farm tractor on a two-lane highway with a no-passing stripe, it sets the pace for the whole process. And just as the farm tractor will gather an accumulation of vehicles behind it, so there will be a pile-up of the intermediate NO_3, until the slower second reaction has time to complete the conversion. The rate of production of product will thus in effect be primarily determined by the rate of this second step.

The implication here is that NO_3 is a comparatively stable intermediate. As such it could be synthesized in the laboratory, and a separate study made of the rate of its reaction with NO. If the kinetics of the two experiments are in agreement, whether starting with NO and O_2 or with NO_3 and NO, the validity of the proposed mechanism is supported.

In other two-step sequences the converse situation may be observed: the first reaction may be by nature a great deal slower than the second. Consider for instance the decomposition of ozone for which the following mechanism might be proposed:

$$\begin{array}{lrcll} \text{Net reaction:} & 2O_3 & \longrightarrow & 3O_2 & \\ \text{Mechanism:} & O_3 & \xrightarrow{k_1} & O_2 + O & \text{(I)} \\ & O + O_3 & \xrightarrow{k_2} & 2O_2, & \text{(II)} \end{array}$$

where $k_1 \ll k_2$.

Here the first step, being slower than the second, is rate-determining. Atomic oxygen is very reactive; it would therefore be a short-lived, "unstable" intermediate. Produced by a slow reaction and consumed by a fast one, it would never be present in quantity. The result is that the concentration of O is maintained at a low but essentially steady value. This situation is called a *steady state* or *stationary state*, a very useful concept in chemical kinetics. If there is such a steady state with respect to oxygen atoms, the rate of change of this species would be zero, that is

$$\frac{d[O]}{dt} = 0 = k_1[O_3] - k_2[O][O_3].$$

The $k_1[O_3]$ term is the rate of appearance of oxygen atoms in the first step of the proposed mechanism; $k_2[O][O_3]$ is the rate of their disappearance in the second step. The above equation can be solved for [O]:

$$k_1[O_3] = k_2[O][O_3]$$
$$[O] = k_1/k_2 .$$

The rate of decomposition of ozone by this mechanism is:

$$R = -\frac{1}{2}\frac{d[O_3]}{dt} = k_1[O_3] + k_2[O_3][O].$$

Substituting the steady-state concentration of atomic oxygen yields:

$$R = k_1[O_3] + k_2[O_3]\frac{k_1}{k_2} = 2k_1[O_3] = k'[O_3].$$

For the proposed mechanism to be a reasonable possibility, the observed rate must be first-order with respect to ozone. The rate would be solely determined by the rate of the very slow first step. Again we see how a complex mechanism may yield a rather simple rate expression which obeys the Law of Mass Action.

The simplifications suggested, rate-determining step and steady state, depend on the rate constants of the successive elementary reactions in a given sequence being very different in magnitude. If the constants are nearly the same, the complexity remains and is further compounded if there are more than two steps or if higher orders or a mixture of orders are involved. In such cases integration of the rate equations becomes excessively difficult, and approximation methods must be used.

Example 56 Propose a mechanism for the following reaction and show that it is consistent with the experimental rate law.

Stoichiometric reaction: $NO_2 + CO \longrightarrow CO_2 + NO$

Rate law:

$$\frac{d[CO_2]}{dt} = k[NO_2]^2$$

Answer: A possible mechanism is

$$2NO_2 \xrightarrow{k_1} NO_3 + NO$$
$$NO_3 + CO \xrightarrow{k_2} NO_2 + CO_2.$$

Using the steady state approach for $[NO_3]$:

$$\frac{d[NO_3]}{dt} = 0 = k_1[NO_2]^2 - k_2[NO_3][CO]$$

$$[NO_3] = \frac{k_1}{k_2}\frac{[NO_2]^2}{[CO]}.$$

The rate of appearance of carbon dioxide is then

$$\frac{d[CO_2]}{dt} = k_2[CO][NO_3] = k_2[CO]\left(\frac{k_1}{k_2}\frac{[NO_2]^2}{[CO]}\right) = k_1[NO_2]^2.$$

Example 57 The amino acid histidine

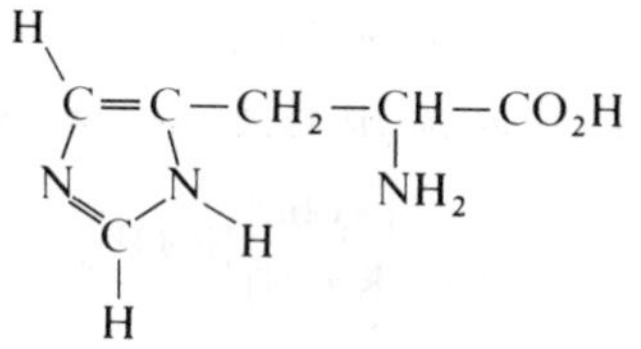

```
H
 \
  C=C—CH2—CH—CO2H
 /   \     |
N     N    NH2
 \\  / \
   C    H
   |
   H
```

is implicated as an active part of enzymes in a number of enzyme-catalyzed reactions such as hydrolysis and the transfer of acyl groups. In the detailed study of such reactions chemists often use a model system: they simplify the reaction so that they can study specific interactions without the complexities caused by large, intricate molecules. One such model reaction used in the study of physiological trans-acetylation involved the use of imidazole instead of a protein containing histidine, and simple alcohols and esters instead of the corresponding complex substrates:

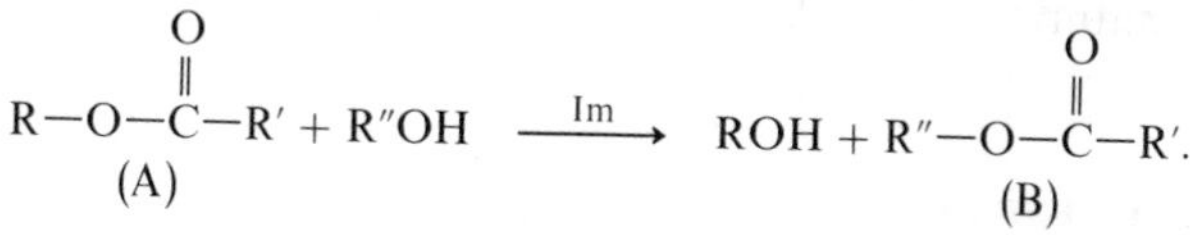

$$\underset{(A)}{R{-}O{-}\overset{\overset{\displaystyle O}{\|}}{C}{-}R'} + R''OH \xrightarrow{\text{Im}} ROH + \underset{(B)}{R''{-}O{-}\overset{\overset{\displaystyle O}{\|}}{C}{-}R'}.$$

The Im written over the reaction arrow means that while imidazole is required for the reaction (as a catalyst; see Chapter 12), it does not appear in the stoichiometric equation; it is not changed in the reaction process. Propose a mechanism which agrees with the experimentally determined rate equation:

$$R = k[A][Im].$$

Answer: One mechanism might be

$$\underset{\text{(A)}}{R{-}O{-}\overset{\overset{\displaystyle O}{\|}}{C}{-}R'} + \underset{\text{(Im)}}{\text{Im}} \xrightarrow[\text{slow}]{k_1} ROH + \underset{\text{(IE)}}{\text{IE}}$$

$$IE + R''OH \xrightarrow[\text{fast}]{k_2} Im + \underset{\text{(B)}}{R''{-}O{-}\overset{\overset{\displaystyle O}{\|}}{C}{-}R'}$$

$$R = \frac{d[B]}{dt} = k_2[IE][R''OH].$$

Using the steady-state method for [IE],

$$\frac{d[IE]}{dt} = 0 = k_1[A][Im] - k_2[IE][R''OH]$$

$$[IE] = \frac{k_1}{k_2}\frac{[A][Im]}{[R''OH]}.$$

Substituting this expression for [IE] into the rate expression:

$$R = k_2 \frac{k_1}{k_2}\frac{[A][Im]}{[R''OH]}[R''OH] = k_1[A][Im].$$

The rate predicted from this mechanism thus agrees with the experimental expression; this is a feasible mechanism. Are there other possibilities?

Radioactive decay series constitute beautiful, if limited, examples of consecutive reactions and steady states. This is because all natural radioactive processes are pure first-order reactions, insensitive to temperature, completely irreversible, and capable of very precise measurement. For example, the 14-step series from ^{238}U to ^{206}Pb is well understood, as are a number of other such series both natural and artificial.

10.7. Chain Reactions

The fourth type of complex mechanism made up of a sequence of elementary reaction steps is actually a special case of the consecutive reaction scheme discussed above. Chain reactions do, however, possess certain characteristics which make their treatment different, while their importance makes treating them necessary.

The classic example of a chain reaction is the $H_2(g) + Br_2(g) \rightarrow 2HBr(g)$ reaction mentioned in Section 9.4. The experimental rate expression for this reaction is

$$R = \frac{1}{2}\frac{d[HBr]}{dt} = \frac{k[H_2][Br_2]^{1/2}}{1 + k'[HBr]/[Br_2]}. \tag{193}$$

As in all cases, we must propose a mechanism from which we derive a theoretical rate expression. If the mechanism is a feasible one, this theoretical expression must agree with the experimental rate equation.

The elementary reaction steps proposed for this reaction are

$$Br_2 \xrightarrow{k_1} 2Br \tag{i}$$

$$Br + H_2 \xrightarrow{k_2} HBr + H \tag{ii}$$

$$H + Br_2 \xrightarrow{k_3} HBr + Br \tag{iii}$$

$$H + HBr \xrightarrow{k_4} H_2 + Br \tag{iv}$$

$$Br + Br \xrightarrow{k_5} Br_2. \tag{v}$$

In chain reactions, elementary steps of type (i) are called *initiation* reactions. Such processes produce from the reactants some highly reactive intermediate species which moves immediately into a subsequent reaction. Elementary steps such as (ii), (iii), and (iv) are called *propagation* reactions. In step (ii) the highly reactive intermediate from step (i) joins with a reactant molecule to produce not only a product molecule but a second reactive intermediate as well. Similarly in step (iii) a product molecule *and* an intermediate are produced. These processes are responsible for the unique nature of chain reactions: each product-producing step also yields some highly reactive "left-over" which is then recycled to participate with reactants in another step which produces another product molecule and an additional reactive intermediate, which reacts with reactant, and so on, resulting in a cascading chain effect. Thus one reactive particle produced in the initiation process may result in thousands of product molecules via these propagation steps.[2]

---

[2] This chain-reaction concept of the chemist (Bodenstein, 1915; Nernst, 1918) is somewhat similar to the later meaning of the term as used by the nuclear physicist, whose "fission chain reaction" may be illustrated by

$$^{235}U + n \rightarrow Ba + Kr + \sim 3n + \text{energy}.$$

Here the neutron, the intermediate which brought about the reaction, is in turn produced by the same reaction, thus sustaining the chain. Since more neutrons are produced than consumed, the uncontrolled result is a branching chain and a nuclear explosion. If an appropriate damper is used to "soak up" the surplus neutrons harmlessly, the atomic bomb becomes a nuclear reactor.

Reaction (iv) does not produce product but merely transforms one reactive intermediate into another. It does not break the chain and is therefore included in the propagation category.

Elementary reaction (v) is a *termination* reaction. This process consumes intermediate and produces none in its place. With the intermediate thus out of circulation, the chain is terminated.

Additional elementary processes of initiation, propagation, and termination can be conceived for this reaction. For instance:

$$\begin{aligned} H_2 &\longrightarrow 2H && \text{(initiation)} \\ Br + HBr &\longrightarrow Br_2 + H && \text{(propagation)} \\ \left.\begin{aligned} H + H &\longrightarrow H_2 \\ H + Br &\longrightarrow HBr \end{aligned}\right\} & && \text{(termination)} \end{aligned}$$

These steps, however, are not found to be a necessary part of the mechanism accounting for the observed rate expression.

We can now write a rate expression based on the proposed mechanism. The product HBr is created in steps (ii) and (iii) and is destroyed in step (iv). The rate is then

$$R = \frac{1}{2}\frac{d[HBr]}{dt} = \frac{1}{2}k_2[H_2][Br] + \frac{1}{2}k_3[Br_2][H] - \frac{1}{2}k_4[HBr][H]. \tag{194}$$

The intermediates H and Br are atomic species and are thus very reactive. They react almost as quickly as they appear. A constant, small concentration is quickly established; the steady-state approximation may therefore be used:

$$\frac{d[H]}{dt} = 0 = k_2[H_2][Br] - k_3[Br_2][H] - k_4[H][HBr]. \tag{195}$$

$$\frac{d[Br]}{dt} = 0 = 2k_1[Br_2] - k_2[H_2][Br] + k_3[Br_2][H] + k_4[H][HBr] - 2k_5[Br]^2.$$

Adding these two steady-state equations:

$$0 = 2k_1[Br_2] - 2k_5[Br]^2$$

$$[Br] = \left(\frac{k_1}{k_5}\right)^{1/2}[Br_2]^{1/2}. \tag{196}$$

Now substitute Eq. (196) into Eq. (195):

$$[H] = \left(\frac{k_1}{k_5}\right)^{1/2} k_2[H_2][Br_2]^{1/2}/(k_3[Br_2] + k_4[HBr]), \tag{197}$$

and substitute both Eq. (196) and Eq. (197) into the rate expression (Eq. 194):

$$R = \frac{1}{2}\frac{d[\mathrm{HBr}]}{dt} = \frac{k_2(k_1/k_5)^{1/2}[\mathrm{H_2}][\mathrm{Br_2}]^{1/2}}{1 + (k_4/k_3)[\mathrm{HBr}]/[\mathrm{Br_2}]}. \qquad (198)$$

This equation is identical with the experimental equation (Eq. 193) if $k = k_2(k_1/k_5)^{1/2}$ and $k' = k_4/k_3$. Our chain mechanism is therefore feasible and may be correct. Although we have not proven that this is the mechanism which actually obtains, it is difficult to propose an alternative which fits the rather complex experimental rate expression equally well.

Example 58 What would be the concentration dependence of the rate of the $H_2 + Br_2 \rightarrow 2HBr$ reaction in the very early stages of reaction if we start with only H_2 and Br_2?

Answer: In the early part of the reaction [HBr] is essentially zero. This leads to a significant simplification of the denominator of Eq. (198):

$$1 + (k_4/k_3)[\mathrm{HBr}]/[\mathrm{Br_2}] \simeq 1$$

so that the initial rate becomes

$$R_0 = k[\mathrm{H_2}][\mathrm{Br_2}]^{1/2}.$$

This is exactly the rate equation which would be obtained by repeating the derivation (Eqs. 194–198) ignoring elementary step (iv).

10.8. Mechanisms Resulting from Combining Opposed, Concurrent, and Consecutive Mechanisms

The chain reaction discussed above is an example of a mechanism which combines several types of elementary sequences. It is made up of consecutive and opposed steps. The steps in opposition are (i) and (v), the initiation and termination steps, and (ii) and (iv), two of the propagation steps.

As a further example of this and as an additional demonstration of the process of constructing a mechanism for a reaction, consider the reaction discussed in Section 9.9: $I^- + OCl^- \rightarrow OI^- + Cl^-$. Based on supplied data we derived the following rate expression (Eq. 188):

$$R = \frac{d[\mathrm{OI^-}]}{dt} = k[\mathrm{I^-}][\mathrm{OCl^-}]/[\mathrm{OH^-}].$$

The presence of $[OH^-]$ in the denominator means that it retards the reaction. Since it is neither a product nor a reactant in the net reaction, there must be a step in which it is a reactant and a step in which it is a product. A possible mechanism is:

$$OCl^- + H_2O \underset{k_2}{\overset{k_1}{\rightleftharpoons}} HOCl + OH^-. \tag{I}$$

$$I^- + HOCl \xrightarrow{k_3} HOI + Cl^-. \tag{II}$$

$$OH^- + HOI \underset{k_5}{\overset{k_4}{\rightleftharpoons}} H_2O + OI^-. \tag{III}$$

The rate of formation of hypoiodite is

$$\frac{d[OI^-]}{dt} = k_4[OH^-][HOI] - k_5[H_2O][OI^-]. \tag{199}$$

Applying the steady-state approximation to [HOI],

$$\frac{d[HOI]}{dt} = 0 = k_3[I^-][HOCl] - k_4[OH^-][HOI] + k_5[H_2O][OI^-]$$

$$[HOI] = \frac{k_3[I^-][HOCl] + k_5[H_2O][OI^-]}{k_4[OH^-]},$$

and substituting into Eq. (199):

$$\frac{d[OI^-]}{dt} = k_4[OH^-]\left(\frac{k_3[I^-][HOCl] + k_5[H_2O][OI^-]}{k_4[OH^-]}\right) - k_5[H_2O][OI^-]$$

$$= k_3[I^-][HOCl]. \tag{200}$$

Now if we assume that the initial equilibrium (I) is rapidly established and is maintained throughout the reaction,

$$K_{eq}^{I} = \frac{[HOCl][OH^-]}{[OCl^-][H_2O]}$$

and

$$[HOCl] = \frac{K^{I}[OCl^-][H_2O]}{[OH^-]}.$$

Substituting this into Eq. (200):

$$\frac{d[OI^-]}{dt} = k_3 K^{I} \frac{[OCl^-][I^-][H_2O]}{[OH^-]}.$$

Since the reaction occurs in aqueous solution, water is in great excess and its concentration does not change appreciably during the course of the reaction. $[H_2O]$ may thus be grouped with the constants:

$$\frac{d[OI^-]}{dt} = (k_3 K^{I}[H_2O])\frac{[OCl^-][I^-]}{[OH]^-}. \tag{201}$$

This expression is identical with the experimental rate expression with $k = k_3 K^{\mathrm{I}}[H_2O]$. The proposed mechanism is then feasible but cannot be said to be proved. In fact, alternative approaches may be proposed, one of which[3] is the following:

$$OCl^- + H_2O \underset{k_2}{\overset{k_1}{\rightleftharpoons}} HOCl + OH^-. \tag{I$'$}$$

$$I^- + HOCl \xrightarrow{k'_3} ICl + OH^-. \tag{II$'$}$$

$$ICl + 2OH^- \xrightarrow{k'_4} OI^- + Cl^- + H_2O. \tag{III$'$}$$

Applying the steady state approximation to [ICl] for this case:

$$\frac{d[ICl]}{dt} = 0 = k'_3[I^-][HOCl] - k'_4[ICl][OH^-]^2$$

$$[ICl] = \frac{k'_3}{k'_4}\frac{[I^-][HOCl]}{[OH^-]^2}. \tag{202}$$

The rate of formation of hypoiodite is

$$\frac{d[OI^-]}{dt} = k'_4[OH^-]^2[ICl] = k'_4[OH^-]^2\left(\frac{k'_3}{k'_4}\frac{[I^-][HOCl]}{[OH]^2}\right)$$

$$= k'_3[I^-][HOCl]. \tag{203}$$

The form of Eq. (203) is identical with that of Eq. (200), which was obtained from the first mechanism we proposed for this reaction. The same arguments used there to obtain Eq. (201) may be applied here to obtain an identical equation:

$$\frac{d[OI^-]}{dt} = (k'_3 K^{\mathrm{I}}[H_2O])\frac{[OCl^-][I^-]}{[OH^-]}. \tag{204}$$

This mechanism fits the observed rate. It too is a feasible mechanism for this important reaction.

Based on the kinetic evidence alone, we could not decide which of these mechanisms is the most likely. Attempts to resolve this uncertainty will be the subject of the next section.

Example 59 A number of lactones have been found to be carcinogenic (cancer-causing) while others are oncolytic (anti-cancer) agents. Involved in one of the proposed explanations of how lactones act is their ability to acylate nucleophilic

[3] Edward L. King, *How Chemical Reactions Occur*, Menlo Park, Calif.: W. A. Benjamin (1963), pp. 74–78.

sites (—OH, $-NH_2$, etc.) of physiological molecules. In a study of the aminolysis of γ-butyrolactone (L) with *n*-pentylamine (A),

$$\text{(lactone ring: } H_2,\ H_2,\ H_2,\ O,\ =O) + RNH_2 \longrightarrow HO(CH_2)_3-\overset{\overset{O}{\|}}{C}-NHR,$$

the rate was found to be

$$R = (1.0 \times 10^{-3})[A]^{3/2}[L],$$

when carried out with ethanol present. Show that the following mechanism is consistent with this observed rate:

$$\underset{(A)}{RNH_2} + \underset{(E)}{CH_3CH_2OH} \overset{K_1}{\rightleftharpoons} \underset{(A^+)}{RNH_3^+} + \underset{(E^-)}{CH_3CH_2O^-}$$

$$\underset{(A)}{2\,RNH_2} \overset{K_2}{\rightleftharpoons} \underset{(A^-)}{RNH^-} + \underset{(A^+)}{RNH_3^+}$$

$$\underset{(L)}{\text{(lactone ring: } H_2,\ H_2,\ H_2,\ O,\ =O)} + \underset{(A^-)}{RNH^-} \xrightarrow[\text{slow}]{k_1} \underset{(F^-)}{{}^-O(CH_2)_3-\overset{\overset{O}{\|}}{C}-NHR}$$

$$\underset{(F^-)}{{}^-O(CH_2)_3-\overset{\overset{O}{\|}}{C}-NHR} + \underset{(A^+)}{RNH_3^+} \xrightarrow[\text{fast}]{k_2} \underset{(F)}{HO-(CH_2)_3-\overset{\overset{O}{\|}}{C}-NHR} + \underset{(A)}{RNH_2}$$

Answer: The rate of the reaction is the rate of appearance of the product amide F:

$$R = \frac{d[F]}{dt} = k_2[F^-][A^+].$$

Since F^- is produced slowly and reacts immediately to F, we may use the steady state method:

$$\frac{d[F^-]}{dt} = k_1[L][A^-] - k_2[F^-][A^+] = 0$$

$$[F^-] = \frac{k_1}{k_2}\frac{[L][A^-]}{[A^+]}.$$

Substituting into the rate expression:

$$R = k_1[A^-][L].$$

Since exactly the same amounts of A^+ and A^- are used up (last two mechanism steps), $[A^+] = [E^-]$ and

$$K_1 = \frac{[A^+][E^-]}{[A][E]} = \frac{[A^+]^2}{[A][E]}; \quad [A^+] = K_1^{1/2}[A]^{1/2}[E]^{1/2}$$

$$K_2 = \frac{[A^-][A^+]}{[A]^2}; \quad [A^+] = K_2\frac{[A]^2}{[A^-]}.$$

Equating these two expressions for $[A^+]$ and solving for $[A^-]$:

$$K_1^{1/2}[A]^{1/2}[E]^{1/2} = K_2[A]^2/[A^-]$$

$$[A^-] = (K_2/K_1^{1/2})[A]^{3/2}/[E]^{1/2}.$$

The rate now becomes

$$R = (k_1K_2/K_1^{1/2}[E]^{1/2})[A]^{3/2}[L] = k[A]^{3/2}[L],$$

which agrees with the experimental expression.

The rate constant k obviously depends on the concentration of ethanol, which is a catalyst in this reaction (see Chapter 12). If this mechanism is correct the rate should also vary with [E] in a manner given by this last equation. Notice that the overall order of this reaction (assuming a constant concentration of ethanol) is 5/2.

10.9. Additional Experimental Evidence

Once a mechanism or several mechanisms have been proposed based on kinetic evidence, additional kinetic experiments may be undertaken to support or disprove the hypothesis. For example, if a proposed mechanism of the type

$$A + B \underset{k_2}{\overset{k_1}{\rightleftharpoons}} C$$

$$C + D \xrightarrow{k_3} F$$

requires that $k_3 \gg k_2$ in order to fit the experimental rate expression, an experimental determination of the relative magnitude of $d[A]/dt$ and $d[F]/dt$ when only C and D are mixed should establish whether or not such a hypothesis is reasonable. Chemical observations of a non-kinetic nature may also provide valuable evidence.

Participation by Solvent Molecules

In some cases a proposed mechanism may require participation by solvent molecules (both mechanisms proposed for the hypochlorite reaction in the previous section, for example). Solvent is usually present in such great excess that its concentration is essentially constant through the entire course of the reaction. In this case the process would appear to be zero-order with respect to solvent. In some cases further experiments may be undertaken in which the amount of solvent is reduced. If at some point its concentration begins to limit the rate, then such a mechanism may be considered reasonable. More on this subject will be presented in Chapter 11.

Proof of Intermediates

Mechanisms in which intermediates are postulated can be accepted with greater confidence if the existence of these intermediate species can be proved. In some cases intermediates may be isolated from the reaction and their identity determined. In some other cases highly reactive intermediates can be "trapped" by "scavengers." For instance, suppose a mechanism is based on chloride ion as an intermediate. An experiment with silver ion in the solution will result in the precipitation of AgCl. Thus the silver ion scavenger would stop the progress of the reaction, tending to confirm the role of the chloride ion as an intermediate.

The effect of an intermediate may also be detected by introducing the intermediate into the reaction mixture from an external source. For instance, some mechanisms for organic reactions propose the existance of free radicals such as CH_3 as highly reactive intermediates. These methyl free radicals can be generated by decomposing azomethane in the reaction mixture. Injecting free radicals from azomethane into a reaction whose mechanism is thought to involve such free radicals should affect the rate in the same way as increasing the concentration of other reaction participants.

Other phenomena may also be used to detect the presence of an intermediate. Free radicals have an unpaired electron. Due to properties of such an electron which we shall discuss later, they can be detected by an electron spin resonance (esr) spectrometer.

All chemical intermediates have characteristic electromagnetic absorption spectra (see Section 9.7). Consequently if an intermediate is present in great enough concentration, its absorption spectrum can be observed with the proper spectrometer and its presence in the reaction mixture verified.

Acquaintance with Chemical Properties

General chemical knowledge may also be a great help in deciding which of several kinetically reasonable reactions is most likely. For example, mechanisms which require ions with the same sign to come together are less likely than those which require the coming together of ions of opposite sign, or an ion and a neutral molecule, or two neutral molecules. For instance a simple one-step mechanism for the hypochlorite reaction

$$OCl^- + I^- \longrightarrow OI^- + Cl^-$$

would be less likely than either of the two proposed mechanisms even if the predicted rate expression based on this one-step mechanism agreed with the experiment (which it does not). A mechanism requiring the production of a charged intermediate in a nonpolar solvent is less reasonable than one in which only neutral molecules are involved. A mechanism which necessitates the breaking of strong molecular bonds under mild conditions should be given less credence than one which accomplishes the same result via the disruption of weaker bonds. Thus a broad knowledge of chemical reactions and the conditions under which they occur can provide valuable hints for constructing mechanisms.

Isotopic Labeling

Isotopic labeling is sometimes used in the examination of possible mechanisms. Consider the following reaction:

$$R{-}O{-}\overset{\overset{\large O}{\|}}{C}{-}CH_3 + H_2O \longrightarrow R{-}OH + HO{-}\overset{\overset{\large O}{\|}}{C}{-}CH_3$$

This process might proceed by breaking the bond between the linking oxygen and the carbonyl ($R{-}O{\nmid}\overset{\overset{\large O}{\|}}{C}{-}CH_3$) or by breaking the bond between the linking oxygen and the alkyl ($R{\nmid}O{-}\overset{\overset{\large O}{\|}}{C}{-}CH_3$). In the first case the oxygen from water would become linked with the acetic acid product. In the second case it would be part of the alcohol. One way of testing the two possibilities is to use water containing the oxygen isotope ^{18}O. When the reaction is finished, the products can be separated and checked for the presence of the heavier isotope. In this case the result is found to be

$$R{-}O{-}\overset{\overset{\large O}{\|}}{C}{-}CH_3 + H_2{}^{18}O \longrightarrow R{-}OH + H^{18}O{-}\overset{\overset{\large O}{\|}}{C}{-}CH_3.$$

Radioactive isotopes may be used in a similar manner.

Analogous Reactions

In some cases the mechanism of a reaction analogous to the one being studied has already been worked out, and we may be tempted to assign a similar mechanism to our own reaction. The blind acceptance of such analogies can be dangerous, but they should serve as a guide in planning further experiments.

In summary, kinetic studies provide a great deal of information concerning the details of how reactions occur. To choose between several possible mechanisms and to provide additional support for likely ones may call for additional studies. These may be both imaginative in concept and ingenious in technique.

Problems

10.1 Both reactions of the aldehyde form of glucose to α- and to β-glucose which we discussed in Section 10.5 are actually reversible:

$$\text{aldehyde} \underset{k_2}{\overset{k_1}{\rightleftharpoons}} \alpha\text{-glucose}$$

$$\text{aldehyde} \underset{k_4}{\overset{k_3}{\rightleftharpoons}} \beta\text{-glucose}$$

where at 25°C

$$k_1 = 1.1 \times 10^2 \text{ sec}^{-1}$$
$$k_2 = 7.6 \times 10^{-3} \text{ sec}^{-1}$$
$$k_3 = 6.6 \times 10^1 \text{ sec}^{-1}$$
$$k_4 = 2.7 \times 10^{-3} \text{ sec}^{-1}.$$

(a) Calculate the equilibrium constant for each step. (b) If 100 g of pure aldehyde is placed in 1.0 liter of water, what will be the concentration of aldehyde, α-glucose and β-glucose at equilibrium?

10.2 In Example 54 we discussed the reversibility of the phosgene reaction:

$$CO + Cl_2 \underset{k_2}{\overset{k_1}{\rightleftharpoons}} COCl_2 .$$

The rate constants k_f and k_r are 7.8×10^1 liter$^{3/2}$ mole$^{-3/2}$ sec^{-1} and 3×10^{-1} liter$^{5/2}$ mole$^{-5/2}$ sec^{-1} respectively. Plot reaction rate versus $[COCl_2]$ from the beginning to equilibrium (a) when 1.0 mole each of CO and Cl_2 are placed in a 1.0-liter container; (b) when 1.0 mole of pure phosgene is placed in a 1.0-liter container.

10.3 The spontaneous loss of a hydrogen molecule by ethane

$$C_2H_6 \longrightarrow C_2H_4 + H_2$$

is found by experiment to be first-order in ethane:

$$\frac{d[C_2H_4]}{dt} = k[C_2H_6].$$

One possible mechanism is simply a first-order decomposition which predicts the correct rate expression. Another much more complicated mechanism is the following:

$$C_2H_6 \longrightarrow 2CH_3$$
$$CH_3 + C_2H_6 \longrightarrow CH_4 + C_2H_5$$
$$C_2H_5 \longrightarrow C_2H_4 + H$$
$$H + C_2H_6 \longrightarrow C_2H_5 + H_2$$
$$H + C_2H_5 \longrightarrow C_2H_6.$$

(a) Show that this mechanism also agrees with the experimental results. (b) Propose an experiment which would distinguish between the two proposed mechanisms.

10.4 Some care must be exercised in the use of the term "mechanism." The following sequence has been proposed as the "mechanism" of a reaction in the enzymatic synthesis of fatty acids

$$\text{stearate} + \text{CoA} + \text{ATP} \longrightarrow \text{stearyl-CoA}$$
$$\text{stearyl-CoA} + O_2 + \text{TPNH} + H^+ \longrightarrow \text{oleyl-CoA} + \text{TPN}^+ + 2H_2O.$$

(a) Comment on the likelihood that these two steps are elementary processes. (b) What do you think is meant by "mechanism" in this case?

10.5 The hydrolysis of cane sugar (sucrose) to glucose and fructose proceeds physiologically via enzyme catalysis. It will also occur in aqueous solution with catalysis by acid:

$$\underset{\text{sucrose}}{C_{12}H_{22}O_{11}} + H_2O \xrightarrow{H^+} \underset{\text{glucose}}{C_6H_{12}O_6} + \underset{\text{fructose}}{C_6H_{12}O_6}.$$

The rate of this reaction with a given concentration of acid is found experimentally to be first-order with respect to the concentration of sucrose. Although the hydrogen ion concentration does not change in the reaction, the rate is a function of its concentration. Propose a mechanism for the hydrolysis of sucrose with acid catalysis which agrees with these observations.

10.6 Show that the following proposed mechanism for the gas-phase decomposition of nitrogen pentoxide is consistent with the observed rate law.

Reaction: $N_2O_5 \rightarrow 2NO_2 + \frac{1}{2}O_2$

Rate expression:

$$\frac{d[O_2]}{dt} = k[N_2O_5]$$

Mechanism:

$$N_2O_5 \underset{k_2}{\overset{k_1}{\rightleftharpoons}} NO_2 + NO_3$$

$$NO_2 + NO_3 \xrightarrow{k_3} NO + NO_2 + O_2$$

$$NO + NO_3 \xrightarrow{k_4} 2NO_2.$$

10.7 Show how the following mechanism for the reaction in which phosgene is formed from carbon monoxide and chlorine may agree with the concentration dependence of the *initial* rate found experimentally:

Reaction: $CO + Cl_2 \rightarrow COCl_2$

Rate expression:

$$\frac{d[COCl_2]}{dt} = k[CO][Cl_2]^{3/2}$$

Mechanism:

$$Cl_2 \underset{k_2}{\overset{k_1}{\rightleftharpoons}} 2Cl$$

$$Cl + CO \underset{k_4}{\overset{k_3}{\rightleftharpoons}} COCl$$

$$COCl + Cl_2 \xrightarrow{k_5} COCl_2 + Cl.$$

10.8 The enzyme chymotrypsin possesses esterase activity, catalyzing reactions of the type

$$R{-}\overset{\overset{\displaystyle O}{\|}}{C}{-}OR' + H_2O \xrightarrow{\text{chym}} R{-}\overset{\overset{\displaystyle O}{\|}}{C}{-}OH + R'OH.$$

Suppose the rate expression is found to be

$$\frac{d[R{-}\overset{\overset{\displaystyle O}{\|}}{C}{-}OH]}{dt} = k\frac{[R{-}\overset{\overset{\displaystyle O}{\|}}{C}{-}OR'][\text{chym}]}{[R'OH]}.$$

Propose a mechanism which is consistent with this observation.

10.9 The experimental rate law for the decomposition of ozone ($2O_3 \rightarrow 3O_2$) is

$$\frac{1}{3}\frac{d[O_2]}{dt} = -\frac{1}{2}\frac{d[O_3]}{dt} = k\frac{[O_3]^2}{[O_2]}.$$

The mechanism is known to involve oxygen atoms. Propose a mechanism consistent with these observations. (The rate law which was derived from the mechanism proposed for this fraction in Section 10.6 does not agree with the experimental rate law. Therefore that mechanism cannot be correct.)

11 / Reaction Rate Theories

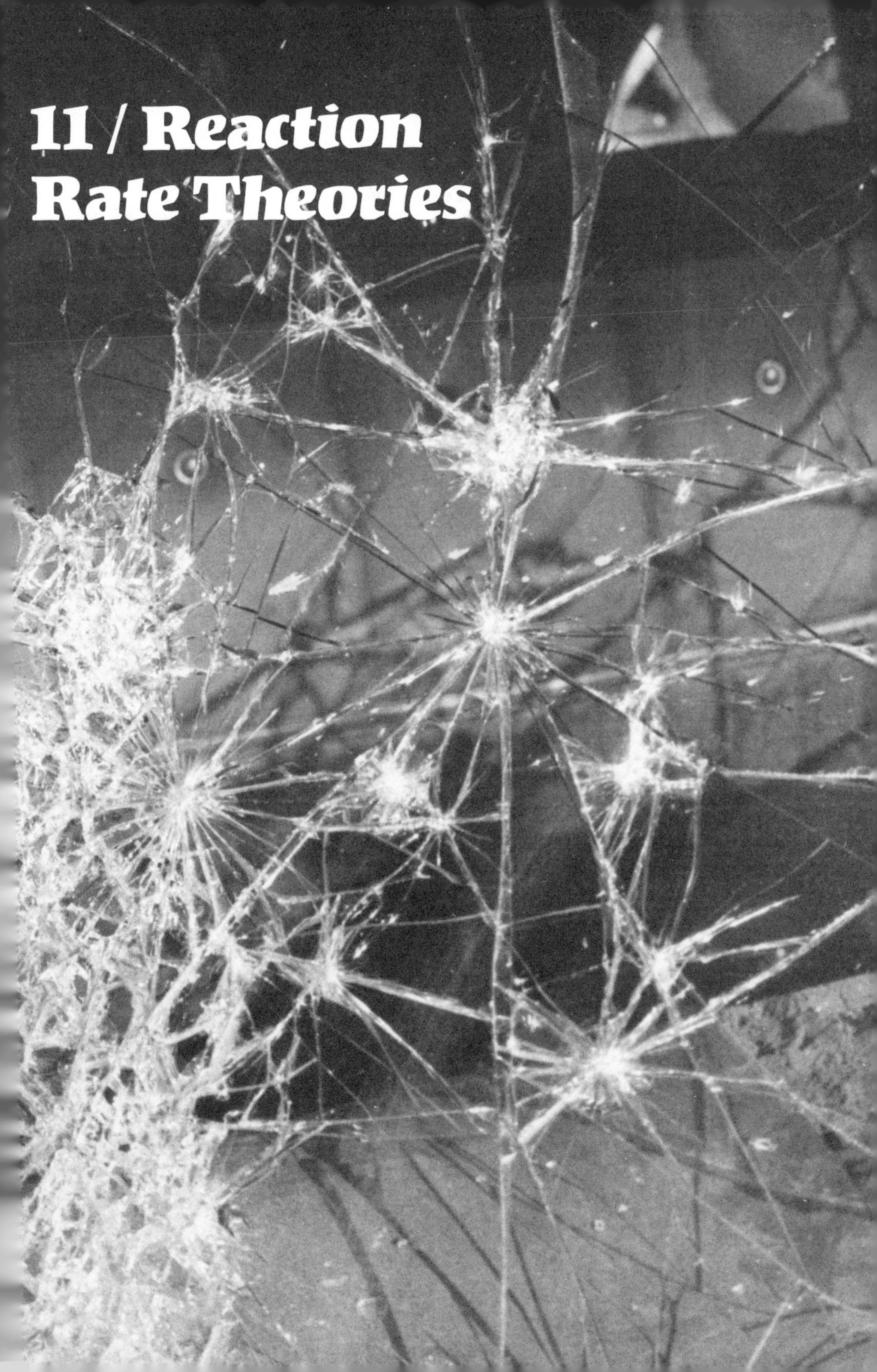

11.1. Dependence of Rate on Temperature

As we have previously mentioned, the rate of a chemical reaction depends on the temperature, the rate generally increasing markedly as the temperature is raised.[1] A time-honored approximate rule of thumb is that the reaction rate will double with each 10° increase in the centigrade temperature. On this basis switching a laboratory reaction from an ice bath to a steam bath (a 100° increase or ten 10° temperature rises), would cause the reaction rate to increase by a factor of 2^{10}, or 1024-fold. While this example does suggest that the influence of temperature is very great, the rule is only approximate and is, if anything, conservative. For many reactions the effect of a 10° rise is closer to a tripling than a doubling. The important point of the rule, however, is that the effect of temperature is *exponential* rather than linear. This observation is formalized with reasonable accuracy by the empirical relationship

$$k = Ae^{-\beta/RT}, \tag{205}$$

where k is the rate constant and A and β are constants which depend on the particular reaction.

Arrhenius, through an analogy with the dependence of the equilibrium constant on temperature,[2] suggested that such a quantitative relationship should be expected. He went further and equated β with E_a which is called the *activation energy*, the minimum amount of energy necessary for a reaction to occur. The constant A is called the *pre-exponential factor*. The slightly modified Eq. (205) is the well-known *Arrhenius equation*:

$$k = Ae^{-E_a/RT}. \tag{206}$$

In logarithmic form we have

$$\ln k = -E_a/RT + \ln A, \tag{207a}$$

and in differential form

$$\frac{d \ln k}{dT} = \frac{E_a}{RT^2}. \tag{208}$$

[1] Photochemical reactions are an interesting exception. The photographer does not need to lengthen the exposure on a cold day, other conditions being equal. Radioactive decay rates are also unaffected by temperature.

[2] This analogy is not difficult to understand. Recall $\Delta G^\circ = -RT \ln K_{eq} = \Delta H^\circ - T\,\Delta S^\circ$; $\ln K = -\Delta H^\circ/RT + \Delta S^\circ/R$; $d \ln K/dT = \Delta H^\circ/RT^2$. Notice the similarity between this equation and Eq. (208) and recall that $K = k_f/k_r$ (Eq. 190). This is by no means a proof; only an analogy.

Assuming constancy of E_a and integrating between two temperatures T_1 and T_2 gives

$$\ln\left(\frac{k_2}{k_1}\right) = -\frac{E_a}{R}\left(\frac{1}{T_2} - \frac{1}{T_1}\right). \tag{209}$$

Converting Eq. (207) to ordinary logarithms gives

$$(\log k) = \frac{-E_a}{2.3R}\left(\frac{1}{T}\right) + \log A. \tag{207b}$$

This form is convenient for evaluating the parameters E_a and A, using rate constant values as measured at a succession of temperatures. When $\log k$ is plotted on the vertical axis against $1/T$ on the horizontal axis, a straight line should result. The slope of this line is $-E_a/2.3R$, and the y-intercept is $\log A$.

Example 60 Calculate the activation energy E_a of a reaction whose rate is doubled by a 10° temperature rise
a) from 100° to 110°K
b) from 200° to 210°K
c) from 300° to 310°K

Answer: Equation (209) is a convenient form for this calculation, which assumes that E_a does not change significantly over a 10° interval.

$$R \ln\left(\frac{k_2}{k_1}\right) = -E_a\left(\frac{1}{T_2} - \frac{1}{T_1}\right)$$

$$R \ln\left(\frac{k_2}{k_1}\right) = E_a\left(\frac{1}{T_1} - \frac{1}{T_2}\right) = E_a\left(\frac{T_2 - T_1}{T_2 T_1}\right)$$

$$E_a = \left\{R \ln\left(\frac{k_2}{k_1}\right)\right\}\left(\frac{T_2 T_1}{T_2 - T_1}\right).$$

When a 10 rise produces a doubling of the rate, this becomes

$$E_a = \left(\frac{R \ln 2}{10}\right) T_2 T_1 = 0.138 T_2 T_1.$$

a) When T_1 is 100°K

$$E_a = (0.138)(110)(100) = 1.5 \text{ kcal.}$$

b) When T_1 is 200°K

$$E_a = (0.138)(210)(200) = 5.9 \text{ kcal.}$$

c) When T_1 is 300°K

$$E_a = (0.138)(310)(300) = 13.0 \text{ kcal.}$$

The approximate nature of the rule which predicts a doubling of the rate with a 10° temperature rise is apparent from these calculations.

Example 61 A good bit of the energy (30–40%) released in metabolic reactions is not used for body functions but is turned into heat. In warm-blooded animals part of this heat can be retained in the animal's body so that its temperature may be maintained at a constant level above that of the surroundings. Cold-blooded animals have no mechanism to retain this excess heat energy. Consequently their bodies assume the temperature of the surroundings. The metabolic processes of warm-blooded animals therefore operate at a fairly constant rate; those of cold-blooded animals vary as the body temperature changes. One example of the results of this variable metabolic rate is the chirping of crickets: a cricket chirps faster as the temperature increases. In fact, a close approximation to the Fahrenheit temperature can be obtained at temperatures near 70°F by counting the number of times a cricket chirps in 15 sec and adding 37. Using this last piece of information, assuming the temperature dependence of the rate of cricket chirps is the same as that for the metabolic reactions, calculate an energy of activation for the metabolic process.

Answer: Employing the relationship °C = 5/9(°F − 32) we first relate the number of chirps in 15 sec (Ch) to the absolute temperature:

$$^\circ\mathrm{K} = \tfrac{5}{9}[Ch + (37 - 32)] + 273$$
$$Ch = \tfrac{9}{5}(^\circ\mathrm{K} - 273) - 5.$$

Now let us compare the rate of chirping at 25°C (298°K) and at 30°C (303°K):

$$Ch_{298} = 40$$
$$Ch_{303} = 49.$$

Relating this to the metabolic rate, we may use the method of Example 45 to solve for E_a:

$$E_a = \left[R \ln\left(\frac{k_2}{k_1}\right)\right]\left(\frac{T_2 T_1}{T_2 - T_1}\right)$$
$$= \left[R \ln\left(\frac{49}{40}\right)\right][(298)(303)/5.0]$$
$$= 7.3\ \mathrm{kcal}.$$

A number of attempts have been made to explain the temperature dependence of the rate constant and to predict its magnitude. These attempts and the theories on which they are based offer significant insight into the problem of how chemical reactions occur. We shall examine two of these: collision theory and transition state theory.

11.2. Collision Theory, Gas Phase

In our previous discussion of chemical reactions we have regarded them as resulting from the coming together of the reacting species in an encounter or collision. Collision theory simply formalizes and quantifies this approach. We shall confine ourselves initially to the gas phase, broadening our discussion to include reactions in solution in the next section.

Collision theory postulates that reaction rate depends on two factors: the frequency with which the reactive species collide and the fraction of these collisions which are effective and result in reaction. If the collisions were 100% efficient and every collision led to reaction, the rate would simply be determined by the collision rate. Since a decrease in collision efficiency would lead to a decrease in rate, the frequency with which reactants collide sets the upper limit to the reaction rate: the collision rate is the maximum reaction rate possible.

The simplest approach to calculating the number of collisions occurring between reactive molecules in a unit volume during a unit time interval considers the molecules to be rigid, hard spheres of radius r which neither attract nor repel each other and whose collisions are elastic. For the bimolecular elementary reaction A + A → Products, a collision occurs whenever the centers of two molecules are a distance $d_A = r_A + r_A = 2r_A$ apart. The number of collisions will be proportional to the relative velocity of the molecules, to the closeness of approach d which constitutes collision and to the "population density" of molecules. If the gas molecules of mass m_A and radius r_A are moving with an average relative velocity $\bar{c}$ in a container in which there are n_A molecules per cm^3 there will be Z_{AA} collisions per cm^3 per sec:

$$Z_{AA} = (1/\sqrt{2})\pi d_A^2 \bar{c} n_A^2 \qquad (210a)^3$$

$$= (1/\sqrt{2})\pi d_A^2 n_A^2 \sqrt{\pi \kappa T/m_A}, \qquad (210b)$$

where κ is Boltzmann's constant.

For the general elementary reaction A + B → Products the number of collisions per cm^3 per sec between A and B is

$$Z_{AB} = \pi d_{AB}^2 \sqrt{8\kappa T/\pi\mu}\, n_A n_B, \qquad (211)$$

where the closeness of approach d_{AB} in this case is equal to $r_A + r_B$, $\mu = m_A m_B/(m_A + m_B)$ is the reduced mass, and n_A and n_B are, respectively, the number of molecules of A and B per cm^3.

[3] For a detailed derivation of Eqs. (210a) and (210b), see G. W. Castellan, *Physical Chemistry*, 2nd ed., Reading, Mass.: Addison-Wesley, p. 693.

Example 62 Which of the following pairs of reactants will have the greatest collision frequency at a given temperature and pressure in the gas phase?

a) Br and Cl

b) CH_3OH and $CH_3-C(=O)-OH$

Answer: In Eq. (211), Z_{AB} can be seen to be proportional to $(r_A + r_B)^2$ and inversely proportional to $\sqrt{\mu}$. Since $r_A + r_B$ is greater for pair (b) and μ is greater for pair (a), the collision frequency is obviously greater for pair (b).

The factors πd^2 and πd_{AB}^2 are called the *collision cross section* σ. In reality molecules attract and repel each other and their collisions are not elastic. Molecules are not hard spheres; they penetrate when they collide. The closest approach of their centers may be less than or greater than the sum of their radii. More exact collision calculations must take these facts into account. One way of doing so is by adjusting the cross section σ.

Note that the collision rate given by Eqs. (210) and (211) is proportional to $\sqrt{T}$. We have postulated that the collision rate Z determines R_{max}, the maximum reaction rate:

$$R_{max} = \pi d_{AB}^2 \sqrt{8\kappa T/\pi\mu}\,[A][B] \tag{212a}$$

$$= C_Z \sqrt{T}\,[A][B]. \tag{212b}$$

For given reactants, C_Z is a constant. This rate is the maximum rate, and rates calculated using this equation are generally much too large. We need to consider the second factor postulated by collision theory: the rate depends on the fraction of collisions which are effective and which result in reaction.

There are several reasons why reaction may not always occur upon collision. The two principal considerations are the energy of the collision and the relative orientation of the molecules when they collide.

In collision theory it is postulated that only if the collision involves at least a minimum amount of energy, the activation energy E_a, will the collision result in reaction. From the *Boltzmann distribution law* of statistical mechanics we can obtain an expression for f, the fraction of collisions which will have at least the requisite energy: $f = e^{-E_a/RT}/(1 - e^{-E_a/RT})$. If $E_a \gg RT$, as is usually the case, this expression reduces to

$$f = e^{-E_a/RT}. \tag{213}$$

Recall from the kinetic theory of gases (Section 1.12) that the translational energy a molecule possesses is directly related to the temperature: the average velocity (and thus the kinetic energy of translation) increases with increasing temperature. At higher temperatures a greater proportion of molecules will qualify (will possess the required high energy) than at lower temperatures. Equation (213) expresses this relationship quantitatively.

To take into account this energy requirement for collision effectiveness, we multiply the number of collisions by that fraction which are energetic enough to result in reaction. The rate would then be

$$R = f \times R_{\max} = R_{\max} e^{-E_a/RT} = C_Z \sqrt{T}\,[\mathrm{A}][\mathrm{B}]e^{-E_a/RT}. \tag{214}$$

For the reaction $O_3 + NO \rightarrow O_2 + NO_2$ the reaction process is depicted schematically in Fig. 11.1. If a collision between the independently moving molecules NO and O_3 is to result in reaction, the collision must be energetic enough for the *activated complex* (or *transition state complex*) NO_4^* to be formed. NO_4^* is not a true reaction intermediate but rather a short-lived species existing at the moment of collision as bonds are being rearranged. For this example,

If the reactant molecules are to come into contact which is intimate enough for such a complex to form, the collision must possess at least an amount of energy E_a.

When the activated complex moves on to products, an amount of energy equal to E_a' (see Fig. 11.1) is released. The net energy change in the reaction is then $E_a - E_a' = \Delta E_f$. This is the energy difference between products and reactants, the thermodynamic energy change for the reaction.

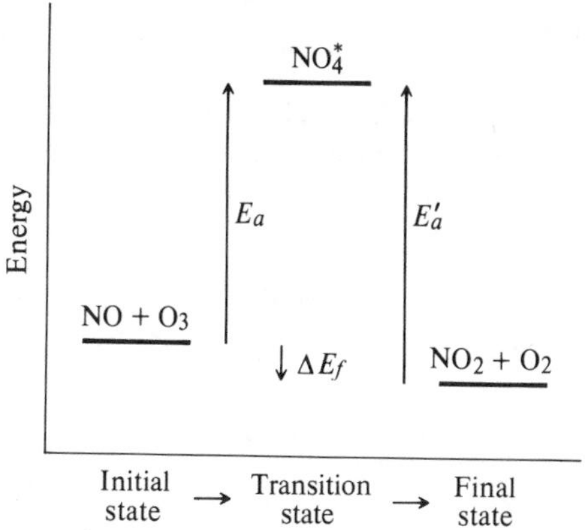

Fig. 11.1. Collision theory energy profile of the reaction $NO + O_3 \rightarrow NO_2 + O_2$.

If the reaction is to proceed in the reverse direction, $NO_2 + O_2 \rightarrow NO + O_3$, the collision between the NO_2 and O_2 molecules must possess at least energy E'_a, and an amount of energy E_a will be released when NO and O_3 are formed from the activated complex. The energy change for this reverse reaction is then $\Delta E_r = E'_a - E_a = -\Delta E_f$.

We must now take into account the effect of molecular orientation on the rate. If the colliding species are uniform spheres, their orientation upon collision is unimportant (Fig. 11.2).

If, however, the molecular species is reasonably complex, certain collisions are much more likely to lead to reaction than others. Consider the following displacement reaction:

$$Br^- + CH_3I \longrightarrow CH_3Br + I^-.$$

It may be pictured as occurring as follows:

$$Br^- + CH_3I \longrightarrow BrCH_3I^-$$
$$BrCH_3I^- \longrightarrow BrCH_3 + I^-.$$

The bromide ion may be regarded as a sphere but CH_3I cannot. There are therefore a number of configurationally different collisions between these two substances, three of which are shown in Fig. 11.3. Only if the collision occurs as in (ii) will there be reaction. In the collisions with incorrect relative arrangement, the two molecules simply bounce away from each other and no reaction occurs. If the relative configuration is correct as in (ii), however, and the collision is energetic enough, the intermediate complex forms and

Fig. 11.2. The collision of spherical reactants indicating the lack of dependence on orientation.

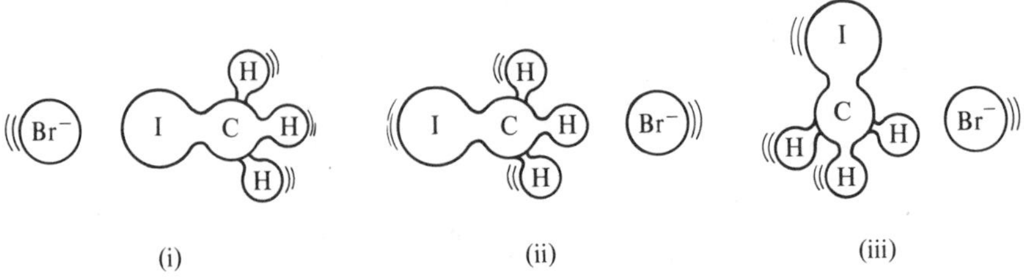

Fig. 11.3. Three possible collision orientations for the reaction $Br^- + CH_3I \rightarrow CH_3Br + I^-$.

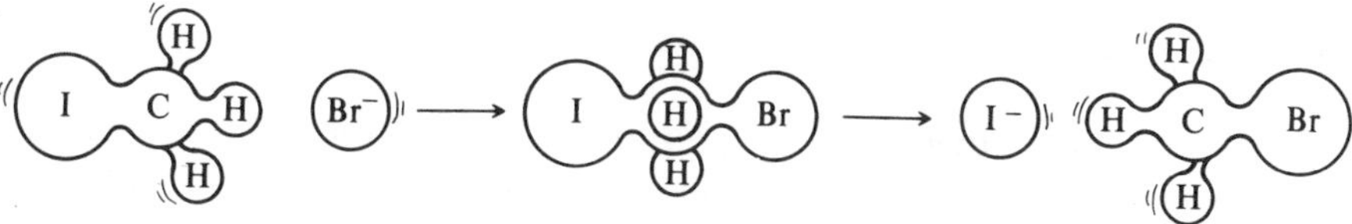

Fig. 11.4. Schematic depiction of the $CH_3I + Br^- \rightarrow CH_3Br + I^-$ reaction.

reacts on to products (Fig. 11.4). Only a fraction of the collisions with sufficient energy will therefore react, due to this configurational or *steric* requirement. This fact is taken into consideration by multiplying by a constant p known as the *steric factor*. The steric factor p is usually less than 1 and consequently predicts a reduced rate:

$$R = pR_{\max} e^{-E_a/RT} \tag{215a}$$

$$= pC_Z\sqrt{T}\, e^{-E_a/RT}[A][B] \tag{215b}$$

Comparison of the rate expression of Eq. (215b) with the second-order rate expression $R = k[A][B]$ yields

$$k = pC_Z\sqrt{T}\, e^{-E_a/RT}. \tag{216}$$

The pre-exponential factor A of the Arrhenius equation (Eq. 206) may obviously be identified with $pC_Z\sqrt{T}$ and is thus a function of temperature.

Notice that as the temperature is raised, the number of collisions increases and the fraction of collisions possessing the requisite energy for reaction increases. Both effects lead to an increase in rate, but the latter, being an exponential, is the more important.

Example 63 Which of the following reactions would have the smallest steric factor in collision theory?

a) $Br + Br \rightarrow Br_2$

b) $CH_3{-}CH_2{-}OH + CH_3{-}\overset{\overset{\displaystyle O}{\|}}{C}{-}OH \longrightarrow CH_3{-}CH_2{-}O{-}\overset{\overset{\displaystyle O}{\|}}{C}{-}CH_3$

c) $O_3 + NO \rightarrow NO_2 + O_2$

d) $CH_4 + Br_2 \rightarrow CH_3Br + HBr$

Answer: No special orientation is required for the atomic collisions in (a). There are orientation requirements in (b), (c), and (d). Both molecules in (d), however, are highly symmetric and there are a number of equivalent effective orientations. The collision in (c) is more restrictive than that in (d), but the symmetry and the relative simplicity of the molecules involved result in more equivalent effective collisions than in (b). The magnitude of p will thus be in the order (a) > (d) > (c) > (b).

11.3. Collision Theory, Reactions in Solution

The collision-theory treatment of liquid-phase solution reactions is somewhat more involved than its reasonably straightforward application to gas-phase processes. Even though there are significant differences between the two situations, the collision approach developed in Section 11.2 will form the basis for our treatment of solution reactions. Modifications will of course be necessary.

In the solution phase a reactant molecule moves through a maze of molecules, most of which are solvent; it undergoes many collisions before colliding with a molecule with which it can react. Unlike the random motion of molecules in the gas phase, the movement of liquid-phase solute molecules is best discussed in terms of *diffusion.* While free to move about, molecules in the liquid phase are closely surrounded by other molecules. In the case of a solute molecule, these close neighbors are usually solvent molecules (Fig. 11.5). A solute molecule spends a reasonably long time "trapped" in the *cage* formed by the solvent. However, there is a continuous jostling of nearest neighbors and a shifting of molecular positions. As a result the solute molecule will frequently escape from one cage into another. Moving from cage to cage in this manner, the solute molecule diffuses through the solution. Occasionally it will slip into a cage in which one of its nearest neighbors is a molecule with which it may react.

In the gas phase, collisions are fairly evenly spaced in time as the gaseous molecules dart randomly about. If upon colliding they possess the right orientation and energy, reaction proceeds; otherwise they bounce away from each other and speed through space until collision occurs with another reactant. In solution, by contrast, a longer time elapses between close approaches of reactants, but once they meet, they are held in close proximity by the solvent cage. If their initial collision does not have the correct orientation or enough energy, they bounce away from each other but immediately collide with the solvent cage and bounce back to collide with each other again. Thus an *encounter* (the coming of reactant molecules into close proximity) lasts for significantly longer than a gas-phase collision, and the cage effect results in a considerable number of collisions occurring during each encounter. Collisions in solution are therefore bunched into encounters, with an interval of time between encounters. One of the results of this is that approximately the same number of collisions between reactant molecules may be expected to occur in the gas phase and in solution.

Reactions with stringent configurational requirements (low pre-exponential factor p) or high energy requirements (large energy of activation E_a) require numerous collisions before reaction is successful. For this type of reaction the rate depends primarily on the number of collisions. Consequently if the reaction

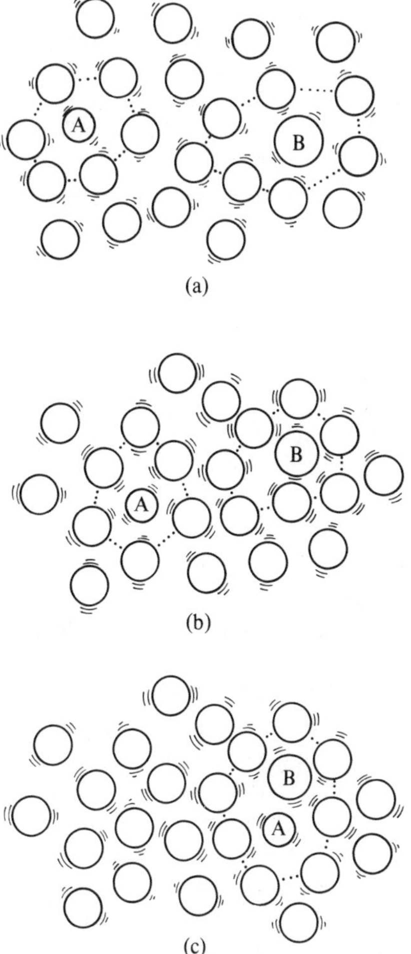

Fig. 11.5. Schematic illustration of the diffusion of reactant molecules (A and B). Solvent cages are indicated by dotted lines.

mechanism is the same in the gas phase and in solution, the rate will be essentially the same in the two phases. The fact that there is a longer time between encounters in solution than between collisions in gases is balanced by the fact that during an encounter in solution there are many collisions within a short time.

A good example of a reaction whose rate constant has close to the same value in the gas phase and in various solvents is the decomposition of

Table 11.1. Rate Constants for the Decomposition of Diacetylperoxide in the Gas Phase and in Various Solvents at 85.2°C (A. Rembaum and M. Szwarc, *Journal of the American Chemical Society*, **76**, 5978, 1954.)

Solvent	$k \times 10^5$, $\sec^{-1}$
gas phase	22.4
isooctane	14.9
cyclohexane	12.7
benzene	16.2
toluene	15.9
acetic acid	13.0
proprionic acid	16.6

diacetyl peroxide ($CH_3COO{-}OCOCH_3$). Table 11.1 contains the values of the rate constant for this reaction.

There are reactions, however, which occur at the first collision or require only a few collisions. Their configurational requirements are minimal and the activation energy is small. The rate of these reactions is then determined not by the number of collisions of a reactant in a given time but rather by how long it takes for one collision to occur. In the gas phase this is the time between collisions; in solution it is the time between encounters. As we have seen, the latter depends on the rate of diffusion of solute through the solution.

The rate with which a given solute diffuses in a given solvent may be determined experimentally by introducing the solute into one location in a container of solvent and observing the time required for its appearance in another part of the container. The diffusion rate is found to be proportional to the difference in solute concentration and inversely proportional to the distance between the two parts of the container; that is, the rate is proportional to the *concentration gradient* of the solution (Fick's First Law, 1855). The proportionality constant is the *diffusion coefficient D*. The magnitude of this constant depends on the ease with which the solute molecule moves through the solvent. This will depend on size (smaller molecules will collide less frequently with the solvent) and on the interactions between solute and solvent (attractive forces will slow a solute's progress).

For the general reaction A + B → Products, the encounter rate is given by

$$Z_{AB} = 4\pi\, d_{AB}(D_A + D_B) n_A n_B, \tag{217}$$

using the neutral hard-sphere model.[4] (Compare with Eq. 211). The rate in this case where each encounter leads to reaction would be equal to the frequency of encounters. Reactions of this sort are said to be *diffusion-controlled* and generally have rate constants in the 10^{10} to 10^{11} range. This maximum rate is modified by introducing a steric term p and an energy term $e^{-E_a/RT}$ for reactions with significant configuration and energy requirements. The result is an equation similar to Eq. (211) for gas-phase reactions; thus it is of the form of the Arrhenius equation.

Reactions between simple ions are primarily diffusion-controlled. For instance the reaction between hydroxide ion OH^- and hydrogen ion H^+ (actually hydronium ion H_3O^+) is diffusion-controlled with a rate constant of 5×10^{10} liter mole^{-1} sec^{-1}.

Example 64 Arrange the following in order of increasing diffusion coefficient in aqueous solution:

a) insulin (molecular weight: 41,000)
b) hemoglobin (molecular weight: 67,000)
c) urea
d) sucrose
e) glucose
f) glycine

Answer: The smaller molecules move fastest through the solvent. Thus, on the basis of molecular size, the diffusion coefficients would be expected to be arranged

glycine > urea > glucose > sucrose > insulin > hemoglobin.

The measured diffusion coefficients of these six substances in aqueous solution at 25°C are:

Substance	*Diffusion coefficient,* cm^2 sec^{-1}
insulin	0.08×10^{-5}
hemoglobin	0.06×10^{-5}
urea	1.2×10^{-5}
sucrose	0.52×10^{-5}
glucose	0.67×10^{-5}
glycine	1.1×10^{-5}

[4] For derivation of this equation see A. W. Adamson, *A Textbook of Physical Chemistry*, New York: Academic Press (1973), p. 696.

Involvement of the solvent in solution reactions affects the rate and even the type of reactions possible. We have mentioned the cage effect, and we have noted the attractive forces between solute and solvent molecules which prolong the period between encounters. These factors result in diffusion control; they also lengthen the encounter so that many collisions may occur in each cage. Moreover, the solute and solvent transfer energy in these collisions, thereby increasing or decreasing the energy supply a reactant solute molecule has available for a subsequent collision with another reactant.

Ions are seldom produced in gas-phase reactions; they are too unstable and/or reactive. But in solution the solvent, if it is a polar species, can stabilize ions through *solvation* and thus make it possible for ions to participate in reactions. The solvent may also function as a catalyst (see Chapter 12), making a reactant more reactive by interacting directly with it. Solvent may also participate in a reaction as a reactant or product. Since it is present in such large excess, any change in its concentration is hard to detect; accordingly, such direct participation is difficult to prove in some cases.

11.4. Transition State Theory

A complete development of transition state theory (also known as *absolute rate theory*) would require the use of rather sophisticated statistical mechanics. We can, however, discuss the main features without a great deal of difficulty.

Transition state theory shares certain similarities with collision theory. Reaction is postulated as occurring when molecules collide with or encounter each other. A transition state complex of relatively high energy is formed; the complex then decays to product. This is pictured schematically in Fig. 11.6 for the general reaction $A + B \rightarrow F + G$. (Compare Fig. 11.1). $AB^{\ddagger}$ is the

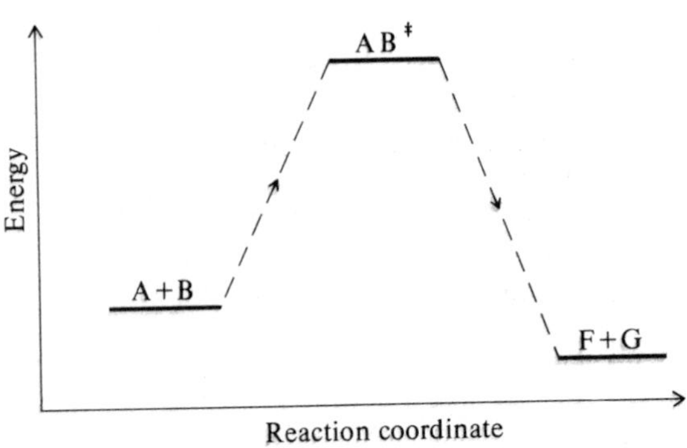

Figure 11.6

activated complex or *transition state complex*. The *reaction coordinate* simply depicts the progress of the reaction from reactants toward products via the transition state complex.

The further postulates of transition state theory depart from those of collision theory. First, the transition state complex is postulated to be in equilibrium with the reactants:

$$A + B \rightleftharpoons AB^{\ddagger}$$

$$K^{\ddagger} = [AB^{\ddagger}]/[A][B] \tag{218}$$

where $K^{\ddagger}$ is the equilibrium constant. Further, the rate of the reaction is postulated to depend only on the rate of decay of $AB^{\ddagger}$ to products:

$$A + B \underset{k_2}{\overset{k_1}{\rightleftharpoons}} AB^{\ddagger} \overset{k_3}{\rightleftharpoons} F + G.$$

The observed rate R is then predicted to be

$$R = k_3[AB^{\ddagger}]. \tag{219}$$

The concentration of $AB^{\ddagger}$ can be obtained from Eq. (218) and thus

$$R = k_3 K^{\ddagger}[A][B]. \tag{220}$$

This agrees with the second-order (bimolecular) rate expression with the observed rate constant k given by

$$k = k_3 K^{\ddagger}. \tag{221}$$

Statistical mechanical considerations predict that the transition state complex will decay with a rate constant

$$k_3 = RT/Nh, \qquad (222)^5$$

where R is the gas constant, N is Avogadro's number, and h is *Planck's constant* with value 6.63×10^{-27} erg-sec. (This last constant will play an important role in our discussion in later chapters.) The observed rate constant is therefore

$$k = (RT/Nh)K^{\ddagger}. \tag{223}$$

Recall from thermodynamics that $\Delta G^{\circ} = -RT \ln K$. Let us define the standard free energy of activation for the process $A + B \rightarrow AB^{\ddagger}$ as $\Delta G^{\ddagger}$. We may then write

$$\Delta G^{\ddagger} = -RT \ln K^{\ddagger} \tag{224a}$$

$$K^{\ddagger} = e^{-\Delta G^{\ddagger}/RT} \tag{224b}$$

---

[5] See G. W. Castellan, *Physical Chemistry*, 2nd ed., Reading, Mass.: Addison-Wesley (1971), pp. 780–782.

and

$$k = (RT/Nh)e^{-\Delta G^{\ddagger}/RT}. \tag{225}$$

Recalling further the relation of free energy to enthalpy and entropy, namely $\Delta G = \Delta H - T\,\Delta S$, we may write

$$k = (RT/Nh)e^{-(\Delta H^{\ddagger}/RT - \Delta S^{\ddagger}/R)}$$

$$k = (RT/Nh)e^{\Delta S^{\ddagger}/R}e^{-\Delta H^{\ddagger}/RT}. \tag{226}$$

Example 65 From the following data calculate E_a and A for the Arrhenius equation and evaluate $\Delta G^{\ddagger}$ at 25°C for the transition state treatment.

t, °C	20	25	30	35
k	3.76×10^4	5.01×10^4	6.61×10^4	8.64×10^4

Answer: To evaluate E_a and A we plot log k versus $1/T$ (Eq. 207b) in Fig. 11.7.

T, °K	293	298	303	308
$1/T$	3.41×10^{-3}	3.36×10^{-3}	3.30×10^{-3}	3.25×10^{-3}
log k	4.58	4.71	4.83	4.94

The slope of the line in Fig. 11.7 is equal to $-E_a/2.3R$. The slope, determined from the graph, is -2.17×10^3. Thus

$$E_a = (2.30)(1.99)(2.17 \times 10^3) = 9.9 \text{ kcal}.$$

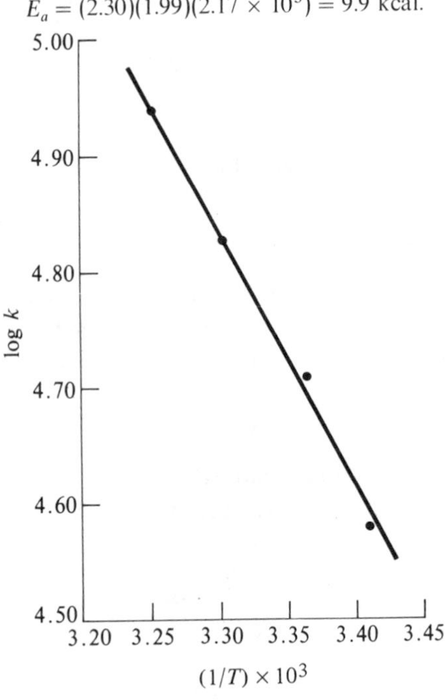

Figure 11.7

The y-intercept is equal to log A, or we could calculate A using Eq. (207b):

$$\log A = \log k + \frac{E_a}{2.3R}\frac{1}{T}$$

$$= 4.94 + \frac{10000}{(2.3)(2)}(3.25 \times 10^{-3})$$

$$= 12.0$$

$$A = 10^{12}.$$

To evaluate $\Delta G^{\ddagger}$ at 25°C we use Eq. (225):

$$e^{-\Delta G^{\ddagger}/RT} = \frac{kNh}{RT}$$

$$\Delta G^{\ddagger} = -RT \ln (kNh/RT)$$

$$= -2(298)(2.3) \log \left[\frac{(5.01 \times 10^4)(6.02 \times 10^{23})(6.63 \times 10^{-27})}{(8.31 \times 10^7)(298)}\right]$$

$$\Delta G^{\ddagger} = 11.0 \text{ kcal}.$$

A comparison of Eq. (226) with the Arrhenius equation and with Eq. (216) for the collision theory reveals some interesting parallels. In transition state theory, the energy of activation E_a is related to the enthalpy of activation $\Delta H^{\ddagger}$. The pre-exponential factor A is now $(RT/Nh)e^{\Delta S^{\ddagger}/R}$. In the collision theory the pre-exponential factor contains a steric factor p. This is related to the entropy of activation $\Delta S^{\ddagger}$. The entropy of activation will be negative for reactions of the type $A + B \rightarrow AB^{\ddagger} \rightarrow$ Products, since $AB^{\ddagger}$ is definitely more organized than A and B. For processes of the type $AB \rightarrow AB^{\ddagger} \rightarrow A + B$ the entropy of activation will generally be positive, since the activated complex will most likely have acquired some of the disorder which will eventually result in its total breakdown into A and B. If in the process $A + B \rightarrow AB^{\ddagger}$ a particular relative collision configuration is required, this will result in a negative $\Delta S^{\ddagger}$ with a large magnitude, since even more ordering than usual is required. The factor $e^{\Delta S^{\ddagger}/R}$ is therefore small, corresponding to a small steric factor p.

In order to use Eq. (225) to make quantitative rate predictions we must calculate $\Delta G^{\ddagger}$. This is done by postulating a structure for the transition state and using the methods of statistical mechanics, a subject beyond the scope of this text.

Figure 11.8, a schematic diagram similar to Fig. 11.6, may be used to describe a reaction in terms of free-energy changes. Even if the net free-energy change in a reaction is negative, if $\Delta G^{\ddagger}$ has a very large positive value, the rate constant will be quite small (Eq. 225) and consequently the reaction may

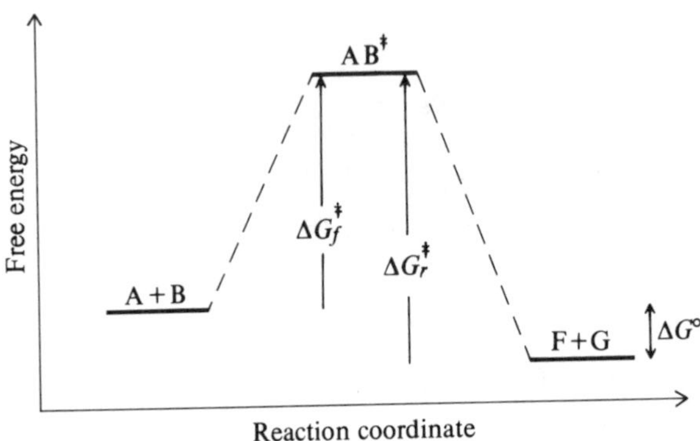

Figure 11.8

be extremely slow. This explains how, while thermodynamics may tell us a reaction is possible because $\Delta G < 0$, the reaction may in fact not occur at all if $\Delta G^{\ddagger}$ is very large.

The free-energy change for the forward reaction is $\Delta G_f^{\circ} = \Delta G_f^{\ddagger} - \Delta G_r^{\ddagger}$. For the reverse reaction: $\Delta G_r^{\circ} = \Delta G_r^{\ddagger} - \Delta G_f^{\ddagger} = -\Delta G_f^{\circ}$. For the forward reaction

$$R_f = k_f[\mathrm{A}][\mathrm{B}] = (RT/Nh)e^{-\Delta G_f^{\ddagger}/RT}[\mathrm{A}][\mathrm{B}],$$

and for the reverse reaction

$$R_r = k_r[\mathrm{F}][\mathrm{G}] = (RT/Nh)e^{-\Delta G_r^{\ddagger}/RT}[\mathrm{F}][\mathrm{G}].$$

Example 66 Show that the kinetic view of chemical equilibrium applied to transition state theory leads to the same equation as the thermodynamic view, $\Delta G^{\circ} = -RT \ln K_{\mathrm{eq}}$.

Answer: In the kinetic approach, reaction is still occurring at equilibrium, but the rate of production of products, R_f, is equal to R_r, the rate of production of reactants, so that there is no net change in the concentration of either:

$$R_f = R_r$$

$$(RT/Nh)e^{-\Delta G_f^{\ddagger}/RT}[\mathrm{A}][\mathrm{B}] = (RT/Nh)e^{-\Delta G_r^{\ddagger}/RT}[\mathrm{F}][\mathrm{G}].$$

Since $K_{\mathrm{eq}} = [\mathrm{F}][\mathrm{G}]/[\mathrm{A}][\mathrm{B}]$,

$$K_{\mathrm{eq}} = \frac{e^{-\Delta G_f^{\ddagger}/RT}}{e^{-\Delta G_r^{\ddagger}/RT}} = e^{-(\Delta G_f^{\ddagger} - \Delta G_r^{\ddagger})/RT} = e^{-\Delta G^{\circ}/RT}$$

$$\ln K_{\mathrm{eq}} = -\Delta G^{\circ}/RT$$

$$\Delta G^{\circ} = -RT \ln K_{\mathrm{eq}}.$$

Problems

11.1 The energy of activation in the isoprene dimerization reaction (Example 51) is 28.9 kcal $mole^{-1}$ and the pre-exponential factor A is 5.3×10^{11} cm^3 $mole^{-1}$ sec^{-1}. Calculate the rate constant of this reaction at 200°C.

11.2 The rate constant for the decomposition of isopropyl nitrite (Problem 9.2) is given as follows as a function of temperature.

Temperature, °C	k, sec^{-1}
150°	1.3×10^{-5}
200°	1.1×10^{-3}
250°	4.2×10^{-2}
300°	1.2

Determine A and E_a.

11.3 The rate constant for the decomposition of diacetyl peroxide in cyclohexane varies with the temperature as follows:

Temperature, °C	k, sec^{-1}
55	0.21×10^{-5}
65	0.97×10^{-5}
75	3.60×10^{-5}
85	17.7×10^{-5}

Determine E_a and A for this reaction.

11.4 Calculate the rate constant at 70°C for the reaction in Problem 11.3.

11.5 Calculate $\Delta G^\ddagger$ for the reaction in Problem 11.3 at 70°C.

11.6 The rate constant for the reaction

$$\text{methylmaleic acid} \longrightarrow \text{methylfumaric acid}$$

at 300°C is 4000 times its value at 250°C. Calculate the energy of activation.

11.7 Diethyl ether and *n*-pentane are approximately the same size. Which should diffuse faster in an aqueous solution?

11.8 The value of $\Delta S^\ddagger$ for the unimolecular decomposition of isopropyl nitrite in the gas phase is small and positive. Can you explain this?

12 / Catalysis

12.1. Introduction

We have discussed in some detail the effects of concentration and of temperature on reaction rates. In many cases a third factor is of considerable importance.

A *catalyst* is a substance which increases the rate of a reaction but is not itself used up; its concentration does not change during the course of the reaction. If it is consumed in one step of a process, it must be regenerated in another step. Similarly, an *inhibitor* is a substance which slows a reaction.

There are two types of catalysts. A *heterogeneous catalyst* exists in a different phase from the reaction it catalyzes. The catalysis by a solid of a reaction between two gases is an example. A *homogeneous catalyst* exists in the same phase as the reaction whose rate it increases. This is illustrated by the catalysis by an acid of the hydrolysis of an ester in dilute water solution. The two situations involve basic differences in mechanism and in applicable theory.

12.2. Theory

Catalysis can be best understood in terms of transition state theory (Section 11.4). A catalyst functions by lowering the free energy of activation. This is illustrated schematically in Fig. 12.1. Notice that the presence of the catalyst alters neither the free energy of the products nor that of the reactants. The total free-energy change for the reaction is therefore the same with or without a catalyst. Since ΔG° is not changed, the equilibrium constant ($\Delta G^\circ = -RT \ln K_{eq}$) is the same with or without the catalyst. A catalyst therefore cannot alter the relative proportion of products and reactants at equilibrium,

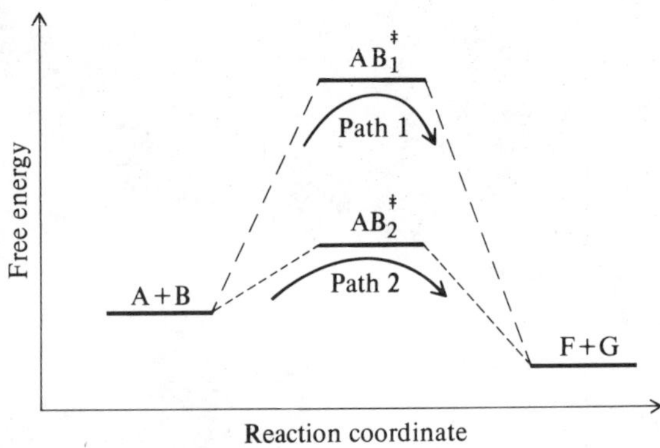

Fig. 12.1. Free-energy profile of reaction. Path 1 is without catalyst; path 2, with catalyst.

and thus it cannot increase the relative amount of product obtainable at equilibrium.

As illustrated in Fig. 12.1, the free energy of activation of both the forward reaction and the reverse reaction is lowered. Equation (225) indicates that the respective rates will therefore increase:

$$k_f = (RT/Nh)e^{-\Delta G_f^\ddagger/RT}. \tag{225}$$

With a lower $\Delta G_f^\ddagger$ it follows that $e^{-\Delta G_f^\ddagger/RT}$ will be larger and the rate constant correspondingly larger. A similar equation, of course, holds for the reverse reaction. Since $K_{\text{eq}} = k_f/k_r$ (Eq. 190),

$$K_{\text{eq}} = \frac{(RT/Nh)e^{-\Delta G_f^\ddagger/RT}}{(RT/Nh)e^{-\Delta G_r^\ddagger/RT}}$$

$$= e^{-(\Delta G_f^\ddagger - \Delta G_r^\ddagger)/RT}$$

$\Delta G_f^\ddagger - \Delta G_r^\ddagger$ is the change in free energy for the reaction under standard conditions, ΔG°, which has the same value with or without a catalyst. Thus K_{eq} has the same value whether a catalyst is present or not:

$$K_{\text{eq}} = e^{-\Delta G^\circ/RT}.$$

It should be clear that although a catalyst does not alter the equilibrium constant or the free energy of the reaction, it does change the rate of the reaction in both forward and reverse directions by altering the free energy of activation. The catalyst may lower $\Delta G^\ddagger$ by decreasing $\Delta H^\ddagger$ and/or by increasing $\Delta S^\ddagger$. But in either case the catalyst (referring again to Fig. 12.1) speeds the reaction by offering an easier route from reactants to products.

In summary then, a catalyst cannot give the experimenter a greater yield than the reaction's equilibrium constant would authorize. Often, however, a catalyst can make feasible an otherwise useless reaction, if the waiting time for this yield is cut from months to minutes.

12.3. Heterogeneous Catalysis

Some reactions are speeded by the addition of a solid catalyst. The effect is pictured as involving the *adsorption* of a reactant on the surface of the catalyst. For example, platinum catalyzes the reaction of hydrogen gas with various substances. The hydrogen molecule is adsorbed onto the surface of the metal, a process which weakens the bond holding the H_2 molecule together. This facilitates the cleavage of that bond and assists the reaction. The catalyst therefore functions primarily by lowering $\Delta H^\ddagger$.

For catalysis to occur on the surface of a solid, the reactant must diffuse to the surface, be adsorbed, and react chemically. The products must then break from the surface and diffuse away into the surrounding phase.

The rate of the overall reaction may be determined by the rate of any or all of these steps.

Catalysis by solids depends on the surface area available for adsorption. Since surface area per gram is larger for small particles than for large aggregates, a solid catalyst is often subjected to processes which increase the surface area and thereby enhance catalytic effectiveness.

12.4. Homogeneous Catalysis by Acids and Bases

One of the most common types of homogeneous catalysis for reactions taking place in liquid solutions is catalysis by acids and bases.

For instance, the hydrolysis of ethylacetate may be catalyzed by acid. A hydrogen ion H^+ may join with ethylacetate (B) to form an ionic complex HB^+. The complex may then react with water to yield the final products and regenerate the H^+ ion.[1]

Reaction:

$$CH_3\overset{\overset{\displaystyle O}{\|}}{C}OCH_2CH_3 + H_2O \xrightarrow{H^+} CH_3\overset{\overset{\displaystyle O}{\|}}{C}OH + CH_3CH_2OH$$

Mechanism:

$$\underset{(B)}{CH_3\overset{\overset{\displaystyle O}{\|}}{C}OCH_2CH_3} + H^+ \underset{}{\overset{K_{eq_1}}{\rightleftharpoons}} \underset{(HB^+)}{CH_3\overset{\overset{\displaystyle O}{\|}}{C}-\overset{\overset{\displaystyle H}{|}}{\underset{+}{O}}-CH_2CH_3} \qquad \text{fast}$$

$$HB^+ + H_2O \overset{K_{eq_2}}{\rightleftharpoons} \underset{(C)}{CH_3\overset{\overset{\displaystyle O^-}{|}}{\underset{\underset{\displaystyle H_2O^+}{|}}{C}}-\overset{\overset{\displaystyle H}{|}}{\underset{+}{O}}-CH_2CH_3} \qquad \text{fast}$$

$$C \xrightarrow{k} CH_3\overset{\overset{\displaystyle O}{\|}}{C}OH + CH_3CH_2OH + H^+ \qquad \text{slow}$$

Rate:

$$\frac{d[CH_3CH_2OH]}{dt} = k[C] \tag{227}$$

[1] Writing H^+ above the arrow of the stoichiometric equation means that it affects the reaction rate even though its concentration is not changed in the process.

By assuming that equilibrium is established quickly in the first two steps and that the last step is much slower, we may write

$$[C] = K_{eq_2}[HB^+][H_2O]$$

and

$$[HB^+] = K_{eq_1}[B][H^+].$$

The rate is thus

$$\begin{aligned} \frac{d[CH_3CH_2OH]}{dt} &= kK_{eq_2}[HB^+][H_2O] \\ &= kK_{eq_1}K_{eq_2}[B][H^+][H_2O] \\ &= k'[B][H^+][H_2O]. \end{aligned} \tag{228}$$

Example 67 The condensation of acetone with itself in the so-called aldol condensation reaction is catalysed by hydroxide ion in aqueous solution:

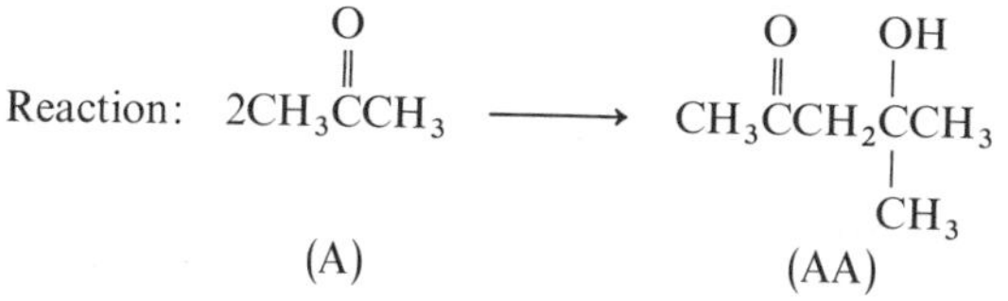

$$\text{Reaction:}\quad 2CH_3\overset{\overset{O}{\|}}{C}CH_3 \;\longrightarrow\; CH_3\overset{\overset{O}{\|}}{C}CH_2\underset{\underset{CH_3}{|}}{\overset{\overset{OH}{|}}{C}}CH_3$$

(A) (AA)

$$\text{Rate:}\quad \frac{d[AA]}{dt} = k[A]^2[OH^-]$$

Propose a mechanism for this reaction.

Answer: One possible mechanism is the following:

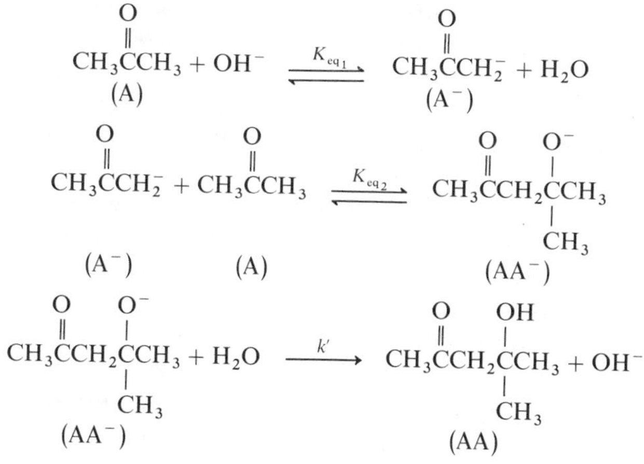

$$\underset{(A)}{CH_3\overset{\overset{O}{\|}}{C}CH_3} + OH^- \underset{}{\overset{K_{eq_1}}{\rightleftharpoons}} \underset{(A^-)}{CH_3\overset{\overset{O}{\|}}{C}CH_2^-} + H_2O$$

$$\underset{(A^-)}{CH_3\overset{\overset{O}{\|}}{C}CH_2^-} + \underset{(A)}{CH_3\overset{\overset{O}{\|}}{C}CH_3} \overset{K_{eq_2}}{\rightleftharpoons} \underset{(AA^-)}{CH_3\overset{\overset{O}{\|}}{C}CH_2\underset{\underset{CH_3}{|}}{\overset{\overset{O^-}{|}}{C}}CH_3}$$

$$\underset{(AA^-)}{CH_3\overset{\overset{O}{\|}}{C}CH_2\underset{\underset{CH_3}{|}}{\overset{\overset{O^-}{|}}{C}}CH_3} + H_2O \overset{k'}{\longrightarrow} \underset{(AA)}{CH_3\overset{\overset{O}{\|}}{C}CH_2\underset{\underset{CH_3}{|}}{\overset{\overset{OH}{|}}{C}}CH_3} + OH^-$$

The rate predicted by this mechanism is

$$\frac{d[AA]}{dt} = k'[AA^-][H_2O].$$

If the first two steps are equilibria:

$$[AA^-] = E_{eq_2}[A][A^-]$$
$$[A^-] = K_{eq_1}[A][OH^-]/[H_2O],$$

and the rate with respect to stable species is

$$\frac{d[AA]}{dt} = k'K_{eq_1}K_{eq_2}[A]^2[OH^-].$$

Thus the rate predicted by this mechanism agrees with the experimentally observed rate.

These are only two of numerous examples of reactions which are catalyzed by acids and/or bases.

12.5. Enzyme Catalysis

Of particular importance is the catalysis of biological reactions by enzymes. An enzyme is in general a large protein molecule of complex structure. Various portions of the enzyme molecule interact with a reactant (called a *substrate*) to form an enzyme-substrate complex. Some of these interactions may be weak, but usually one particular section of the enzyme molecule interacts strongly with the substrate and is thus the site where reaction occurs. This part of the enzyme is known as the *active site*. Enzymes are usually very specific, catalyzing only one type of reaction. For instance, leucine aminopeptidase catalyzes the hydrolysis of the peptide linkage of L-leucylglycine, whereas chymotrypsin catalyzes the hydrolysis of peptide bonds of aromatic or large aliphatic amino acids.

Not only the enzyme's chemical make-up but also its spatial arrangement (conformation) is important. An enzyme and its interaction with a substrate are represented schematically in Fig. 12.2. The complexity of the enzyme and the substrate is reduced in this figure by indicating only those chemical groups which are important for the formation of the enzyme-substrate complex. The remainder of these molecules is simply represented by ribbons. If the enzyme were spatially oriented in any other way (in a straight line for

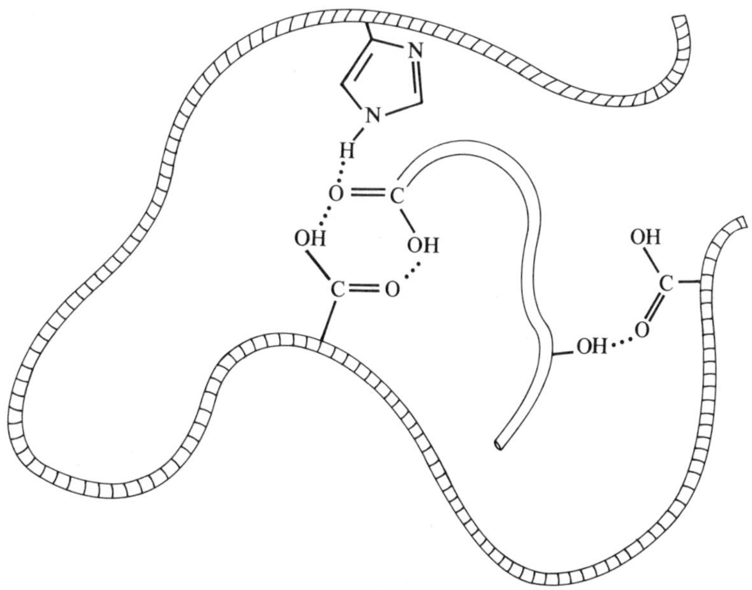

Fig. 12.2. Schematic representation of the conformation of an enzyme-substrate complex (the enzyme is the long striped portion, the substrate the short open part).

example) the substrate would not "fit." The analogy to a lock (enzyme) and key (substrate) is appropriate here.

The kinetic behavior of enzyme-catalyzed reactions can be generalized by a treatment initially developed by L. Michaelis and M. L. Menten in 1913. The following general mechanism is postulated:

$$\mathrm{E + S} \underset{k_2}{\overset{k_1}{\rightleftharpoons}} \mathrm{ES} \tag{229}$$

$$\mathrm{ES} \xrightarrow{k_3} \mathrm{P + E}, \tag{230}$$

where E is the enzyme, S the substrate, ES an enzyme-substrate complex, and P the product. Using the terminology of enzyme catalysis, v is the *initial* rate (see Section 9.9). When a high initial concentration of substrate is used, the initial rate of the enzyme-catalyzed reaction is found experimentally to be zero-order in [S], but when the initial concentration of substrate is low the experimental order with respect to [S] is 1. This is an example of *biphasic* kinetics and indicates that the actual mechanism is complex. Figure 12.3 illustrates this behavior. The rate of the reaction when zero-order obtains is called v_{max}.

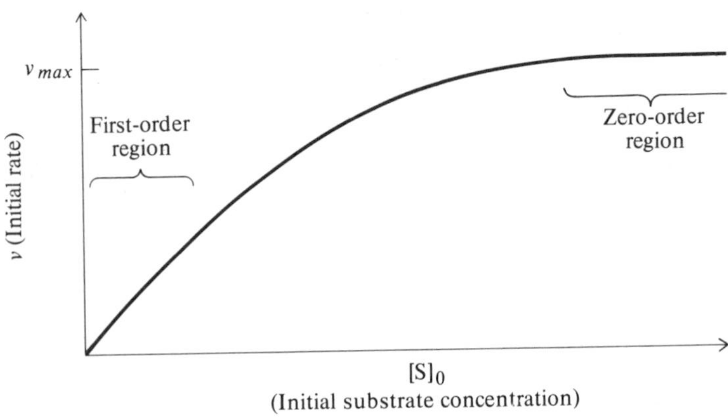

Figure 12.3

In the Michaelis-Menten treatment it is assumed that the formation of ES is fast relative to the step which produces product and that the steady-state method may be applied to [ES]:

$$\frac{d[\mathrm{ES}]}{dt} = 0 = k_1[\mathrm{E}][\mathrm{S}] - k_2[\mathrm{ES}] - k_3[\mathrm{ES}]$$

$$[\mathrm{ES}] = k_1[\mathrm{E}][\mathrm{S}]/(k_2 + k_3). \tag{231}$$

It is in general difficult to determine at any moment the concentration of either the enzyme or the enzyme-substrate complex. Instead we may know $[\mathrm{E}]_T$, the total enzyme concentration (calculated from the amount put in at the beginning). At any instant, then,

$$[\mathrm{E}]_T = [\mathrm{E}] + [\mathrm{ES}]$$

or

$$[\mathrm{E}] = [\mathrm{E}]_T - [\mathrm{ES}]. \tag{232}$$

Substituting Eq. (232) into Eq. (231) and rearranging:

$$[\mathrm{ES}] = k_1[\mathrm{E}]_T[\mathrm{S}]/(k_2 + k_3 + k_1[\mathrm{S}]).$$

The rate of the enzyme-catalyzed reaction is $d[P]/dt = k_3[\text{ES}]$ from the mechanism. The initial rate is then

$$v = k_3 k_1[\text{E}]_T[\text{S}]_0/(k_2 + k_3 + k_1[\text{S}]_0)$$

$$= \frac{k_3[\text{E}]_T[\text{S}]_0}{(k_2 + k_3)/k_1 + [\text{S}]_0}$$

$$= k_3[\text{E}]_T[\text{S}]_0/(K_M + [\text{S}]_0). \tag{233}$$

K_M is called the Michaelis-Menten constant. It and k_3 are characteristic constants for a particular enzyme-catalyzed reaction.

This rate equation can be seen to fit the experimental observations of biphasic behavior. At low $[\text{S}]_0$ (when $[\text{S}]_0 \ll K_M$):

$$v = \frac{k_3}{K_M}[\text{E}]_T[\text{S}]_0. \tag{234}$$

At high $[\text{S}]_0$ (when $[\text{S}]_0 \gg K_M$):

$$v = k_3[\text{E}]_T. \tag{235}$$

Determining $[\text{E}]_T$ can be quite difficult and may prove impossible. If we know the number of moles of enzyme used we can calculate $[\text{E}]_T$ but this requires a pure enzyme of known molecular weight. Many enzymes are virtually impossible to prepare pure, and the molecular weights of many are not accurately known. Luckily we can get around these problems.

The rate given by Eq. (235) is the rate when the order with respect to $[\text{S}]_0$ is zero. This is the maximum rate v_{max} referred to earlier (see Fig. 12.3). This initial rate can be measured for any enzyme-catalyzed reaction, even those for which $[\text{E}]_T$ is not accurately known. We could simply increase the substrate concentration until the initial rate no longer depends on its concentration; the rate at this point is v_{max}. Notice that in this case $[\text{E}]_T = v_{\text{max}}/k_3$ (Eq. 235). If we substitute this into Eq. (233) we obtain the *Michaelis-Menten equation*:

$$v = v_{\text{max}}[\text{S}]_0/(K_M + [\text{S}]_0). \tag{236}$$

This equation and its modifications are of great value in the study of enzyme-catalyzed reactions. One such modification is the *Lineweaver-Burk equation*:

$$\frac{1}{v} = \frac{K_M}{v_{\text{max}}[\text{S}]_0} + \frac{1}{v_{\text{max}}} \tag{237}$$

If the reaction being studied follows the Michaelis-Menten development exactly, a plot of $1/v$ versus $1/[\text{S}]_0$ will be a straight line with slope K_M/v_{max} and with an intercept on the $1/v$ axis of $1/v_{\text{max}}$. The extrapolated intercept on the $1/[\text{S}]$ axis is equal to $-1/K_M$. Such plots are used to determine the characteristic constants K_M and v_{max}. Figure 12.4 is an example of a Lineweaver-Burk plot.

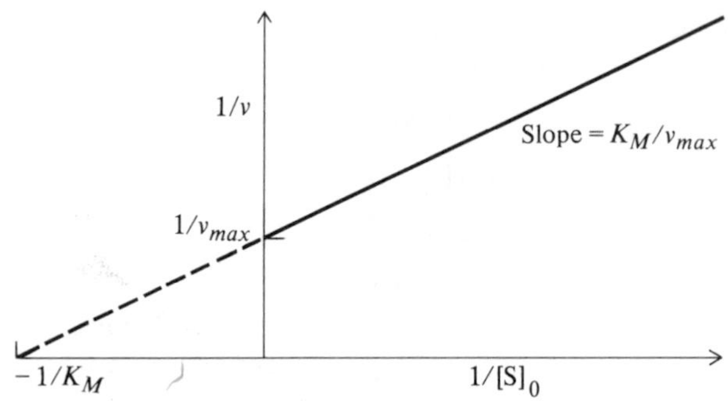

Fig. 12.4. A typical Lineweaver-Burk plot.

Example 68 The enzyme carbonic anhydrase catalyzes the reaction

$$CO_2(aq) + H_2O \longrightarrow HCO_3^- + H^+$$

In a typical experiment the following data were obtained:

Initial CO_2 concentration, moles liter^{-1} × 10^3	Initial rate, moles liter^{-1} sec^{-1} × 10^3
1.0	.02
2.5	.05
5.0	.08
7.5	.10
10.0	.12
25.0	.17
50.0	.19
100.0	.21

Determine the values of the Michaelis-Menten constant and of v_{max}.

Answer: This is done most easily by converting the data to reciprocals and using a Lineweaver-Burk plot.

$1/[S]_0$, liters mole^{-1}	$(1/v) \times 10^{-3}$, liter-sec mole^{-1}
1000	50.00
400	20.00
200	12.50
130	10.00
100	8.33
40	5.88
20	5.26
10	4.76

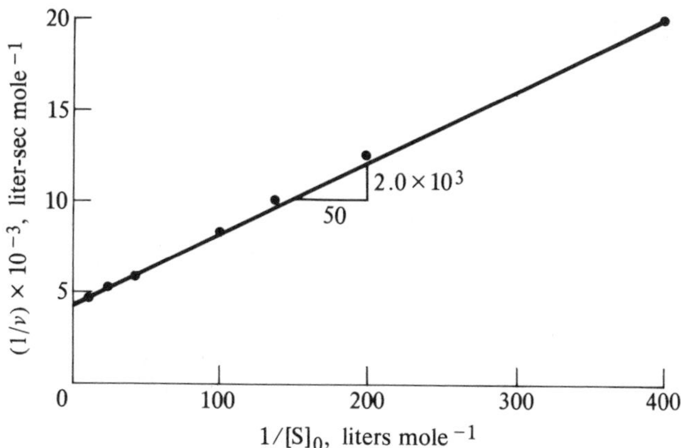

Figure 12.5

Figure 12.5 is a plot of $1/[\mathrm{S}]_0$ versus $1/v$. From the Lineweaver-Burk equation

$$\frac{1}{v} = \frac{K_M}{v_{\max}[\mathrm{S}]_0} + \frac{1}{v_{\max}}$$

we see that the intercept when $1/[\mathrm{S}]_0 = 0$ is $1/(v_{\max} \times 10^{-3}) \simeq 4.3$. Therefore $v_{\max}$ is 0.23×10^{-3} mole liter^{-1} sec^{-1}. The slope of the line in Fig. 12.5 is $K_M/v_{\max} = 2.0 \times 10^3/50 = 40$. With the value of $v_{\max}$ solved from the intercept, this gives us a value of K_M equal to 9.2×10^{-3} mole liter^{-1}.

The accepted values for these constants appear in H. Devoe and G. B. Kistiakowsky, *Journal of the American Chemical Society*, **83**, 274 (1961). They are $v_{\max} = 0.23 \times 10^{-3}$ and $K_M = 9.1 \times 10^{-3}$.

The mechanism of action of some medicinal agents—and some poisons—involves *inhibition*. In one type of inhibition, the inhibitor molecules compete with the substrate for positions at the active site of the enzyme. (An example is the inhibition of succinic dehydrogenase by malonate.) The process is assumed to be

$$\mathrm{E} + \mathrm{S} \underset{k_2}{\overset{k_1}{\rightleftharpoons}} \mathrm{ES} \xrightarrow{k_3} \mathrm{P} + \mathrm{E}$$

$$\mathrm{E} + \mathrm{I} \overset{K_\mathrm{I}}{\rightleftharpoons} \mathrm{EI}.$$

It is postulated that the inhibtor I reacts with enzyme to form an EI complex and that this process rapidly reaches equilibrium with an equilibrium constant K_I. The kinetic behavior of such an inhibited enzyme-catalyzed reaction can be readily developed. The total enzyme concentration is now

$$[\mathrm{E}]_T = [\mathrm{E}] + [\mathrm{ES}] + [\mathrm{EI}]$$
$$[\mathrm{E}]_T = [\mathrm{E}] + [\mathrm{ES}] + K_\mathrm{I}[\mathrm{E}][\mathrm{I}],$$

and the instantaneous enzyme concentration is

$$[E] = ([E]_T - [ES])/(1 + K_I[I]).$$

Substituting this last expression into Eq. (231), rearranging, and recalling that $K_M = (k_2 + k_3)/k_1$:

$$[ES] = \frac{k_1[S][E]_T}{k_1[S] + (k_2 + k_3)(1 + K_I[I])}$$

$$[ES] = \frac{[S][E]_T}{[S] + K_M(1 + K_I[I])}.$$

The rate of product formation is still $d[P]/dt = k_3[ES]$ so that the initial rate for this *competitive inhibition* is

$$v = \frac{v_{max}[S]_0}{[S]_0 + K_M(1 + K_I[I]_0)}. \tag{238}$$

The Lineweaver-Burk equation for this situation is

$$\frac{1}{v} = \frac{K_M}{[S]_0 v_{max}}(1 + K_I[I]_0) + \frac{1}{v_{max}}. \tag{239}$$

A plot of $1/v$ versus $1/[S]_0$ would be a straight line with slope $(K_M/v_{max}) \times (1 + K_I[I]_0)$ and an intercept on the ordinate of $1/v_{max}$.

There are several other types of inhibitions which distrupt or modify enzyme-catalyzed reactions. Each of these has its own characteristic rate equation similar to Eq. (239). The study of the kinetic behavior of enzyme-inhibitor reactions can suggest the probable mechanism of physiological inhibition, information which is vitally important in molecular pharmacology.

Problems

12.1 In Example 68 we determined K_M and v_{max} for the reaction

$$CO_2 + H_2O \longrightarrow HCO_3^- + H^+$$

catalyzed by carbonic anhydrase. This enzyme also catalyzes the reverse process

$$HCO_3^- + H^+ \longrightarrow CO_2 + H_2O.$$

From the following data determine K_M and v_{max} for this latter process at pH 7.0.

Initial concentration of $[HCO_3^-]$, moles liter^{-1}	Initial rate, moles liter^{-1} sec^{-1}
1×10^{-3}	0.57×10^{-5}
2×10^{-3}	1.07×10^{-5}
5×10^{-3}	2.22×10^{-5}
10×10^{-3}	3.50×10^{-5}

12.2 Suppose the products formed in an enzyme-catalyzed reaction can react with the enzyme to produce ES:

$$\mathrm{E} + \mathrm{S} \underset{k_2}{\overset{k_1}{\rightleftharpoons}} \mathrm{ES} \underset{k_4}{\overset{k_3}{\rightleftharpoons}} \mathrm{E} + \mathrm{P}$$

How will this modify the Michaelis-Menten equation (Eq. 236)?

12.3 The dephosphorylation of ATP is catalyzed by the enzyme myosin. From the following data determine K_M and v_{max}.

Initial concentration of ATP, moles liter^{-1}	Initial rate, moles liter^{-1} sec^{-1}
2×10^{-3}	0.51×10^{-5}
4×10^{-3}	0.67×10^{-5}
6×10^{-3}	0.76×10^{-5}
8×10^{-3}	0.81×10^{-5}
10×10^{-3}	0.83×10^{-5}

12.4 (a) Explain how the energy of activation in the Arrhenius equation may be determined for the reaction

$$\mathrm{ES} \xrightarrow{k_3} \mathrm{P} + \mathrm{E}$$

by measuring v_{max} at several temperatures. (b) Is it *necessary* for the energy of activation of an enzyme-catalyzed reaction to be less than that for the reaction with no catalyst?

12.5 The rate of pharmacological action of drugs may be treated in a manner analogous to enzyme kinetics. The drug D is assumed to react reversibly with a physiological receptor R to form a drug-recepter complex DR. The rate of response is then assumed to be proportional to the concentration of this complex.

$$\mathrm{D} + \mathrm{R} \underset{k_2}{\overset{k_1}{\rightleftharpoons}} \mathrm{DR} \xrightarrow{k_3} \text{response}$$

(A. J. Clark, in *Heftner's Handbuch der Experimentellen Pharmakologie*, Vol. IV, Berlin: Springer, 1937.) Show that the response rate v may be given by

$$v = v_{max} D_t / \{(k_2/k_1) + D_t\},$$

where D_t is the total drug concentration.

12.6 Listed below are several competitive inhibitors of carboxypeptidase (H. Neurath and G. W. Schwert, *Chemical Reviews*, **46**, 69 (1950).) Which of these most retards the rate of a carboxypeptidase-catalyzed reaction?

Inhibitor	K_1 (millimolar)
D-histidine	0.05
p-nitrophenylacetic acid	0.4
γ-phenylbutyric acid	0.9
β-phenylpropionic acid	16.7

12.7 In some cases of inhibition (for example the inhibition of rat liver amidase by hexanoate ion), the inhibitor forms a complex with the enzyme-substrate complex as well as with the free enzyme:

$$E + S \underset{k_2}{\overset{k_1}{\rightleftharpoons}} ES \xrightarrow{k_3} P$$

$$E + I \overset{K_I}{\rightleftharpoons} EI$$

$$ES + I \overset{K_I'}{\rightleftharpoons} ESI.$$

Show that the equation for the reciprocal of the expected initial rate in this case is

$$\frac{1}{v} = \frac{K_M}{[S]_0\, v_{max}}(1 + K_I[I]_0) + \frac{1}{v_{max}}(1 + K_I'[I]_0).$$

12.8 In *noncompetitive inhibition*, $K_I = K_I'$ (see Problem 12.7). What is the expression for the reciprocal of the initial rate for this special case? (The inhibition of fumarase by thiocyanate is noncompetitive.)

12.9 In the inhibition of cytochrome oxidase by azide, the inhibitor combines with the enzyme-substrate complex but not with the free enzyme. What is the expression for the reciprocal of the initial rate for this case (called, for lack of a better name, *uncompetitive inhibition*)?

PART THREE

Quantum Chemistry

13 / Wave Mechanics: Basic Theory

13.1. Introduction

The methods of thermodynamics and kinetics provide considerable information concerning chemical processes. The simplified microscopic models proposed for explaining the observed behavior can help us visualize how molecules undergo transformations. In our further inquiry into the details of chemical reactivity, we need now to look at the molecule itself. How are the parts of a molecule held together? What properties of molecular bonds make some molecules more reactive than others? What is the most accurate and helpful way of visualizing a molecule? These are only a few of many questions which might be asked. The answers can provide a basis for understanding why a given reaction is accompanied by a particular energy and entropy change, and why one mechanism should be accepted while another is rejected.

The answers to such questions as those posed above and, indeed, our current understanding of the structure and behavior of molecules come from *quantum chemistry*, a science ultimately based on the theory of *quantum mechanics* or *wave mechanics*. Our purpose at this point is to develop an appreciation of that theory and its application. Doing so will equip us with a microscopic model which will help us visualize molecular phenomena and allow us to deal quantitatively with them. The net result should be a more satisfying and more useful knowledge of chemical behavior.

13.2. The Development of Wave Mechanics

Isaac Newton (1642–1727) developed a set of three postulates which, taken together, apparently accounted for ("explained") the dynamic behavior of the physical world. If the position of and the forces acting on any chunk of matter were known, its future could be accurately predicted. These postulates have come to be known as the laws of *Newtonian* or *classical mechanics.*

These laws, however, have not proved to be as unassailable as the laws of thermodynamics. Although they work well when macroscopic systems are treated, seemingly insurmountable problems appear in the treatment of microscopic phenomena. Efforts to develop an accurate quantitative method for dealing with the dynamics of microscopic matter led to the emergence in the first quarter of the twentieth century of a new approach to such systems, a new mechanics.

The first major break with classical mechanics came with the proposal in 1900 by Max Planck that energy is absorbed or given off by matter in small packets called *quanta.* A system cannot therefore possess any energy it might choose; energy is *quantized.* This is illustrated in Fig. 13.1.

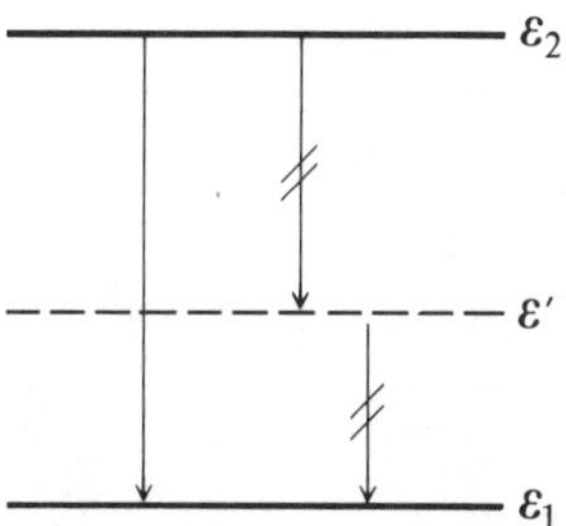

Fig. 13.1. Here ε_1 and ε_2 are the energies of two allowed states of a microscopic system. When the system changes from state 2 to state 1, a quantum of energy equal to $\varepsilon_2 - \varepsilon_1$ is released. Since ε' corresponds to a state which is not allowed for this system, transitions between the allowed state and this state cannot occur; quanta of energy $\varepsilon_2 - \varepsilon'$ or $\varepsilon' - \varepsilon_1$ will not be observed.

Example 69 In photosynthesis by plants and in other photochemical reactions the energy needed to reach an activated state is obtained via the absorption of quanta of radiation, or *photons*. As we shall see in Chapter 18, the energy ε of a single photon is given by $\varepsilon = h\nu = hc/\lambda$ where h is Planck's constant, ν is the frequency of the radiation, λ is the radiation's wavelength, and c is the speed of light (3.00×10^{10} cm sec^{-1}). If the energy of activation for a certain photochemical reaction is 10.0 kcal mole^{-1}, what must be the wavelength of the photon which when absorbed by a reactant molecule could elevate it to the transition state?

Answer: $$10 \text{ kcal mole}^{-1} = \frac{(10000 \text{ cal mole}^{-1})(4.18 \times 10^7 \text{ ergs cal}^{-1})}{6.02 \times 10^{23} \text{ molecules mole}^{-1}}$$

$$= 6.9 \times 10^{-13} \text{ erg molecule}^{-1}$$

The activation energy per molecule is thus 7×10^{-13} erg, and

$$\lambda = hc/\varepsilon = \frac{(6.6 \times 10^{-27} \text{ erg-sec})(3.00 \times 10^{10} \text{ cm sec}^{-1})}{(6.9 \times 10^{-13} \text{ erg})}$$

$$= 2.9 \times 10^{-4} \text{ cm} = 2.9 \times 10^3 \text{ m}\mu \text{ (millimicrons).}$$

The subsequent work of Albert Einstein supported and extended Planck's quantum ideas. Niels Bohr successfully treated the hydrogen atom (the simplest atom, having only one electron and one proton) using quantum ideas, but the attempt to extend his approach to other atoms was not particularly successful.

A major breakthrough came in 1923 with the revolutionary suggestion by Louis de Broglie that moving particles might possess certain properties commonly associated with waves. Included would be a wavelength, predicted as $\lambda = h/mv$, where h is Planck's constant (6.626×10^{-27} erg-sec), m is the mass of the particle, and v is its velocity. This concept was confirmed by C. Davisson and L. H. Germer who observed that a stream of electrons (particles) is *diffracted* (a wave phenomenon) as it passes through a slit.[1]

Example 70[2] Calculate the wavelength of

a) a baseball of mass 5.25 oz delivered with a velocity of 100 mph by Sandy Koufax in his prime or by Nolan Ryan.

b) an electron of mass 9.11×10^{-28} g with velocity 10^8 cm sec^{-1}.

Answer: The predicted wave length is $\lambda = h/mv$.

a) 5.25 oz = 148.84 g, 100 mph = 4470 cm sec^{-1}

$$\begin{aligned}\lambda &= 6.63 \times 10^{-27} \text{ erg-sec}/(148.8 \text{ g})(4470 \text{ cm sec}^{-1}) \\ &= 10.0 \times 10^{-21} \text{ cm.}\end{aligned}$$

This value is much too small to be observed.

b) $\lambda = 6.63 \times 10^{-27}$ erg-sec/$(9.11 \times 10^{-28}$ g$)(1.00 \times 10^8$ cm sec$^{-1})$
$= 7.3 \times 10^{-8}$ cm.

This value is large enough to detect.

In 1926 Erwin Schrödinger combined the developing ideas of quantized systems and the wave nature of particles into a set of postulates which form the core of wave mechanics.[3] This body of theory has led to a new understanding of the microscopic world.

13.3. Wave Mechanics, Basic Postulates

The theory of wave mechanics is based on a set of postulates which cannot be derived. They can be proved correct only in the sense of comparing our observations and measurements of the physical world with the predictions made using the postulates. If the theory proves useful in organizing and predicting the behavior of a part of the physical world, if it provides models which help us visualize and understand this behavior, then we may pronounce

---

[1] See Walter J. Moore, *Physical Chemistry*, 4th ed., Englewood Cliffs, N.J.: Prentice-Hall (1972), p. 596.

[2] Ignoring this example and all subsequent parts of Chapters 13–18 which are shaded will lead to a briefer and less mathematically difficult treatment of subjects related to quantum chemistry.

[3] Werner Heisenberg simultaneously and independently proposed an alternative but equivalent set of postulates which are the basis of *matrix mechanics*. The wave mechanical theory of Schrödinger is somewhat the easier to work with and consequently will form the basis of our discussion.

it a good theory and use it with a confidence which increases with the continuing absence of exceptions.

The postulates of wave mechanics may be presented in several ways. Some of the complex math will be avoided here; only those features essential to the level of our discussion will be given.

It is postulated that a *wave function* is associated with each system of particles and that the behavior of this wave function (and thus of the system of particles) is given by the *Schrödinger equation.*

The value of the postulated wave function $\Psi(x_1, x_2, \ldots, t)$ associated with a system of particles depends on the position of each of the particles and on the time. The position $\boldsymbol{x}$ is the vector position. Mathematically it is written $\boldsymbol{x} = \hat{\imath}x + \hat{\jmath}y + \hat{k}z$ (see Section 1.12). In other words, the vector position is a function of the x, the y, and the z coordinates.

It is further proposed that the time dependence of the wave function (that is, how Ψ changes with the passage of time) is given by

$$-\frac{h}{2\pi i}\frac{\partial\Psi}{\partial t} = E\Psi, \tag{240}$$

where i is $\sqrt{-1}$ and E is the total energy of the system. This equation is known as the *time-dependent Schrödinger equation.*

When we are dealing with microscopic states which are *conservative,* whose total energy is by definition not a function of time (energy is neither being added to nor removed from the system so that there can be no change in its total energy as time passes), a significant simplification can be made in Eq. (240). For such systems the wave function may be written

$$\Psi(\boldsymbol{x}_1, \boldsymbol{x}_2, \ldots, t) = e^{-2\pi iEt/h}\psi(\boldsymbol{x}_1, \boldsymbol{x}_2, \ldots). \tag{241}$$

The wave function ψ now depends only upon the positions of the particles of the system, not on the time. Equation (240) may now be written

$$\mathscr{H}\psi = E\psi, \tag{242}$$

which is known as the *time-independent Schrödinger equation.* $\mathscr{H}$ is the quantum mechanical *Hamiltonian operator*. Its exact form will depend on the particular system, but there are general rules for its construction. First we write the kinetic energy as a function of *momentum.* For a system of only one particle with mass m and velocity v this would be $\frac{1}{2}mv^2 = p^2/2m$, where $p = mv$ is the momentum. We then replace p^2 with $-(h/2\pi)^2 \times (\partial^2/\partial x^2 + \partial^2/\partial y^2 + \partial^2/\partial z^2)$, which is often symbolized $-(h/2\pi)^2\nabla^2$. To this we add the potential energy V. Hence in this particular case the Hamiltonian operator would be

$$\mathscr{H} = \frac{-(h/2\pi)^2\nabla^2}{2m} + V = -(h^2/8\pi^2 m)\nabla^2 + V,$$

and $\mathscr{H}\psi = E\psi$ becomes

$$-(h^2/8\pi^2 m)\nabla^2\psi + V\psi = E\psi$$

$$-(h^2/8\pi^2 m)\nabla^2\psi + (V - E)\psi = 0. \tag{243}$$

Although this process may seem rather complicated, it becomes more clear as we apply it to particular systems. The solutions of this second-order differential equation will provide us with the wave functions associated with each particular system.

The Schrödinger equation for all systems will be similar to Eq. (243), containing the second derivatives of the wave function with respect to position ($\nabla^2\psi$), the potential energy (V) and the total energy of the system (E). This equation resembles the wave equation of classical mechanics. It is a second-order differential equation (that is, it contains second derivatives). Of course this is not the first time we have encountered differential equations which describe the behavior of physical systems. The rate equations of chemical kinetics are good examples.

Since we shall be primarily concerned with conservative systems, our efforts will be directed toward the solution of Eq. (242).

Example 71 Write the Schrödinger equation for

a) a system consisting of a single particle of mass m moving only in the x direction in a container in which the potential energy V is (i) zero, (ii) a function of position, $V = (x/L)$ erg, where L is the length of the container.
b) a particle with a charge of +1 coulomb moving in a one-dimensional container of length L which has a +1-coulomb charge attached to the end of the container.

Answer: a) The kinetic energy part of all the Hamiltonian operators dealing with a single particle will be the same: $-(h^2/8\pi^2 m)(d^2/dx^2)$. The potential energy parts will be different. (i) $V = 0$; therefore $\mathscr{H} = -(h^2/8\pi^2 m)(d^2/dx^2) + 0$, and the Schrödinger equation is $-(h^2/8\pi^2 m)(d^2\psi/dx^2) = E\psi$. (ii) $V = x/L$; the total Hamiltonian operator is $\mathscr{H} = -(h^2/8\pi^2 m)(d^2/dx^2) + x/L$. The Schrödinger equation is $-(h^2/8\pi^2 m)(d^2\psi/dx^2) + (x/L)\psi = E\psi$.

b) The potential energy in this case will be the result of electrostatic repulsion between the two positive charges: $V = q_1 q_2/r$, where r is the distance between the two charges q_1 and q_2. If we situate the container so that one end is at $x = 0$, the other end will be at L. If now the fixed charge is attached to the end at $x = 0$, then $r = x$, the position of the particle in the box. Now $V = q_1 q_2/x$. The units of V, however, are $\text{coul}^2\ \text{cm}^{-1}$, while the "kinetic energy" part of the Hamiltonian is in terms of ergs. To get both in ergs we multiply q in coulombs by 3×10^9. This gives q in electrostatic units (esu) and V in ergs if x is in centimeters.

$$V = (1\ \text{coul})(1\ \text{coul})/x = (3 \times 10^9)^2/x = (9 \times 10^{18}/x)\ \text{erg}.$$

The Schrödinger equation is similar to part (ii):

$$-(h^2/8\pi^2 m)(d^2\psi/dx^2) + (9 \times 10^8/x)\psi = E\psi.$$

Thus we postulate a wave function associated with every system of particles. We further postulate an equation which describes the behavior of this wave function and relates it to the total energy of the particles. Let us now examine a particular case to see how these postulates are applied in practice.

13.4. Translational Motion, One Dimension

The simplest application of wave mechanics of importance to us is the system consisting of a single gaseous particle (an atom) in a container. The atom is free to move about in this box, its velocity being related to the temperature (see Section 1.12). In order to simplify the problem somewhat, we shall confine our atom to a one-dimensional box (for example, a wire) of length L. (The more general, three-dimensional case will be discussed in the next section.)

We first set up the Schrödinger equation for this system; then we solve the equation to obtain the wave functions and the energy associated with each wave function which quantum mechanics predicts for this system.

If there are no electrical charges, the potential energy within the box will be zero. The particle's only energy will be its kinetic translational energy: $mv^2/2 = p^2/2m$. The Hamiltonian operator is thus (see Example 71a, i)

$$\mathscr{H} = -(h^2/8\pi^2 m)\nabla^2 + V = -(h^2/8\pi^2 m)\frac{d^2}{dx^2} + 0.$$

Here ∇^2 is d^2/dx^2 because we are dealing with only one dimension. Since the particle is confined to the wire, the wave function will be equal to zero outside the box. A requirement of all wave functions is that they be *continuous*. One result of this requirement is that the wave function must also have a value of zero at each end of the wire. We conveniently set up our coordinate system so that one end of the wire is located at $x = 0$ and the other end at $x = L$. We may then write the *boundary conditions*:

$$\psi(0) = \psi(L) = 0.$$

Thus the Schrödinger equation ($H\psi = E\psi$) for this system is

$$-\frac{h^2}{8\pi^2 m}\frac{d^2\psi}{dx^2} = E\psi. \tag{244}$$

The next step is the solution of this equation.

The function ψ we seek must be such that when it is differentiated twice we obtain a constant (E) multiplied by the function. The functions $A \sin (ax)$ and $B \cos (bx)$ (A, B, a, and b being constants) are of this type. We can eliminate the cosine function by using the boundary condition $\psi(0) = 0$: $A \sin (a \cdot 0) = 0$, but $B \cos (b \cdot 0) = B$. If ψ is indeed $A \sin (ax)$, the boundary condition $\psi(L) = 0$ must also be obeyed, and $A \sin (aL) = 0$. This will be true whenever aL is equal to 0, π, 2π, etc., whenever aL is equal to an integer multiple of π: $aL = n\pi$. The constant a is therefore equal to $n\pi/L$ and our proposed solution is

$$\psi = A \sin\left(\frac{n\pi}{L}x\right). \tag{245}$$

This solution (Eq. 245) is actually a set of functions: each different value of the integer n results in a different wave function. The function corresponding to a particular value of n will be designated ψ_n.

Let us now substitute this wave function into Eq. (244) to make sure it is a correct solution, and perhaps to obtain further information:

$$-\frac{h^2}{8\pi^2 m}\frac{d^2}{dx^2}\left[A \sin\left(\frac{n\pi}{L}x\right)\right] = E\left[A \sin\left(\frac{n\pi}{L}x\right)\right]$$

$$-\frac{h^2}{8\pi^2 m}\left(\frac{n\pi}{L}\right)\frac{d}{dx}\left[A \cos\left(\frac{n\pi}{L}x\right)\right] = E\left[A \sin\left(\frac{n\pi}{L}x\right)\right]$$

$$\frac{h^2}{8\pi^2 m}\left(\frac{n^2\pi^2}{L^2}\right)\left[A \sin\left(\frac{n\pi}{L}x\right)\right] = E\left[A \sin\left(\frac{n\pi}{L}x\right)\right]$$

$$\frac{n^2h^2}{8mL^2}\psi = E\psi. \tag{246}$$

We see that this wave function (Eq. 245) is indeed a solution of the Schrödinger equation (Eq. 244).

We must now determine the value of the constant A. The quantity $\psi(x)\psi^*(x)\,dx$ may be interpreted as the *probability* of finding the particle within an infinitesimal range dx of x (in other words, the probability of finding it at x). If the particle is in the one-dimensional box, the probability of finding it somewhere between $x = 0$ and $x = L$ must of course be unity. Therefore

$$\int_0^L \psi_n(x)\psi_n(x)^*\,dx = 1. \tag{247}$$

ψ^* is the *complex conjugate* of ψ. If ψ is a *complex number* (that is, if it contains $\sqrt{-1} = i$), ψ^* is obtained by replacing i with $-i$ each time it occurs in ψ. For example:

$$\begin{aligned} \text{if } \psi &= e^{iax}, & \psi^* &= e^{-iax} \\ &= A - iB, & &= A + iB \\ &= Ax, & &= Ax. \end{aligned}$$

Notice that when the function is *real* (does not contain i), $\psi^* = \psi$.

We may now use Eq. (247) to solve for A:

$$\int_0^L \left(A \sin\frac{n\pi x}{L}\right)\left(A \sin\frac{n\pi x}{L}\right) dx = 1$$

$$A^2 \int_0^L \sin^2\left(\frac{n\pi x}{L}\right) dx = 1$$

$$A^2\left(\frac{L}{2}\right) = 1$$

$$A = \sqrt{2/L}.$$

This procedure of determining A so that the total probability is 1 (Eq. 247) is called *normalization*.

The wave functions for a particle moving in a one-dimensional box are therefore

$$\psi_n = \sqrt{2/L}\,\sin\left(\frac{n\pi}{L}x\right). \tag{248}$$

Example 72 A particle is in a one-dimensional box of length L. If its wave function has quantum number $n = 1$, calculate the probability of finding this particle somewhere between 0 and $\frac{1}{4}L$.

Answer: $P = \int_0^{L/4} \left(\sqrt{\frac{2}{L}} \sin \frac{\pi}{L} x\right)^2 dx$

$$= \frac{2}{L} \int_0^{L/4} \sin^2 \left(\frac{\pi}{L} x\right) dx.$$

If $y = \frac{\pi}{L} x$, then $dy = \frac{\pi}{L} dx$ and when $x = \frac{L}{4}$, $y = \frac{\pi}{4}$. Substituting:

$$P = \left(\frac{2}{L}\right)\left(\frac{L}{\pi}\right) \int_0^{\pi/4} \sin^2 y \, dy$$

$$P = \frac{2}{\pi} \left(\tfrac{1}{2}y - \tfrac{1}{4} \sin 2y\right)\Big|_0^{\pi/4}$$

$$= \frac{2}{\pi}\left[\left(\frac{\pi}{8} - \frac{1}{4} \sin \frac{\pi}{2}\right) - \left(0 - \frac{1}{4} \sin 0\right)\right]$$

$$= 0.091.$$

The energy of a particle with wave function ψ_n may be obtained from Eq. (246). It is equal to

$$E_n = \frac{n^2 h^2}{8mL^2}. \tag{249}$$

The energy depends on the mass of the particle and on the length of the wire. It is also *quantized*; the particle may possess only those translational energies given by Eq. (249) where n is a positive integer. Figure 13.2 gives the wave functions with their energies for the first few values of n. There is a different wave function with its own associated energy for each value of n. If no energy is being added or taken out of the system (that is, if the system is conservative) no other translational energies are possible. This integer n is called a *quantum number*. Its value determines the wave function and the energy. We shall find that other quantum numbers will also appear in the solutions to the wave equations for other situations. It is through their presence that we have quantized energy levels with some energies allowed while others are forbidden (see Fig. 13.1). Accounting for this distinctive feature of microscopic systems, which is easily verified by observation, has been one of the great successes of wave mechanics.

Also given in Fig. 13.2 is the probability $\psi\psi^*$ as a function of the position x for each wave function. The particle is more likely to be found at positions where $\psi\psi^*$ is large.

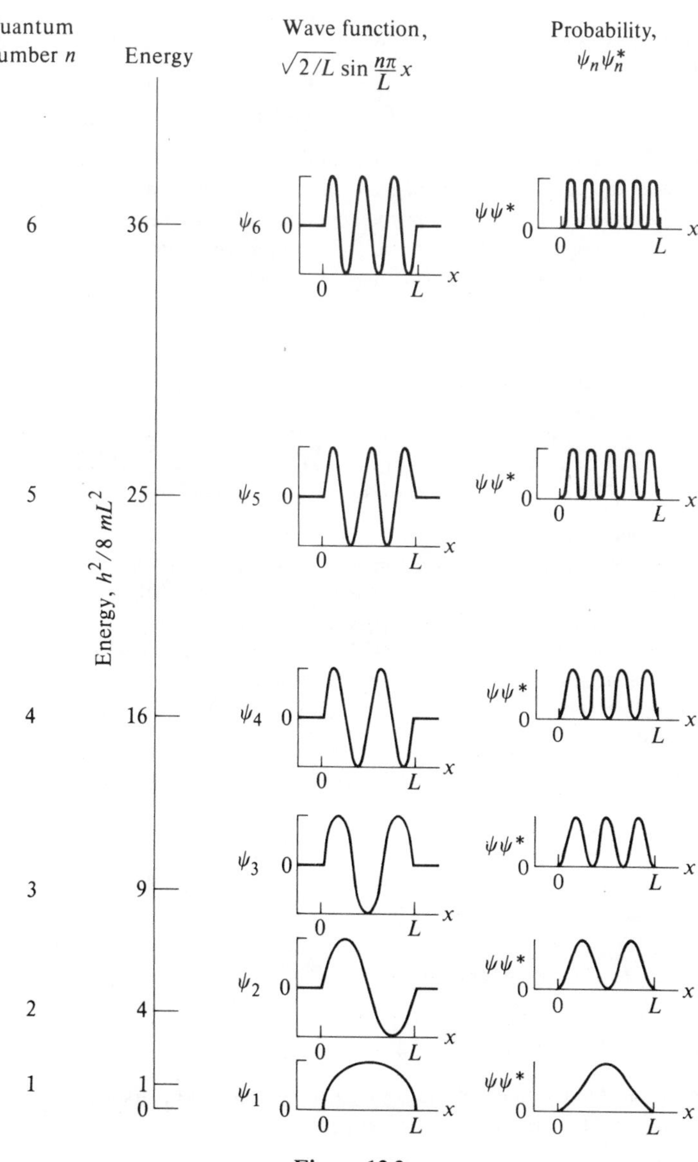
Quantum number n
Energy
Wave function, $\sqrt{2/L}\,\sin\frac{n\pi}{L}x$
Probability, $\psi_n\psi_n^*$
Energy, $h^2/8\,mL^2$
6
36
ψ_6
5
25
ψ_5
4
16
ψ_4
3
9
ψ_3
2
4
ψ_2
1
1
0
ψ_1
$\psi\psi^*$
0
L
x

Figure 13.2

Example 73 Conjugated molecules with alternating single and double bonds are of the type

$$R{-}CH{=}CH{-}CH{=}CH{-}R'.$$

We shall discuss the bonding in such molecules in Chapter 15. Some of the electrons in the conjugated system may be treated approximately by considering them to be particles in a one-dimensional container. The container is the conjugated portion of the molecule. One electron from each carbon atom of the conjugated system may be considered to be involved. For vitamin A_1,

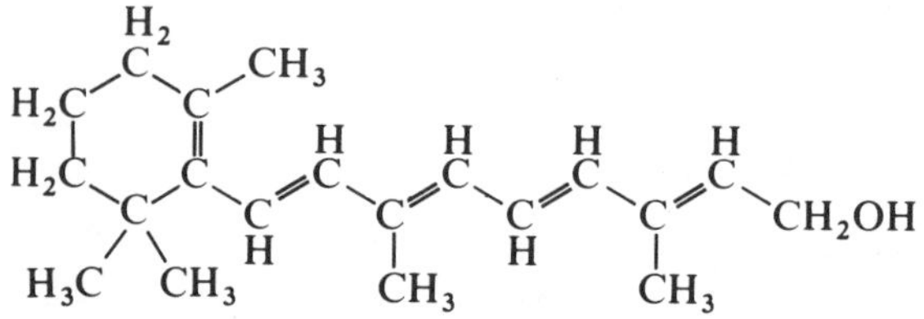

```
          H2
          C      CH3
   H2C        C
    |         ||       H        H        H        H
   H2C        C        C        C        C        C
          C        C        C        C        C        CH2OH
         / \       H        |        H        |
      H3C   CH3            CH3               CH3
```

using the average length 1.40 Å (1.40×10^{-8} cm) for each bond, double and single, the container stretches over the carbon atoms and would be 1.40×10^{-7} cm long.[4] There are ten electrons in this container, one from each carbon atom. Only two electrons may occupy a single energy level (see Section 14.4).

a) Calculate the energy of each of the six quantum states of lowest energy.
b) Assuming that electrons will occupy the lowest energy levels possible in their most stable state (the *ground state* or *unexcited state*), calculate the energy of that state.

Answer: a) Using Eq. (249), the energy levels for the conjugated portion of vitamin A_1 will be

$$En = \frac{n^2(6.63 \times 10^{-27}\ \text{erg-sec})^2}{8(9.11 \times 10^{-28}\ \text{g})(1.40 \times 10^{-7}\ \text{cm})^2}$$
$$= n^2(3.1 \times 10^{-13}\ \text{erg}).$$

The six states of lowest energy are:

n	1	2	3	4	5	6
E_n, erg $\times 10^{13}$	3.1	12.4	27.9	49.6	77.5	111.6

b) Two of the ten electrons of the conjugated system will be associated with each of the five lowest energy levels. The total energy will then be

$$E = 2E_1 + 2E_2 + 2E_3 + 2E_4 + 2E_5$$
$$= 3.41 \times 10^{-11}\ \text{erg}.$$

[4] Although there are only nine bonds in the conjugated system, the use of $10(1.40 \times 10^{-8}$ cm) results in a longer container, allowing the electrons to move about the terminal carbons of the conjugated chain rather than picturing them as not being able to go beyond the centers of these atoms.

13.5. Translational Motion, Three Dimensions

The results for one dimension can be extended to the more useful case of three dimensions. For a particle moving in a rectangular container of dimensions a, b, and c, the wave functions obtained from the Schrödinger equation are

$$\psi_{n_x\, n_y\, n_z} = \sqrt{8/abc}\,\sin\left(\frac{n_x \pi}{a} x\right)\sin\left(\frac{n_y \pi}{b} y\right)\sin\left(\frac{n_z \pi}{c} z\right). \tag{250}$$

The corresponding energies are

$$E_{n_x\, n_y\, n_z} = \frac{h^2}{8m}\left(\frac{n_x^2}{a^2} + \frac{n_y^2}{b^2} + \frac{n_z^2}{c^2}\right). \tag{251}$$

The wave functions and their associated energies possess a set of three quantum numbers n_x, n_y, and n_z, each of which may have any positive, nonzero integer value.

If the container is cubic, then $a = b = c$ and the energy is

$$E_{n_x\, n_y\, n_z} = \frac{h^2}{8ma^2}(n_x^2 + n_y^2 + n_z^2). \tag{252}$$

In this special case there may be more than one wave function with the same energy. For instance the six different wave functions with (n_x, n_y, n_z) equal to (1, 2, 3), (1, 3, 2), (3, 2, 1), (3, 1, 2), (2, 1, 3), and (2, 3, 1), respectively, all have the same energy, $14(h^2/8ma^2)$. This is because the sum of the squared quantum numbers is identical in all six cases. The number of different wave functions which have the same energy is called the *degeneracy*. The energy level $14(h^2/8ma^2)$ is six-fold degenerate.

Example 74 The electrons in metals move about freely. They may be treated approximately like particles in a three-dimensional container of size equal to that of the piece of metal. Give the degeneracy of the energy level in a cube of solid copper with the following quantum numbers:

a) $n_x = 1$, $n_y = 1$, $n_z = 1$

b) $n_x = 1$, $n_y = 1$, $n_z = 2$

Answer: a) $E = (h^2/8ma^2)(1^2 + 1^2 + 1^2)$

$= 3(h^2/8ma^2)$

No other set of quantum numbers (that is, no other wave function) will result in this energy. This energy level is nondegenerate.

b) $E = (h^2/8ma^2)(1^2 + 1^2 + 2^2)$

$= 6(h^2/8ma^2)$

The wave functions with (n_x, n_y, n_z) equal respectively to (2, 1, 1) and (1, 2, 1) will also possess this value of energy. This energy level is therefore three-fold degenerate.

Example 75 Consider an oxygen molecule in a cubic container with a volume of 3.0 liters (approximately equal to that of adult lungs).

a) Calculate the oxygen molecule's average kinetic energy $\bar{\varepsilon}$ at 37°C assuming it behaves ideally.

b) What translational quantum state with $n_x = n_y = n_z$ corresponds to this energy?

Answer: a) $\bar{\varepsilon} = (3/2)kT$ from Eqs. (19) and (22).

$$= (3/2)(1.38 \times 10^{-16} \text{ erg deg}^{-1})(310°) = 6.42 \times 10^{-14} \text{ erg}$$

b) $E_{n_x\, n_y\, n_z} = (h^2/8ma^2)(n_x^2 + n_y^2 + n_z^2)$

$$a^3 = 3.0 \text{ liters} = 3000 \text{ cm}^3$$

$$a = 14.4 \text{ cm}$$

$$(n_x^2 + n_y^2 + n_z^2) = \frac{(6.42 \times 10^{-14} \text{ erg})(8)(14.4 \text{ cm})^2}{(6.63 \times 10^{-27} \text{ erg-sec})^2} \times \frac{32 \text{ g mole}^{-1}}{6 \times 10^{23} \text{ molecules mole}^{-1}}.$$

$$= 1.29 \times 10^{20}$$

If $n_x = n_y = n_z = n$,

$$3n^2 = 1.29 \times 10^{20}$$

$$n = 6.6 \times 10^9.$$

This rather large integer results from the fact that the translational energy levels in a container of this size are extremely closely spaced. The thermal energy available to such a molecule is thus enough to allow the molecule to be in a quantum state far above the ground state (corresponding to $n_x = n_y = n_z = 1$ and thus possessing the least amount of translational energy).

The wave mechanical treatment of a particle moving about in a container therefore predicts that the particle has associated with it a wave function which must be one of a set of wave functions (Eq. 250) which are the solutions of the Schrödinger equation. The specific wave function is determined by the values of the three quantum numbers n_x, n_y, and n_z. The particle possesses translational kinetic energy of magnitude given by Eq. (251), again dependent upon the values of the three quantum numbers. It may change its wave function and its energy only if it absorbs or emits energy of an amount exactly equal to the energy difference between its former state and its new state (as in Fig. 13.1). It is interesting to calculate this energy difference for several cases.

Example 76 Determine the energy required for a transition from the $n_x = n_y = n_z = 1$ to the $n_x = n_y = 1$, $n_z = 2$ state for

a) an argon atom (atomic weight = 39.95) in a cubic container with a 1.0-cm side.

b) an electron (mass = 9.11×10^{-28} g) in a cubic hole of a crystal with a 10^{-8}-cm edge.

Answer: a) The atomic weight of argon is 39.95 g. Therefore the mass of one argon atom is 39.95 divided by Avogadro's number, the number of atoms in a gram-atom. The energy of the lower energy state (Eq. 252) is:

$$E_{1,1,1} = \frac{(6.63 \times 10^{-27})^2 \text{ erg}^2 \text{ sec}^2}{8[39.95/(6.02 \times 10^{23})](1)^2 \text{ g-cm}^2}(1^2 + 1^2 + 1^2)$$
$$= (8.3 \times 10^{-32})(3) \text{ erg}$$
$$= 2.5 \times 10^{-31} \text{ erg.}$$

The energy of the higher state is

$$E_{1,1,2} = (8.3 \times 10^{-32})(1^2 + 1^2 + 2^2)$$
$$= (8.3 \times 10^{-32})(6)$$
$$= 5.0 \times 10^{-31} \text{ erg.}$$

The energy absorbed by the argon atom in changing its translational energy from the quantum state (1, 1, 1) to (1, 1, 2) is then

$$E_{1,1,2} - E_{1,1,1} = 2.5 \times 10^{-31} \text{ erg.}$$

b) The energies of the quantum states in the second case are

$$E_{1,1,1} = \frac{(6.63 \times 10^{-27})^2 \text{ erg}^2 \text{ sec}^2}{8(9.11 \times 10^{-28} \text{ g})(10^{-8})^2 \text{ cm}^2} \times 3$$
$$= (6.0 \times 10^{-11})(3)$$
$$= 1.8 \times 10^{-10} \text{ erg}$$

$$E_{1,1,2} = (6.0 \times 10^{-11})(6) = 3.6 \times 10^{-10} \text{ erg,}$$

and the energy absorbed by the electron is

$$E_{1,1,2} - E_{1,1,1} = 1.8 \times 10^{-10} \text{ erg.}$$

For the case of atoms and molecules in containers of usual size, the energy levels are very close together (Example 76a). When this energy gap is compared with kT, the thermal energy available from collisions with other molecules and with the walls of the container ($\sim 5 \times 10^{-14}$ ergs at room temperature), it is apparent that the energy of this gap is small indeed and may readily be obtained in collisions. In fact we cannot measure such minute differences in energy and if we determine the energy distribution of particles in a macroscopic box, we obtain a seemingly continuous plot; the energy steps due to quantization are simply too small to detect. As a result, the translation of particles in a macroscopic container may be treated equally well by classical mechanics. This is an example of the *correspondence principle*: when dealing with macroscopic systems quantum mechanics and classical mechanics must predict the same results.

On the other hand the microscopic situation in Example 76b has a significantly large separation of energies. We can detect the quantized energy levels in this case. For most cases that we need to consider, however, our molecules will be in macroscopic containers. We can therefore conveniently use the classical viewpoint in dealing with their translational energy.

Problems

13.1 Calculate the wavelength of an average helium gas molecule in the Goodyear blimp at (a) 0°C, (b) 25°C, (c) 100°C.

13.2 Calculate the probability of finding an electron with $n = 1$ in a wire of length L somewhere between $\frac{1}{4}L$ and $\frac{1}{2}L$ assuming it behaves as a particle in a one-dimensional box.

13.3 Calculate the wavelength of a 3-g bullet travelling at 300 m sec^{-1}.

13.4 Show that $\psi(x) = Ae^{iax} + Be^{-ibx}$ is a suitable wave function for the particle in a one-dimensional box.

13.5 The range of wavelengths which are effective in photosynthesis is from about 425 to 700 millimicrons. (a) Calculate the energy available from each quantum for radiation within this range. (b) To produce adenosine triphosphate (ATP), 7 kcal mole^{-1} are required. How many molecules of ATP could be produced using the energy of one photon of green light ($\lambda = 500$ millimicrons) if the process were 100% efficient?

13.6 Repeat the calculations of Example 75 for CO_2 at 10°K.

13.7 Vitamin A_2 has one more double bond in its conjugated system than does vitamin A_1 (see Example 73). Its structure is

```
        H
        C     CH3
   HC //  \  /
     |     C
     |     ||      H       H       H       H
   H2C     C       C       C       C       C
      \   / \    //  \   //  \   //  \   //  \
        C     C         C       C       C        CH2OH
       / \    H         |       H       |
   H3C    CH3          CH3             CH3
```

(a) What is the length of the one-dimensional container in vitamin A_2? (b) How many electrons are involved in its conjugated system? (c) Will its E_n for a given value of n be less than, greater than, or the same as that for vitamin A_1? (d) Calculate the total energy of the electrons in the conjugated portion of vitamin A_2.

13.8 (a) By analogy with the one-dimensional and three-dimensional cases, write the expressions for the wave functions and energies of a particle in a two-dimensional container. (b) The six conjugated electrons of benzene may be treated approximately as particles in a circular one-dimensional box or even more crudely as particles in a two-dimensional box of dimensions equal to the molecular dimensions. Calculate the energy of these electrons if we assume them to be particles in a square two-dimensional container with the length of a side equal to 3.25 Å. (c) Repeat part (b) for anthracene using the dimensions 3.5 × 10.0 Å.

13.9 The limit of visability for the human eye is a candle at fourteen miles. This amounts to 6 photons of light reaching the eye each second. How much energy and how much power does this correspond to if the wavelength is 550 millimicrons?

14 / Atoms

14.1. Introduction

A working picture of what an atom is and how it may be expected to behave is essential to any understanding of chemical systems. Since all molecules, no matter how complex, are made up of atoms, our approach to molecular structure and behavior will be largely based on atomic concepts. It is therefore necessary for us to develop as clear an understanding of atomic structure as possible before examining molecular structure.

In Chapters 1 and 13 we dealt with the translational motion and energy of atoms, treating them as point particles. As we have seen (Section 13.5), if these atoms are confined to a macroscopic container the quantum mechanical results are the same as those of classical mechanics.

We know, however, that atoms are not simply single particles. In 1911, E. Rutherford proposed a model for the atom based on an analogy with the solar system. Corresponding to the sun was a heavy inner core called the *nucleus*, made up of positively charged particles called *protons* and uncharged *neutrons*, while much lighter, negatively charged particles called *electrons* revolved about the nucleus in a manner analogous to the planets. This simple model has been superseded by a more complex quantum mechanical model, but the basic conception of a nuclear atom with extranuclear electrons remains.

The total energy of such an atom is the sum of the kinetic energy of the nucleus and that of the electrons, plus the potential energy due to the interaction of the positive nucleus with the negative electrons and of the electrons with each other. To this we must add nuclear energy due to forces and interactions within the nucleus. Since nuclear energy is of a magnitude far beyond the reach of ordinary chemical processes, we may consider the nucleus as a single uncomplicated particle that remains unchanged in chemical reactions.

14.2. Hydrogen-like Wave Functions: The Schrödinger Equation

We shall confine our attention initially to atomic and ionic cases which involve only two particles: the nucleus and a single electron. The hydrogen atom is one example; its nucleus, composed of a single proton ($Z = 1$), has associated with it a single outer electron. Other examples include the helium ion He^+ and the lithium ion Li^{+2}, whose nuclei are associated with only one electron. Atoms and ions of this type are described as *hydrogen-like*.

Let us now attempt to develop a quantum mechanical model for these hydrogen-like atoms and ions by determining the wave functions associated with them. We shall proceed as in the previous chapter: first develop the proper Schrödinger equation for the hydrogen-like atom, then solve this equation for the appropriate wave functions and their associated energies.

The Hamiltonian operator for this system may be constructed by the procedure outlined in Section 13.3. The kinetic energy of the nucleus and of the single electron may be written as $(p^2/2m)_N$ and $(p^2/2m)_E$ respectively. These are transformed into $-(h^2/8\pi^2 m_N)\nabla_N^2$ and $-(h^2/8\pi^2 m_E)\nabla_E^2$ respectively. The potential energy due to the interaction of the positive nucleus (charge = $+Ze$, where e is the charge on one proton, equal to 1.60×10^{-19} coulomb or 4.8×10^{10} esu) with the negative electron (charge = $-e$) is given by

$$V = (Ze)(-e)/r = -Ze^2/r,$$

where r is the instantaneous distance between the nucleus and the electron. The Hamiltonian operator is therefore

$$\mathscr{H} = -(h^2/8\pi^2 m_N)\nabla_N^2 - (h^2/8\pi^2 m_E)\nabla_E^2 - Ze^2/r. \tag{253}$$

The complexity of the differential equation which results when this operator is substituted into the Schrödinger equation (Eq. 242) may be reduced somewhat. Without going into the steps involved, the result is a separation of the Hamiltonian into two parts:

$$\mathscr{H} = \mathscr{H}_{\text{trans}} + \mathscr{H}_{\text{int}} \tag{254}$$

$$\mathscr{H}_{\text{trans}} = -[h^2/8\pi^2(m_N + m_E)]\nabla^2 \tag{255}$$

$$\mathscr{H}_{\text{int}} = -(h^2/8\pi^2\mu)\nabla^2 - Ze^2/r. \tag{256}$$

The first of these, $\mathscr{H}_{\text{trans}}$, is called the *translational Hamiltonian* operator since it is identical with the Hamiltonian operator for a particle of mass equal to that of the atom, moving with zero potential energy in a three-dimensional container. The second, $\mathscr{H}_{\text{int}}$, is called the *internal Hamiltonian* operator since it is a function only of the internal atomic coordinates, the relative positions of the nucleus and electron. Since the mass of the nucleus is so much larger than that of an electron, the *reduced mass* $\mu = m_N m_E/(m_N + m_E) \simeq m_N m_E/m_N = m_E$.

As a result of this separation of the Hamiltonian operator

$$\psi = \psi_{\text{trans}}\psi_{\text{int}}, \tag{257}$$

$$E = E_{\text{trans}} + E_{\text{int}}, \tag{258}$$

and the Schrödinger equation itself may be separated into two equations:

$$\mathscr{H}_{\text{trans}}\psi_{\text{trans}} = E_{\text{trans}}\psi_{\text{trans}} \tag{259}$$

$$\mathscr{H}_{\text{int}}\psi_{\text{int}} = E_{\text{int}}\psi_{\text{int}} \tag{260}$$

The total wave function of a hydrogen-like ion is thus the product of a wave function associated with the translation of the ion and a wave function associated with the relative position and movement of the electron and nucleus. The former is the translational wave function ψ_{trans} and the latter the internal wave function ψ_{int} (Eq. 257). The total energy of the ion is the sum of the translational energy and the internal energy (Eq. 258).

14.3 The Translational Wave Functions

First consider the Schrödinger equation for translation (obtained by substituting Eq. 255 into Eq. 259):

$$[-h^2/8\pi^2(m_N + m_E)]\nabla^2\psi_{\text{trans}} = E_{\text{trans}}\psi_{\text{trans}}. \tag{261}$$

The solutions for this differential equation have already been given in Section 13.5. The wave functions and the allowed energies are the same as those for a particle in a three-dimensional container:

$$\psi_{n_x\, n_y\, n_z} = (8/abc)^{1/2} \sin\left(\frac{n_x \pi}{a} x\right) \sin\left(\frac{n_y \pi}{b} y\right) \sin\left(\frac{n_z \pi}{c} z\right) \tag{249}$$

$$E_{n_x\, n_y\, n_z} = \frac{h^2}{8m}\left(\frac{n_x^2}{a^2} + \frac{n_y^2}{b^2} + \frac{n_z^2}{c^2}\right), \tag{250}$$

where the mass m is now the total mass $(m_N + m_E)$. In a box of ordinary size these relationships predict the same behavior as does classical mechanics. In that situation, they do not add significantly to our understanding of the atom.

14.4. Evaluation of the Internal Wave Functions: Separation of Variables

The internal Schrödinger equation is of considerably more interest since it concerns itself with internal variables. By using spherical coordinates[1] we may further separate this equation mathematically into three equations. The internal wave function is then a product of three functions. One is a function of the radial coordinate r, one is a function of the angular coordinate θ, and one is only a function of ϕ:

$$\psi_{\text{int}} = \psi(r)\psi(\phi)\psi(\theta). \tag{262}$$

[1] Spherical coordinates are illustrated in Fig. 14.1.

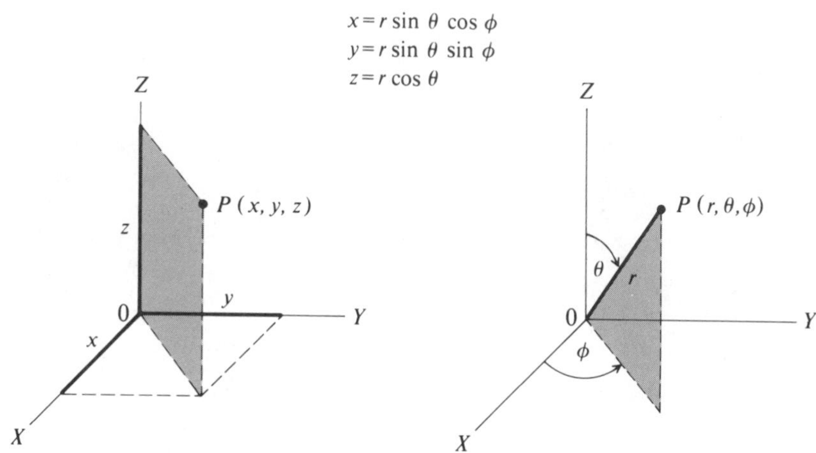

Fig. 14.1. Spherical coordinates. Adapted from M. W. Hanna, *Quantum Mechanics in Chemistry*, 2nd ed. (1969), W. A. Benjamin, Menlo Park, California.

The three equations into which the internal Schrödinger equation can be separated are

$$\frac{1}{\psi(\phi)}\frac{d^2\psi(\phi)}{d\phi^2} = -m^2 \tag{263}$$

$$\frac{1}{\sin\theta}\frac{d}{d\theta}\sin\theta\frac{d\psi(\theta)}{d\theta} - \frac{m^2}{\sin^2\theta}\psi(\theta) + g\psi(\theta) = 0 \tag{264}$$

$$\frac{1}{r^2}\frac{d}{dr}r^2\frac{d\psi(r)}{dr} + \left[\frac{8\mu\pi^2}{h^2}\left(E_{\text{int}} + \frac{Ze^2}{r}\right) - \frac{g}{r^2}\right]\psi(r) = 0. \tag{265}$$

Note that each of these three differential equations is a function of only one variable, either θ, ϕ, or r. Complicated as these equations may appear, they do have exact solutions.

14.5. The Angular Wave Functions: $\psi(\theta)$ and $\psi(\phi)$

It may be easily confirmed that the solutions of the ϕ equation (Eq. 263) are

$$\psi(\phi) = e^{im\phi}, \tag{266}$$

where m must be an integer ($m = 0, \pm 1, \pm 2, \ldots$).

The solutions of Eq. (264) are more difficult to obtain, but their general form is

$$\psi(\theta) = \frac{(-1)^{\ell}}{2^{\ell}\ell!}\sqrt{\frac{2\ell+1}{2}\frac{(\ell-|m|)!}{(\ell+|m|)!}}\sin^{|m|}\theta\,\frac{d^{\ell+|m|}\sin^{2\ell}\theta}{d(\cos\theta)^{\ell+|m|}}, \tag{267}$$

where ℓ is a positive integer ($\ell = 0, 1, 2, \ldots$). The *absolute value* of m (its magnitude ignoring its sign, written $|m|$) must be less than or equal to ℓ; thus $|m| \leq \ell$. The notation $m!$ is called "m factorial," and it means $m(m-1)(m-2)(m-3)\cdots(m-m+2)(m-m+1)$. Thus $5! = 5 \times 4 \times 3 \times 2 \times 1 = 120$ and $0! = 1$.

Example 77 What are all the possible ϕ and θ wave functions for $\ell = 1$?

Answer: When $\ell = 1$, then $|m| \leq 1$ and it follows that m may have the values 0, $+1$, or -1. First consider the case when $\ell = 1$ and $m = 0$. Then $\psi(\phi)$ is simply $e^{-0} = 1$. The θ function is

$$\psi(\theta) = \frac{(-1)^1}{2^1 1!}\sqrt{\frac{2+1}{2}\frac{(1-0)!}{(1+0)!}}(\sin^0\theta)\frac{d^{1+0}\sin^2\theta}{d(\cos\theta)^{1+0}}$$

$$= -\frac{1}{2}\sqrt{\frac{3}{2}\frac{1}{1}}(1)\frac{d\sin^2\theta}{d\cos\theta}$$

$$= -\frac{1}{2}\sqrt{\frac{3}{2}}\frac{d(1-\cos^2\theta)}{d\cos\theta} = -\frac{1}{2}\sqrt{\frac{3}{2}}(-2\cos\theta)$$

$$\psi(\theta) = \sqrt{\frac{3}{2}}\cos\theta.$$

For the case when $\ell = 1$ and $m = 1$, we find that $\psi(\phi) = e^{i\phi}$, and

$$\psi(\theta) = -\frac{1}{2}\sqrt{\frac{3}{2}\frac{(1-1)!}{(1+1)!}}(\sin^1\theta)\frac{d^{1+1}\sin^2\theta}{d(\cos\theta)^{1+1}}$$

$$= -\frac{1}{2}\sqrt{\frac{3}{4}}(\sin\theta)\frac{d^2\sin^2\theta}{d(\cos\theta)^2}$$

$$= -\frac{1}{2}\sqrt{\frac{3}{4}}(\sin\theta)\frac{d^2(1-\cos^2\theta)}{d(\cos\theta)^2}$$

$$\psi(\theta) = -\frac{1}{2}\sqrt{\frac{3}{4}}(\sin\theta)(-2) = \sqrt{\frac{3}{4}}\sin\theta.$$

For the third case, when $\ell = 1$ and $m = -1$, $\psi(\phi) = e^{-i\phi}$. Notice, however, that in $\psi(\theta)$ each time m appears it is $|m|$. Therefore $\psi(\theta)$ will be the same for both the $+1$ and the -1 values of m.

Table 14.1 gives the $\psi(\phi)$ and $\psi(\theta)$ functions for the cases of usual interest. These functions depend on the quantum numbers ℓ and m where $-\ell \leq m \leq \ell$.

Table 14.1. The Normalized Angular Wave Functions $\psi(\theta)$ and $\psi(\phi)$ for Hydrogen-like Atoms and Ions

Orbital	ℓ	m	$\psi(\phi)$	$\psi(\theta)$
s	0	0	$(2\pi)^{-1/2}$	$\sqrt{2}/2$
p_0	1	0	$(2\pi)^{-1/2}$	$(\sqrt{6}/2)\cos\theta$
$p_{\pm 1}$	1	± 1	$(2\pi)^{-1/2}e^{\pm i\phi}$	$(\sqrt{3}/2)\sin\theta$
d_0	2	0	$(2\pi)^{-1/2}$	$(\sqrt{10}/4)(3\cos^2\theta - 1)$
$d_{\pm 1}$	2	± 1	$(2\pi)^{-1/2}e^{\pm i\phi}$	$(\sqrt{15}/2)\sin\theta\cos\theta$
$d_{\pm 2}$	2	± 2	$(2\pi)^{-1/2}e^{\pm 2i\phi}$	$(\sqrt{15}/4)\sin^2\theta$

14.6. The Radial Wave Functions: $\psi(r)$

The solution of the $\psi(r)$ equation (Eq. 265) is again rather complicated, but it too has exact solutions:

$$\psi(r) = \left(\frac{Z}{0.53}\right)^{3/2} e^{-(rZ/0.53n)}\left(\frac{2rZ}{0.53n}\right)^{\ell} \sum_{j=0}^{k} a_j\left(\frac{2rz}{0.53n}\right)^{j}, \tag{268}$$

where the a_j are constants, ℓ is the same quantum number which appeared in $\psi(\theta)$, and k is a positive integer which relates ℓ to a second quantum number n by $k + \ell + 1 = n$. This requires that

$$\ell + 1 \leq n. \tag{269}$$

Example 78 Give all the possible *radial wave functions* $\psi(r)$ when $n = 2$ for $He^{+}(Z = 2)$.

Answer: When $n = 2$, then ℓ may have the values 0 or 1 (Eq. 269). Since $\psi(r)$ depends on both n and ℓ, there are two different radial wave functions.

First consider $n = 2$, $\ell = 0$. In this case k must be 1 and Eq. (268) gives

$$\begin{aligned}\psi(r) &= (2/0.53)^{3/2}e^{-(2r/1.06)}(4r/1.06)^0[a_0(4r/1.06)^0 + a_1(4r/1.06)^1].\\ &= (2/0.53)^{3/2}e^{-(2r/1.06)}[a_0 + (4r/1.06)a_1].\end{aligned}$$

For the second case, when $n = 2$ and $\ell = 1$, $k = 0$ and

$$\psi(r) = \left(\frac{2}{0.53}\right)^{3/2} e^{-(2r/1.06)}\left(\frac{4r}{1.06}\right)a_0.$$

Table 14.2 contains the radial wave functions $\psi(r)$ for values of the quantum number n from 1 through 3. The functions also depend on the quantum number ℓ ($\ell \leq n - 1$).

Table 14.2. Normalized Radial Wave Functions $\psi(r)$ for Hydrogen-like Atoms and Ions ($\rho = 2Zr/0.53n$)

n	ℓ	$\psi(r)$
1	0	$2(Z/0.53)^{3/2}e^{-\rho/2}$
2	0	$(1/\sqrt{8})(Z/0.53)^{3/2}(2-\rho)e^{-\rho/2}$
2	1	$(1/\sqrt{24})(Z/0.53)^{3/2}\rho e^{-\rho/2}$
3	0	$(1/\sqrt{243})(Z/0.53)^{3/2}(6-6\rho+\rho^2)e^{-\rho/2}$
3	1	$(1/\sqrt{486})(Z/0.53)^{3/2}(4\rho-\rho^2)e^{-\rho/2}$
3	2	$(1/\sqrt{2430})(Z/0.53)^{3/2}\rho^2 e^{-\rho/2}$

14.7. Electronic Energies

In the solution of the radial Schrödinger equation (Eq. 265) we obtain E_{int}, the energy of an electron with a particular associated wave function. This energy depends only on the quantum number n and is given by

$$E_n = -(2\pi^2\mu e^4 Z^2/h^2 n^2. \quad (270)$$

Example 79 Calculate the electronic energies of the $n = 1$ and $n = 2$ states for each of the following and determine the difference in these energies for each.
a) a hydrogen atom
b) a helium +1 ion

Answer: Let us first evaluate the constants of Eq. (270):

$$\begin{aligned} E_n &= -[2\pi^2(9.1 \times 10^{-28}\text{ g})(4.8 \times 10^{-10}\text{ esu})^4/(6.6 \times 10^{-27}\text{ erg-sec})^2](Z^2/n^2) \\ &= -(2.2 \times 10^{-11}\text{ erg})(Z^2/n^2) \end{aligned}$$

(One esu of charge equals 1 $\text{cm}^{3/2}$ $\text{g}^{1/2}$ sec^{-1} and 1 erg = 1 g-cm^2 sec^{-2}.)

a) For hydrogen, $Z = 1$:

$$\begin{aligned} E_1 &= -(2.2 \times 10^{-11}\text{ erg})(1^2/1^2) = -2.2 \times 10^{-11}\text{ erg} \\ E_2 &= -2.2 \times 10^{-11}(1^2/2^2) = -5.5 \times 10^{-12}\text{ erg} \\ \Delta E &= E_2 - E_1 = 1.65 \times 10^{-11}\text{ erg.} \end{aligned}$$

b) For helium, $Z = 2$:

$$E_1 = -(2.2 \times 10^{-11})(2^2/1^2) = -8.8 \times 10^{-11} \text{ erg}$$
$$E_2 = -2.2 \times 10^{-11}(2^2/2^2) = -2.2 \times 10^{-11} \text{ erg}$$
$$\Delta E = E_2 - E_1 = 6.6 \times 10^{-11} \text{ erg.}$$

For the hydrogen atom and ions with only one electron, all wave functions with a particular value of Z and n will have the same energy regardless of the value of ℓ and m. These energy levels are therefore degenerate.

Example 80 Determine the degeneracy of the energy levels corresponding to
a) $n = 1$
b) $n = 2$
c) $n = 3$

Answer: a) When $n = 1$, ℓ and m must be zero ($0 \leq \ell < n$, $|m| \leq \ell$). There is only one wave function with $n = 1$. This level is nondegenerate.
b) With $n = 2$, ℓ can have values 1 or 0. For $\ell = 0$, m must also be 0, but for $\ell = 1$, m may have the values -1, 0, or $+1$. Therefore the four wave functions with the following quantum numbers all have the same energy: $(n, \ell, m) = (2, 0, 0)$, $(2, 1, -1)$, $(2, 1, 0)$, $(2, 1, 1)$. The degeneracy is four.
c) With similar reasoning, wave functions with the following sets of quantum numbers all have the same energy: $(n, \ell, m) = (3, 0, 0)$, $(3, 1, -1)$, $(3, 1, 0)$, $(3, 1, 1)$, $(3, 2, -2)$, $(3, 2, -1)$, $(3, 2, 0)$, $(3, 2, 1)$, $(3, 2, 2)$. This energy level is therefore 9-fold degenerate.

14.8. The Total Atomic Electronic Wave Function; Orbitals

The total internal wave function for a hydrogen-like atom or ion will be the product of the radial and angular wave functions:

$$\psi_{n,\ell,m} = \psi(r)\psi(\theta)\psi(\phi). \tag{271}$$

We now have a set of wave functions to associate with an atom or ion which has a single electron. Each function with a unique set of quantum numbers n, ℓ, and m is called an *orbital*.

The quantum number n is the *principal quantum number*. As Eq. (270) indicates, it determines the internal energy. It only appears in the radial part of the wave function and will determine the *radial distribution* or size of the orbital (how far from the nucleus an electron with this particular associated wave function is likely to be found). The quantum number ℓ is called the *azimuthal quantum number*, and m is the *magnetic quantum number*. These determine the orbital's shape and orientation. A shorthand convention has been adopted to simplify somewhat our orbital notation. When $\ell = 0$, the orbital is called an *s orbital*; $\ell = 1$ for a *p orbital*; $\ell = 2$ for a *d orbital*. A coefficient gives the value of n and a subscript the value of m. We write

$$\begin{aligned} \psi_{n,\ell,m} = \psi_{1,0,0} &\equiv 1s_0 = 1s \\ \psi_{2,0,0} &\equiv 2s \\ \psi_{2,1,0} &\equiv 2p_0 \\ \psi_{2,1,1} &\equiv 2p_{+1} \\ \psi_{2,1,-1} &\equiv 2p_{-1} \\ &\vdots \\ \psi_{3,2,-2} &\equiv 3d_{-2}. \end{aligned}$$

In our subsequent discussion we will employ this notation.

14.9. The Appearance of Radial Wave Functions and Probabilities

What do these orbitals look like? It is difficult to picture both the radial and angular parts of the wave functions simultaneously. It is more convenient and clearer to plot these parts separately; we must remember, however, that the total wave function is the product of these parts.

Consider first the radial functions. Figure 14.2 contains $\psi(r)$ as a function of r (the distance of the electron from the nucleus) for the cases of usual interest. Notice that some of these functions are positive at some values of r and negative at others. The positiveness or negativeness is simply the mathematical sign of the function and has *nothing* to do with electrical charges. The importance of this sign will become clear in our discussion of molecules. There are also values of r where $\psi = 0$; these points are called *nodes*.

Since $\psi\psi^* \, d\tau$ may be interpreted as the probability of finding the particle (the electron) within a given infinitesimal volume element $d\tau$ (essentially the probability of finding it at a point; see Section 13.4), then $\psi(r)\psi^*(r)d\tau$ would be the probability of finding the electron with associated wave function $\psi(r)$ *at a particular point* at a distance r from the nucleus. Of more interest, however, is the probability of finding the electron *at any point* at a distance r from the nucleus; in other words, the probability of finding the electron on the surface of a sphere with radius r centered on the nucleus. For this reason we multiply the radial probability function by $4\pi r^2$, the area of a sphere with radius r.

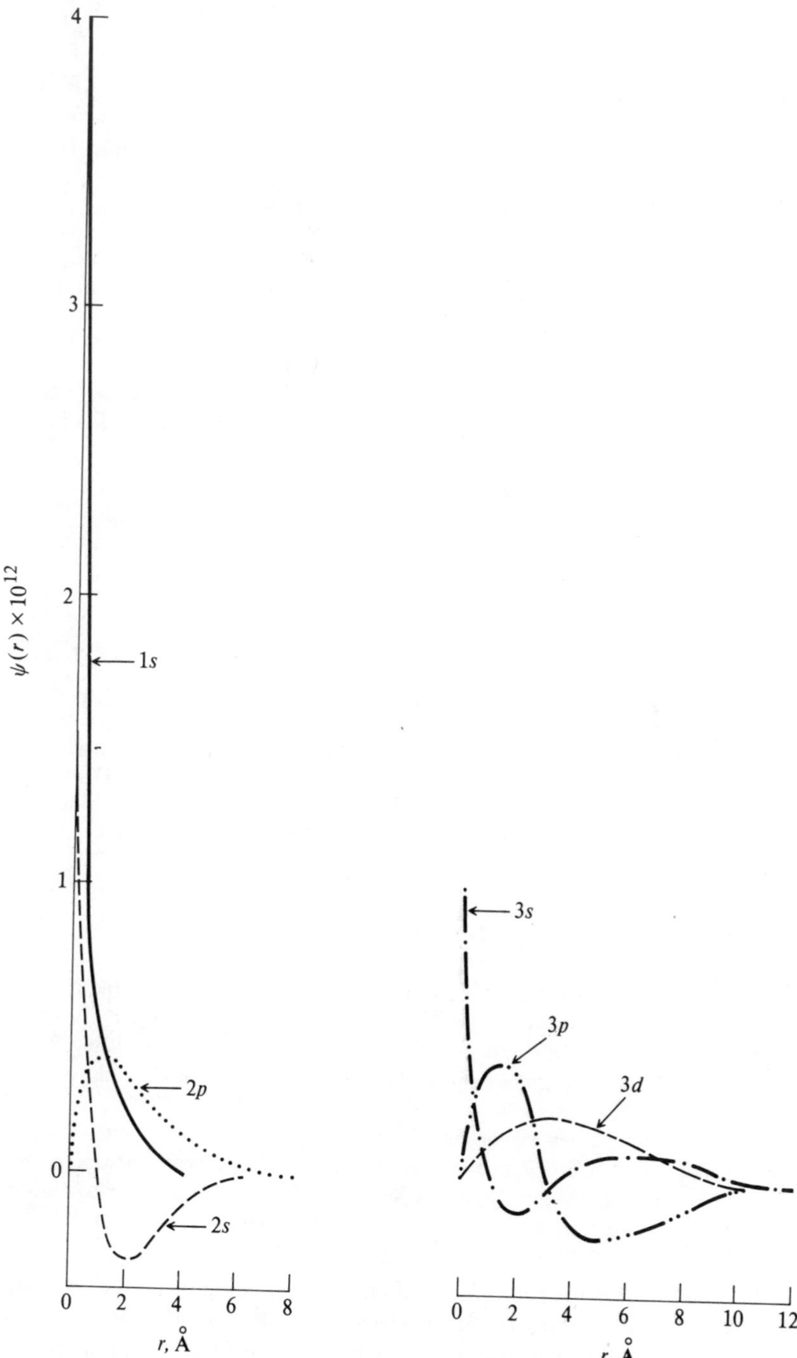

Fig. 14.2. The radial functions for hydrogen-like atoms and ions.

Example 81 The reason for multiplying the radial probability function by $4\pi r^2$ should become clearer if we consider an analogy.

a) How many postage stamps, each covering an area of 5.46 cm^2, may be glued to the surface of

1) a baseball with a diameter of 7.46 cm?

2) a basketball with a diameter of 23.95 cm?

b) Suppose that these balls are not completely covered and that the stamps are randomly distributed over their surfaces but with the stipulation that the probability of finding a stamp on any particular 5.46 cm^2 area of either ball is 0.1.

1) On which ball will there be more stamps?

2) For which ball is the probability of finding stamps greater?

Answer: a) The area A of the surface of a sphere of diameter d (radius r) is $4\pi r^2 = \pi d^2$.

1) $A = \pi(7.46)^2 = 174.8 \text{ cm}^2$

The number of stamps is equal to the area of the baseball divided by the area covered by a single stamp:

$$174.8/5.46 = 32.$$

2) For the basketball, $A = 1802.0 \text{ cm}^2$ and the number of stamps is

$$1802.0/5.46 = 330.$$

b) A probability of 0.1 means that if we check any ten 5.46-cm^2 areas at random we will, on the average, find one stamp. Thus one tenth of the surface of each ball is covered, $(0.1)(4\pi r^2)$. The number of stamps is then the area of the surface covered by stamps divided by the area occupied by a single stamp:

$$(0.1)(4\pi r^2)/5.46. \qquad (272)$$

1) With the larger radius r, the basketball obviously has more stamps than the baseball.

2) Note that the probability *per cm*^{2} is 0.1/5.46 in both cases. To obtain the total probability on the surface, we multiply this probability per unit area by the total area of the surface. Although the probability per unit area is the same for both balls, the actual stamp-finding probability is greater for the basketball, since it has a larger radius and thus a larger surface area.

Hence if we know the probability (of finding an electron, for instance) at a particular point at a distance r from the center of a sphere (that is, from the nucleus), we should multiply this by $4\pi r^2$ to obtain the total probability for all those points which are at a distance r from the center.

If $\psi(r)\psi^*(r)\,dr$ is the probability of an electron being present at a particular point whose distance from the nucleus is r, then $4\pi r^2\psi(r)\psi^*(r)\,dr$ is the probability of finding it somewhere on the surface of a sphere with radius r centered on the nucleus, the total probability, that is, of finding it at a

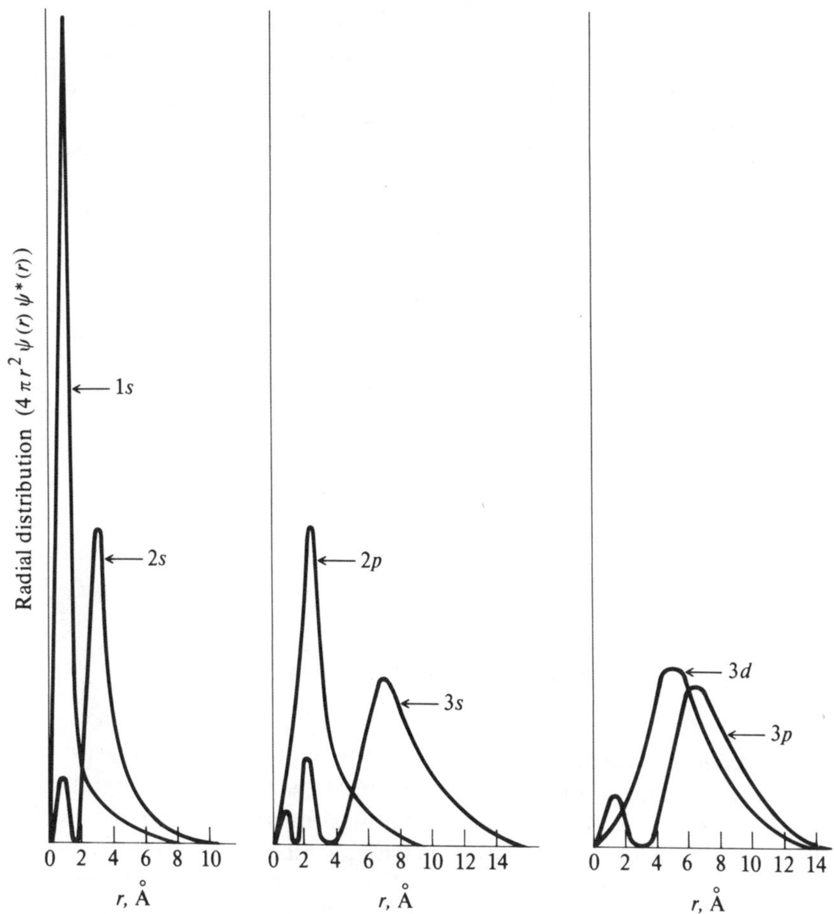

Fig. 14.3. Radial distribution for hydrogen-like atoms and ions.

distance r from the nucleus. This function is called the *radial distribution function.* Figure 14.3 contains plots of this function for various orbitals. The electron is more likely to be found at a distance r where the radial distribution function is large. Note that while this function is always positive there are nodes where the probability is zero; the electron spends no time at these points.

14.10. The Appearance of Angular Functions and Probabilities

Now let us consider the angular functions. For the 1*s*, 2*s*, and in fact all the *s* orbitals, the angular wave functions are constant; there is no angular dependence (see Table 14.1). Therefore the angular wave function $\psi(\theta)\psi(\phi)$

has the same positive value in all directions, regardless of the value of θ or ϕ. A plot of this function in three-dimensional space is a sphere with a positive value in all directions. Figure 14.4 is a two-dimensional representation of such a three-dimensional plot. The length of a line drawn from the origin (that is, the nucleus) to the surface is the magnitude of the angular wave function at that value of θ and ϕ.

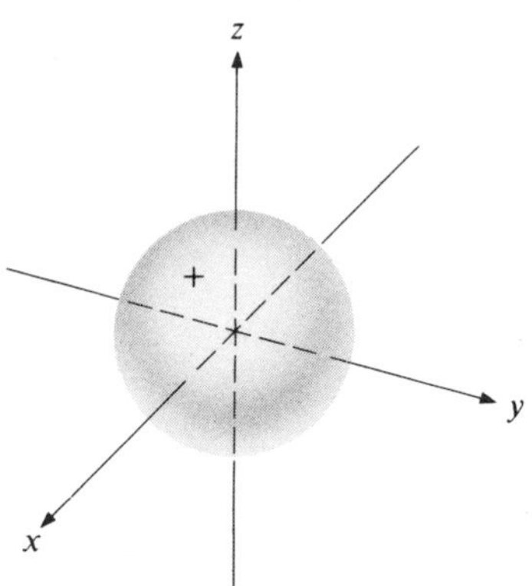

Fig. 14.4. The angular part of an *s* orbital.

The *angular probability* may be obtained by plotting $\psi(\theta)\psi(\phi)\psi^*(\theta)\psi^*(\phi)$ versus θ and ϕ. Since $\psi(\theta)\psi(\phi)$ for the *s* orbitals has the same magnitude at all values of θ and ϕ, the angular probability function is also a constant; the probability of finding the electron is the same at all angles. Such a probability plot of an *s* orbital is essentially identical to Fig. 14.4. The length of a line drawn from the origin to the surface of the sphere in this case is proportional to the probability of finding the electron at that particular angle. Obviously there is equal probability at all angles (that is, in all directions.) An electron in an *s* orbital (one which has an *s* wave function associated with it) spends most of its time at a distance from the origin determined by the radial distribution function for the particular *s* orbital. However, it divides its time equally among all the various positions located at this most likely distance from the origin. In other words it can be found with equal probability at all locations on the surface of a sphere of radius equal to the most probable radius.

The angular functions for all p_0 orbitals ($2p_0$, $3p_0$, etc.) are the same, as are those for all p_+ orbitals and those for all p_- orbitals. Consider first the p_0 angular function. Table 14.1 gives this function as $(3/2)^{1/2} \cos \theta$. There is no ϕ dependence since $e^{im\phi} = e^0 = 1$.

Example 82 Plot the angular part of a p_0 orbital. Do the same for the angular probability.

Answer: First determine the value of the angular function at various values of θ (Table 14.3). Then plot these values, remembering that along the $+z$-axis, $\theta = 0$; along the $-z$-axis, $\theta = 180°$; along $+x$, $\theta = 90°$; along $-x$, $\theta = 90°$ (see Fig. 14.1). For a given angle, the distance from the origin to the corresponding point plotted in Fig. 14.5(a) is equal to the magnitude of $\psi(\theta)\psi(\phi)$. These points are connected to form the plot of the function in the xz plane (Fig. 14.5b). The function is positive from a θ value of 0° to 90° and negative from 90° to 180°, this being indicated by positive signs in the *lobe* of the orbital extending along $+z$ and negative signs in the lobe along $-z$.

Since there is no ϕ dependence, the value of the function for a given θ will be the same for all values of ϕ (Fig. 14.5c).

Table 14.3. Values of $\psi(\theta)$ and $\psi^2(\theta)$ for p_0

θ, deg	$\psi(\theta) = (3/2)^{1/2} \cos\theta$	$\psi^2(\theta) = (3/2)\cos^2\theta$
0	1.23	1.50
15	1.20	1.44
30	1.08	1.17
45	0.88	.77
60	0.62	.38
75	0.32	.10
90	0.00	.00
105	−0.32	.10
120	−0.62	.38
135	−0.88	.77
150	−1.08	1.17
165	−1.20	1.44
180	−1.23	1.50

The angular probability is plotted similarly. The values of $\psi^2(\theta)$ appear in Table 14.3, and this function is plotted for the xz plane in Fig. 14.6(a). The distance from the origin to the curve is a measure of the relative probability at that angle. Figure 14.6(b) is a schematic representation of the three-dimensional angular probability.

Notice that there is a *nodal plane* in the p_0 function; the function has a zero value everywhere in the xy plane.

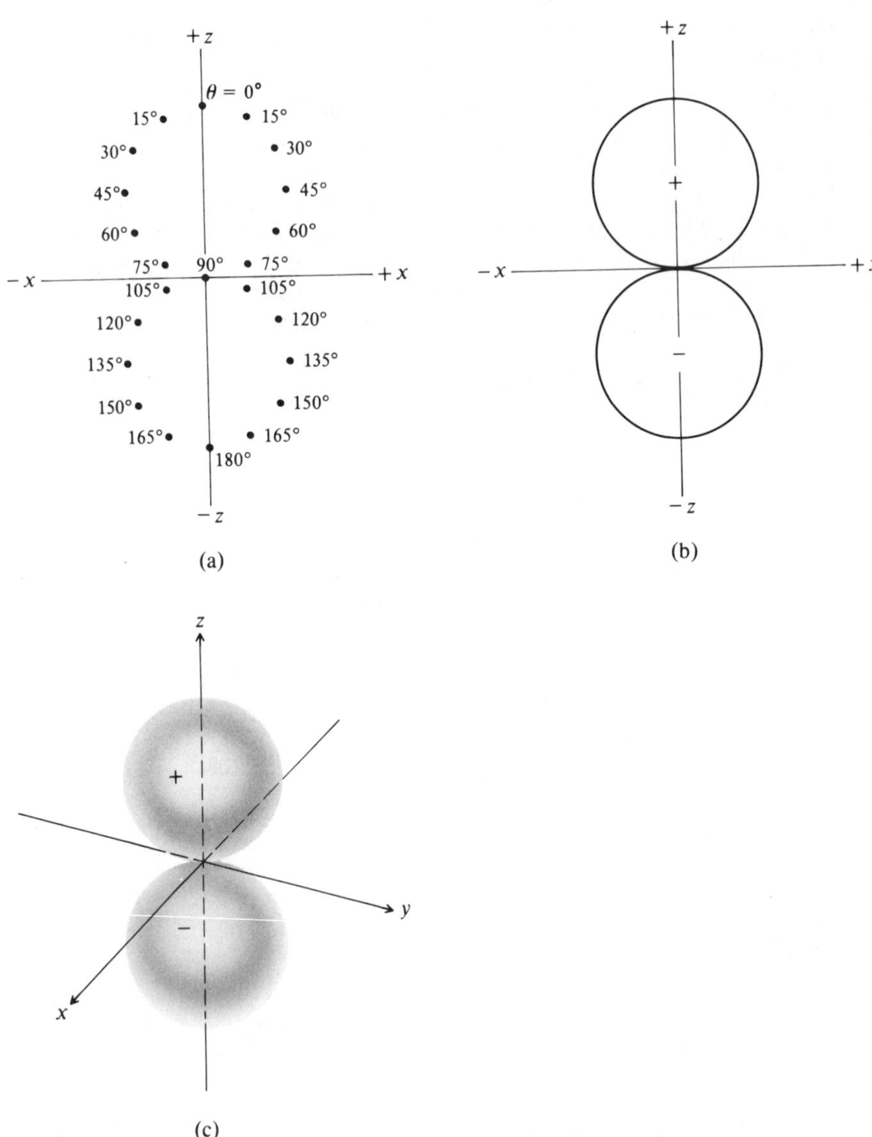

Fig. 14.5. (a) Values of the p_0 angular wave function for selected values of θ. (b) Plot of the angular wave function of a p_0 orbital in the xz plane. (c) Angular part of a p_0 orbital.

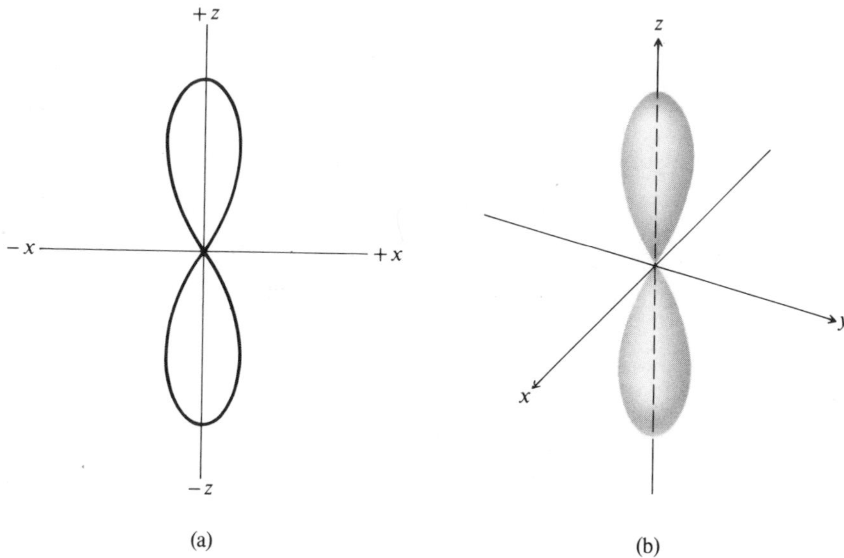

Fig. 14.6. (a) Plot of the angular probability for p_0 in the xz plane. (b) Angular probability of p_0 orbital. Adapted from G. W. Castellan, *Physical Chemistry*, 2nd ed. (1971), Addison-Wesley, Reading, Mass.

Let us now consider the p_+ and p_- functions. From Table 14.1,

$$p_+: \quad \psi(\theta)\psi(\phi) = (3/8\pi)^{1/2} \sin \theta e^{i\phi}$$

$$p_-: \quad \psi(\theta)\psi(\phi) = (3/8\pi)^{1/2} \sin \theta e^{-i\phi}.$$

The θ dependence of both these orbitals is the same, but the ϕ dependence is different. Because $\psi(\phi)$ is complex (it contains $i = \sqrt{-1}$) it is virtually impossible to plot these angular functions.

In the treatment of waves in classical mechanics, if there are two wave functions which are both solutions to the wave equation and are degenerate, then a *linear combination* of these wave functions is also a solution which possesses the same energy. This general property of waves holds in quantum mechanics as well. By taking linear combinations of the angular functions of p_+ and p_- (which are degenerate, that is they have the same energy, if n is the same for both) we obtain new wave functions which are correct solutions of the Schrödinger equation for hydrogen-like atoms and ions. These new functions prove more convenient and more useful for chemical applications than the p_+ and p_- functions from which they are formed.

The two linear combinations are

$$p_x = (p_+ + p_-)/2 \tag{273}$$

$$p_y = (p_+ - p_-)/2i. \tag{274}$$

The angular parts of these functions may be written as follows:

$$\begin{aligned} p_x\colon \quad \psi(\theta)\psi(\phi) &= \sqrt{3/8\pi}\,(\sin\theta e^{i\phi} + \sin\theta e^{-i\phi})/2 \\ &= \sqrt{3/8\pi}\,\sin\theta[(e^{i\phi} + e^{-i\phi})/2] \\ p_y\colon \quad \psi(\theta)\psi(\phi) &= \sqrt{3/8\pi}\,(\sin\theta e^{i\phi} - \sin\theta e^{-i\phi})/2i \\ &= \sqrt{3/8\pi}\,\sin\theta[(e^{i\phi} - e^{-i\phi})/2i]. \end{aligned}$$

The sine and the cosine may be defined in terms of complex exponentials:

$$\cos\phi = (e^{i\phi} + e^{-i\phi})/2$$

$$\sin\phi = (e^{i\phi} - e^{-i\phi})/2i,$$

so that our new p functions are

$$p_x\colon \quad \psi(\theta)\psi(\phi) = \sqrt{3/8\pi}\,\sin\theta\cos\phi \tag{275}$$

$$p_y\colon \quad \psi(\theta)\psi(\phi) = \sqrt{3/8\pi}\,\sin\theta\sin\phi. \tag{276}$$

These functions and their probability functions may now be plotted in a manner similar to the angular parts of p_0. These plots are given in Fig. 14.7. Notice that the positive lobe of p_x extends along the positive x-axis, whereas the positive lobe of p_y is along the positive y-axis. Recalling that the positive lobe of p_0 extends along $+z$, this orbital is often referred to as the p_z orbital so that we now have a set of three mutually perpendicular p orbitals p_x, p_y,

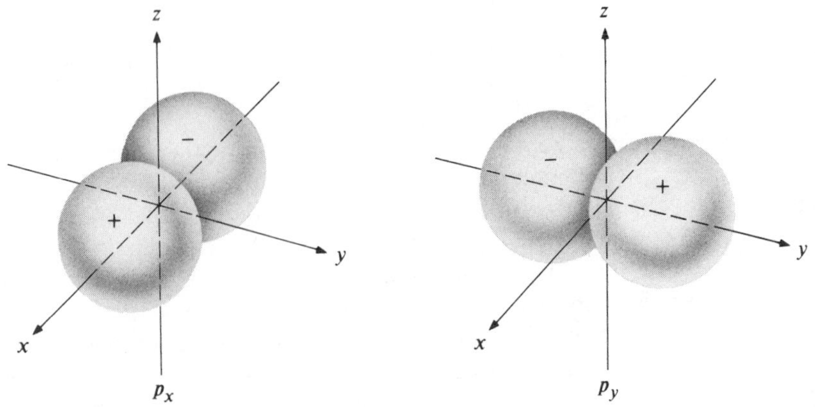

Fig. 14.7. Angular part of p_x and p_y orbitals.

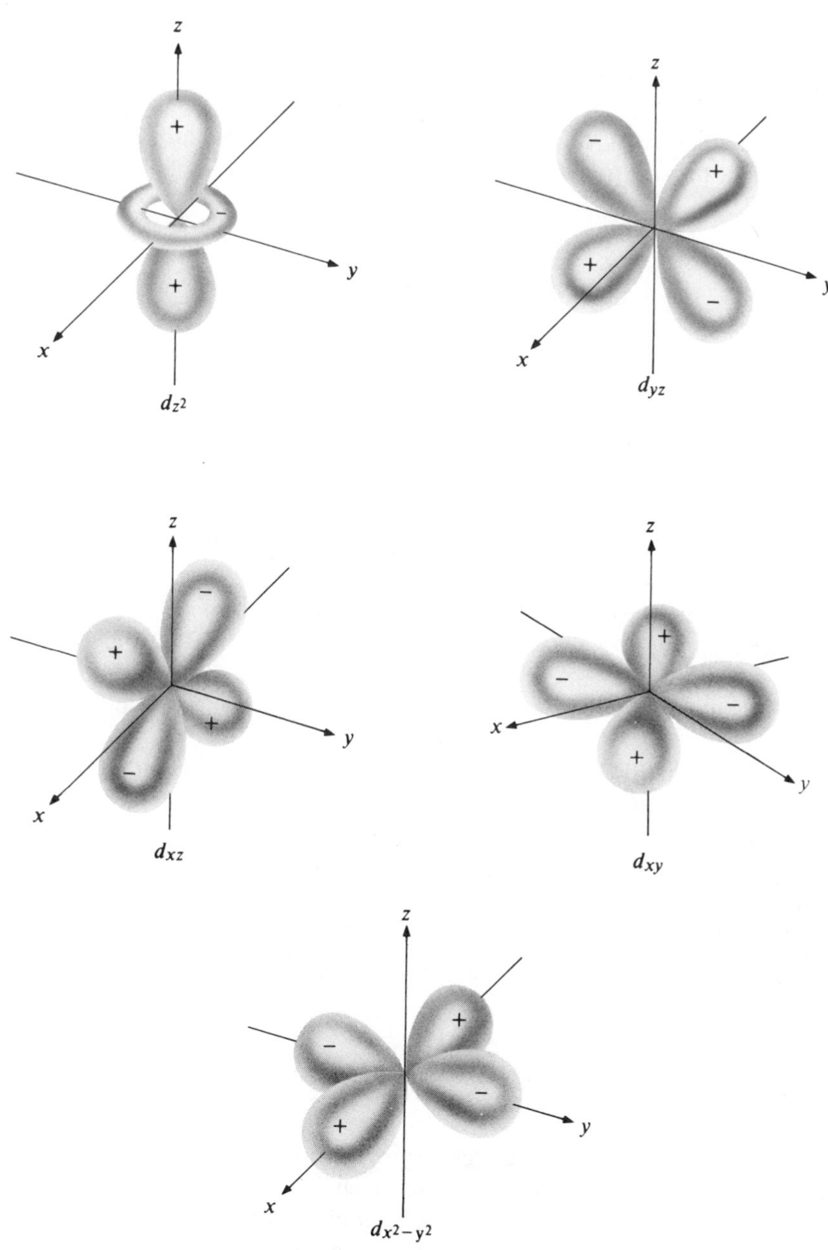

Fig. 14.8. The radial parts of the five d orbitals.

and p_z. The angular probability functions for p_x and p_y are similar to the plots of the functions (Fig. 14.7) except that they are more elongated along their axes and all lobes are positive. The relationship between these two types of plots can be seen by comparing the angular part of p_z (that is, p_0 in Fig. 14.5) with its probability (Fig. 14.6).

The angular parts of the d orbitals may be treated similarly, and the following set of *real* (not containing $\sqrt{-1}$) orbitals is obtained:

$$d_{z^2} = d_0: \quad \psi(\theta)\psi(\phi) = 3\cos^2\theta - 1 \tag{277}$$

$$d_{xz} = \frac{1}{2}(d_{+1} + d_{-1}): \quad \psi(\theta)\psi(\phi) = \sin\theta\cos\theta\cos\phi \tag{278}$$

$$d_{yz} = \frac{1}{2i}(d_{+1} - d_{-1}): \quad \psi(\theta)\psi(\phi) = \sin\theta\cos\theta\sin\phi \tag{279}$$

$$d_{xy} = \frac{1}{2i}(d_{+2} - d_{-2}): \quad \psi(\theta)\psi(\phi) = \sin^2\theta\sin 2\phi \tag{280}$$

$$d_{x^2-y^2} = \frac{1}{2}(d_{+2} + d_{-2}): \quad \psi(\theta)\psi(\phi) = \sin^2\theta\cos 2\phi. \tag{281}$$

Figure 14.8 contains schematic drawings of these five d functions. Plots of the corresponding angular probability functions bear a relationship to these drawings which is similar to the relationship between the p functions and their probabilities; that is, they are elongated along their axes and all lobes are positive.

We have now discussed in some detail the radial and angular functions and probabilities of the s, p, and d one-electron orbitals for hydrogen-like atoms and ions. It must always be remembered that the total atomic orbital is the product of the angular and radial parts; the total probability is a product of these two probabilities. Along with their hybrids (see Section 15.8), the s, p, and d orbitals will form the basis for our discussion of atoms with more than one electron, and for our development of the structure of molecules.

14.11. Atoms with More than One Electron

If we now consider the next simplest atom, helium, insurmountable problems confront us. Helium has a nucleus with a charge of $+2e$, balanced by two electrons. The Schrödinger equation will contain kinetic energy terms for each of these three particles. There will also be potential energy terms for the attraction between the nucleus and electron 1, and between the nucleus and electron 2, as well as for the repulsion between electrons 1 and 2. The resulting equation has no known exact solution. Neither are there exact solutions for any other atom with more than one electron.

Though exact solutions do not exist, we can obtain functions which are at least approximately correct—and quite useful. This may be done by systematically proposing reasonable functions, then testing them in turn to see which is best. The *variation principle* provides the basis for deciding between alternative wave functions: the function whose calculated energy is closest to the experimental value is the most nearly correct wave function.

In applying the variation principle, the energy corresponding to each proposed function is calculated using $\mathscr{H}\psi = E\psi$, employing the correct Hamiltonian operator.

For our purposes one of the most useful procedures for developing good wave functions for atoms with more than one electron is that proposed by J. C. Slater. The following discussion is based on his approach.

In an atom with a single electron, there is only one potential energy term, the direct attractive interaction of nucleus and electron. In a more complex atom an electron not only feels the attraction of the nucleus but also the repulsion of all the other electrons present. The net effect is that a given electron is attracted toward the nucleus with reduced force because of the repulsion of the other electrons. It is as if the electron is *shielded* from the full attractive charge of the nucleus by the presence of the other electrons; the *effective nuclear charge* is reduced from $+Ze$ to $+Z'e$. The effective atomic number Z' is defined to be

$$Z' = Z - \sigma, \tag{282}$$

where σ is the *shielding constant*. The value of σ (and hence of Z') will differ for different atoms and for different electrons within the same atom. Those electrons that are close-in, with few electrons between them and the nucleus, will have a smaller σ and larger Z' than will those farther out, where numerous shielding electrons intervene.

With this in mind, each electron in a complex atom is treated separately. The total Hamiltonian is approximated to be

$$\mathscr{H} = \sum_j \mathscr{H}_j, \tag{283}$$

the sum containing a one-electron Hamiltonian for each electron:

$$\mathscr{H}_j = -(h^2/8\pi^2\mu)\nabla_j^2 - \frac{Z'e^2}{r_j}. \tag{284}$$

Notice that the total Hamiltonian will have a kinetic energy term for each electron, together with a potential energy term containing the attractive interaction between the electron and the nucleus. There is no explicit term for the repulsion between electrons, but this is taken care of by the use of Z' in the attractive term. These one-electron Hamiltonians are now used in Schrödinger equations ($\mathscr{H}_j\psi_j = E_j\psi_j$).

The wave functions which result from this approach are the hydrogen-like atomic orbitals which we have previously derived, but with Z' *instead of* Z. Using the variation principle the best Z' may be obtained.

The total electronic wave function for an atom with n electrons is

$$\psi = \prod_{j=1}^{n} \psi_j; \tag{285}$$

that is, the product[2] of one-electron hydrogen-like wave functions, one function for each of the n electrons of the atom. The total atomic energy is

$$E = \sum_{j=1}^{n} E_j, \tag{286}$$

where E_j is given by Eq. (270) but with Z' substituted for Z.

Each electron is therefore considered to have associated with it one of the hydrogen-like wave functions with Z' instead of Z. This function has a set of quantum numbers n, ℓ and m. In the lowest energy state, the *ground state*, in which an atom will spend most of its time, the electrons will distribute themselves among the atomic orbitals in such a way that the total energy (Eq. 286) is a minimum. One way of doing this might be for all the electrons to be in the orbital of lowest energy, the $1s$ orbital. Nothing we have discussed to this point would forbid this state of affairs. It is not observed to occur, however. Instead the atom behaves as if a given orbital can accommodate a maximum of two electrons. This observation is summarized in the *Pauli exclusion principle* (Wolfgang Pauli, 1924). For our purposes this postulate may best be stated: no two electrons in the same atom can have the same set of four quantum numbers n, ℓ, m, and m_s.

We have discussed the first three of these quantum numbers, but not the fourth. This quantum number m_s is called the *spin quantum* number. In addition to its *spatial wave function* $\psi(r)\psi(\theta)\psi(\phi)$, an electron also possesses *by postulate* a *spin wave function*. There are only two possible spin wave functions for an electron: one with spin quantum number $m_s = \frac{1}{2}$, represented by α, and one with spin quantum number $m_s = -\frac{1}{2}$, represented by β.

---

[2] The $\prod$ notation for the product is similar to the sum notation Σ. The product $\phi_1\phi_2\phi_3\phi_4$ may be abbreviated $\prod_{j=1}^{4}\phi_j$, as the sum $\phi_1 + \phi_2 + \phi_3 + \phi_4$ is written $\sum_{j=1}^{4}\phi_j$.

One consequence of the exclusion principle is the requirement that there be no more than two electrons in the same orbital. Each spatial wave function, that is, each orbital, is characterized by a set of quantum numbers n, ℓ, and m. All electrons in a particular orbital will therefore have the same set of values for n, ℓ, and m. Thus each electron in the orbital must have a different value of m_s because the exclusion principle forbids all four of their quantum numbers matching. Since their spatial quantum numbers are identical, their spin quantum numbers must be different. With only two possible values for m_s, only two electrons may be in the same orbital. For example, the two electrons with $1s$ wave functions would have quantum numbers $n = 1$, $\ell = 0$, $m = 0$, $m_s = \frac{1}{2}$ and $n = 1$, $\ell = 0$, $m = 0$, $m_s = -\frac{1}{2}$, respectively.

To determine the orbitals associated with the electrons in the ground state of a complex atom we use the *aufbau principle*. Beginning with the orbital of lowest energy, we assign two electrons (one with $m_s = \frac{1}{2}$, the other with $m_s = -\frac{1}{2}$); we then assign two more to the orbital of next-lowest energy, and so forth, placing two electrons in each spatial orbital until we have assigned every electron to an appropriate orbital.

In the case of atoms and ions with only one electron, all wave functions with the same principal quantum number n are degenerate (Section 14.7). This is not true for atoms with more than one electron. Due to the shielding effect, Z' is different for orbitals with the same value of n but with different ℓ values. Thus the $2s$ orbital energy is less than that of the $2p$. Orbitals with the same values of n and ℓ but different m values are still degenerate. When shielding is taken into consideration, the energy order of the occupied ground state orbitals follows the approximate pattern:

$$1s < 2s < 2p < 3s < 3p < 4s < 3d < 4p < 5s < 4d \tag{287}$$

As we shall see, the actual order varies somewhat with atomic number.

Example 83 The assignment of electrons to orbitals for a complex atom is called its *configuration*. Give the ground state configuration for boron and for bromine.

Answer: Boron has five electrons. Two will be in the $1s$ orbital, two in $2s$, and the fifth in the next lowest orbital $2p$. This may be written $1s^2 2s^2 2p$.

Bromine has 35 electrons. Its ground state configuration is

$$1s^2 2s^2 2p^6 3s^2 3p^6 3d^{10} 4s^2 4p^5.$$

Notice that we have written the degenerate orbitals with a single symbol: $2p^6$ instead of $2p_x^2 2p_y^2 2p_z^2$ or $2p_0^2 2p_+^2 2p_-^2$. There are three degenerate p orbitals (thus 6 electrons when all are occupied) and five degenerate d orbitals (10 electrons).

Using the orbital energy order given in Eq. (287) and the aufbau principle, we can construct the chart of atomic configurations (periodic table) given in Table 14.4. Exceptions to this order do occur, the first problem being

Table 14.4. Electronic Configurations of the Atoms

Z	Element	1s	2s	2p	3s	3p	3d	4s	4p
1	H	1							
2	He	2							
3	Li	2	1						
4	Be	2	2						
5	B	2	2	1					
6	C	2	2	2					
7	N	2	2	3					
8	O	2	2	4					
9	F	2	2	5					
10	Ne	2	2	6					
11	Na	Neon shell			1				
12	Mg				2				
13	Al				2	1			
14	Si				2	2			
15	P				2	3			
16	S				2	4			
17	Cl				2	5			
18	Ar				2	6			
19	K	Argon shell						1	
20	Ca							2	
21	Sc						1	2	
22	Ti						2	2	
23	V						3	2	
24	Cr						5	1	
25	Mn						5	2	
26	Fe						6	2	
27	Co						7	2	
28	Ni						8	2	
29	Cu						10	1	
30	Zn						10	2	
31	Ga						10	2	1
32	Ge						10	2	2
33	As						10	2	3
34	Se						10	2	4
35	Br						10	2	5
36	Kr						10	2	6

Table 14.4 (*Continued*)

Z	Element		$4d$	$4f$	$5s$	$5p$	$5d$	$5f$	$6s$	$6p$
37	Rb	Krypton shell			1					
38	Sr				2					
39	Y		1		2					
40	Zr		2		2					
41	Nb		4		1					
42	Mo		5		1					
43	Tc		6		1					
44	Ru		7		1					
45	Rh		8		1					
46	Pd		10							
47	Ag		10		1					
48	Cd		10		2					
49	In		10		2	1				
50	Sn		10		2	2				
51	Sb		10		2	3				
52	Te		10		2	4				
53	I		10		2	5				
54	Xe		10		2	6				
55	Cs	Xenon shell							1	
56	Ba								2	
57	La						1		2	
58	Ce			2					2	
59	Pr			3					2	
60	Nd			4					2	
61	Pm			5					2	
62	Sm			6					2	
63	Eu			7					2	
64	Gd			7			1		2	
65	Tb			9					2	
66	Dy			10					2	
67	Ho			11					2	
68	Er			12					2	
69	Tm			13					2	
70	Yb			14					2	
71	Lu			14			1		2	
72	Hf			14			2		2	
73	Ta			14			3		2	
74	W			14			4		2	
75	Re			14			5		2	
76	Os			14			6		2	
77	Ir			14			9			
78	Pt			14			9		1	

Table 14.4 (*Continued*)

Z	Element		4f	5s	5p	5d	5f	6s	6p	6d	7s
79	Au	Xenon shell	14			10		1			
80	Hg		14			10		2			
81	Tl		14			10		2	1		
82	Pb		14			10		2	2		
83	Bi		14			10		2	3		
84	Po		14			10		2	4		
85	At		14			10		2	5		
86	Rn		14			10		2	6		
87	Fr	Radon shell									1
88	Ra										2
89	Ac									1	2
90	Th									2	2
91	Pa						2			1	2
92	U						3			1	2
93	Np						4			1	2
94	Pu						5			1	2
95	Am						6			1	2
96	Cm						7			1	2
97	Bk						8			1	2
98	Cf						9			1	2

encountered with chromium (Cr) with 24 electrons. Instead of continuing the sequence with a fourth electron in a 3*d* orbital, leaving two electrons in the 4*s* wave function, the atom finds it more favorable energetically to have five electrons with 3*d* functions and only one with 4*s*. This results from the fact that electrons repel each other. Placing two in the same orbital confines them to the same region of space; in different orbitals they move in different regions, are able to get further apart, and consequently reduce their mutual repulsion. The difference in the energy between 4*s* and 3*d* is greater for calcium, scandium, titanium and vanadium than this repulsion energy. For each of these the lowest energy state is obtained with two electrons in the 4*s* orbital. When chromium is reached, however, the energy difference between the 4*s* and 3*d* orbitals has been so reduced that the mutual repulsion of the two electrons in 4*s* is significant: the lowest energy state for Cr is found to be that with one electron in 4*s* and one in each of the five 3*d* orbitals. No orbital has more than one electron, and electron repulsion is at a minimum. Similar explanations account for the exceptional configurations of other heavier atoms.

14.12. Applications

The mathematical and physical model of the atom which results from the wave mechanical treatment may be used to rationalize, to organize, to explain, and even to predict the behavior of atomic systems. In many cases we are also able to calculate various observed properties of atoms.

One example of the use of this model is the relative size of atoms. Two effects are most important. For a given atom the radial distribution functions (see Fig. 14.3) clearly indicate that the distance of greatest probability is larger

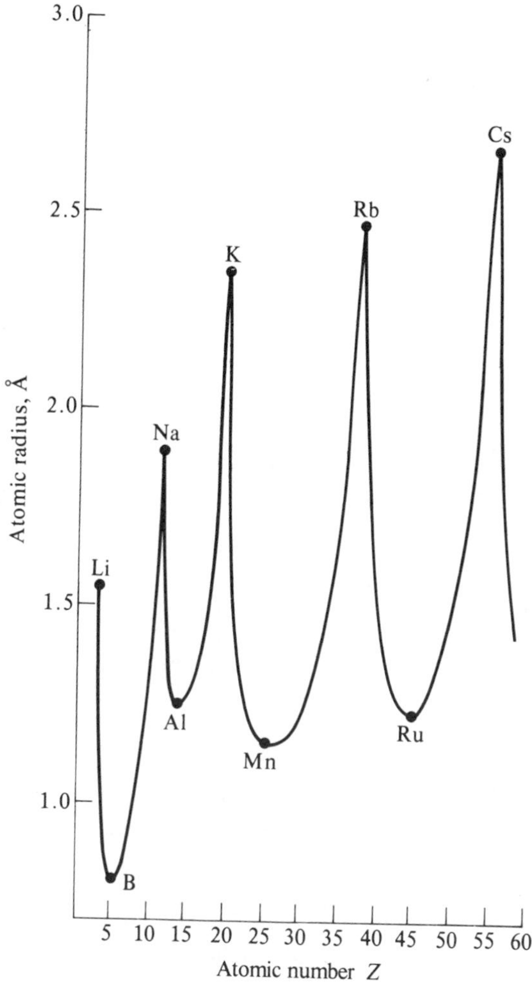

Fig. 14.9. Atomic sizes.

for $n = 2$ than for $n = 1$, for $n = 3$ than for $n = 2$, etc. The larger the principal quantum number n, the greater will be the electron's average distance from the nucleus. We may therefore expect that an atom with an occupied 2s orbital will be bigger (that is the most probable distance of its electrons from the nucleus will be larger) than one with only 1s electrons.

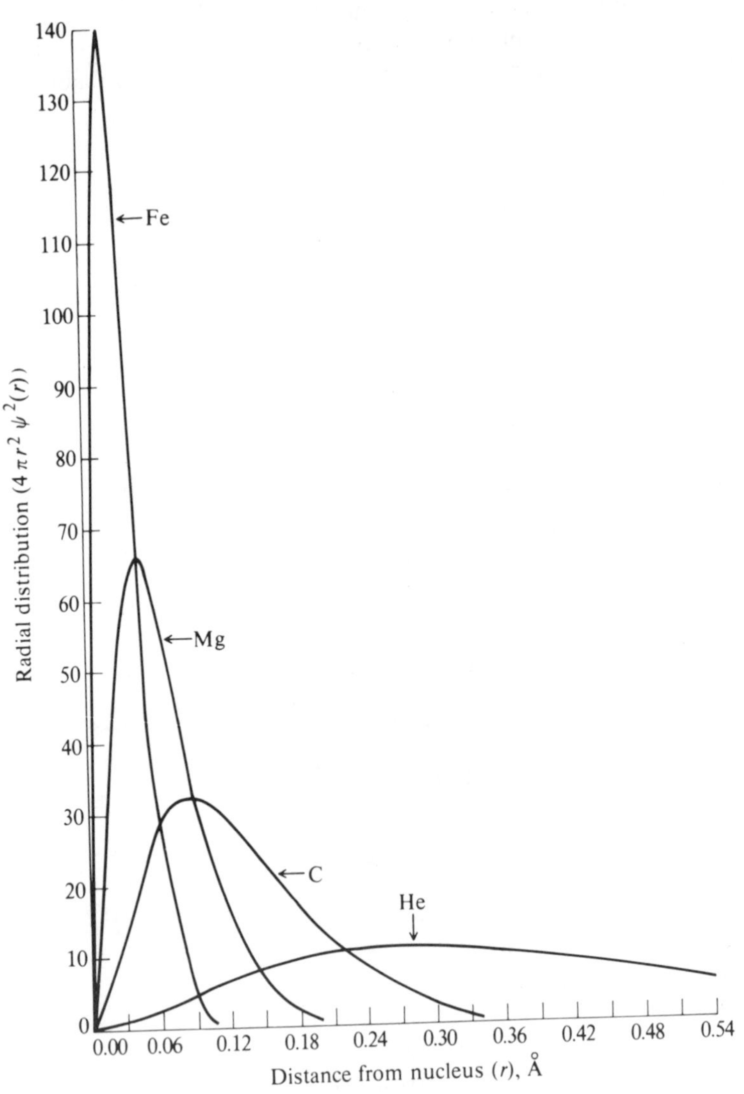

Fig. 14.10. Radial distribution of a 1s electron for selected atoms.

A look at Fig. 14.9 quickly reveals that this is not the only factor involved. The second major effect is the nuclear charge, or rather the effective nuclear charge. An electron in a $1s$ orbital of iron spends its time much closer to the nucleus than one in a $1s$ orbital of helium. This is apparent from the distribution function. The substitution of a larger Z into this function results in a maximum probability value nearer the nucleus. Figure 14.10 illustrates this for helium ($Z = 2$), carbon ($Z = 6$), magnesium ($Z = 12$), and iron ($Z = 26$). (Note that the maximum in the probability distribution curve not only occurs at a lower r value as Z increases but also increases in sharpness). A higher nuclear charge results in a greater attractive force between nucleus and electron, pulling the electron in closer. Similar statements may be made for electrons in other orbitals: an electron in a given orbital will be located nearer the nucleus in an atom of larger Z (and therefore larger Z').

On this basis we can explain the relative atomic sizes given in Fig. 14.9. The order Li < Na < K < Rb < Cs is due to the fact that in each succeeding case the outer electron is in an s orbital of higher n. An electron with a larger value of n is expected to be farther away from the nucleus *unless* the increased nuclear charge is great enough to produce the opposite effect. Although sodium has a Z value of 11 while that of lithium is only 3, this value is significantly decreased for the $3s$ electron (to a Z' of approximately 2.2) due to the shielding effect of electrons in lower orbitals.

In going from lithium ($Z = 3$) to berylium ($Z = 4$) the radius decreases. A second electron is added to the $2s$ orbital and the nuclear charge is increased by one. Except for the added nuclear charge, the radial distribution function will be the same. The additional nuclear charge pulls both electrons in the $2s$ orbital of berylium in closer, resulting in a smaller atomic radius. Although the presence of another $2s$ electron does result in a greater amount of shielding for a $2s$ electron in berylium than in lithium, this increase in σ is not large enough to offset the increase in Z.

Similar arguments may be used to explain other size relationships in the periodic chart.

Example 84 The first *ionization potential* of an atom or ion is the minimum amount of energy necessary to take one electron completely away from the ground state orbital of highest energy. For helium this would be a $1s$ electron; for boron a $2p$ electron; for chlorine a $3p$ electron, etc. This value is an experimental measure of the energy of an electron in this particular atomic environment. Rationalize the relative order of first ionization potentials given in Fig. 14.11.

Answer: The ionization potential is the change in energy in removing an electron from the atom, the difference between a final state in which the electron is

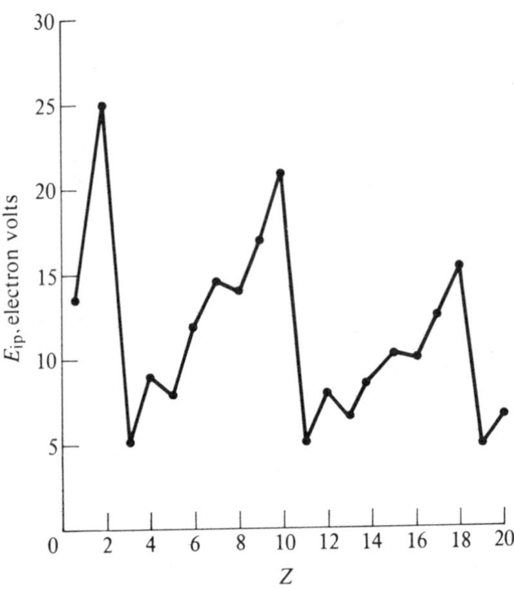

Fig. 14.11. First ionization potentials for the elements.

completely separated from the atom and is not attracted by the nucleus and an initial state in which it is bound by its attraction to the nucleus. Using Eq. (270):

$$\begin{aligned} E_{ip} &= E_{final} - E_{initial} \\ &= 0 - \left[-(2\mu\pi^2 Z'^2 e^4/h^2)\frac{1}{n^2}\right] \\ &= (2\mu\pi^2 Z'^2 e^4/h^2)\frac{1}{n^2}. \end{aligned} \tag{288}$$

It is clear, then, that variation of ionization potential among various atoms is primarily due to differences in Z' and in the quantum number n.

As examples of how the relative order can be rationalized, let us compare (1) neon and argon, (2) neon and sodium, and (3) sodium and magnesium.

1) Although argon has a larger nuclear charge than neon, much of the difference is cancelled due to shielding by the $3s$ electrons and the other five $3p$ electrons. As a result, Z' for the $3p$ electron which is removed from argon is not a great deal larger than Z' for the $2p$ electron removed from neon. Since $n = 3$ for argon and $n = 2$ for neon, Z'/n is actually less for argon than for neon and this leads to a greater ionization potential for neon.

2) In sodium it is the $3s$ electron which is removed in the first ionization. This electron is shielded from the nucleus by the $1s$, $2s$, and $2p$ electrons, all of which

are nearer the nucleus than it is. In neon a $2p$ electron is removed. It is shielded by $1s$, $2s$, and $2p$ electrons, but the $2s$ and the other $2p$ electrons are at approximately the same distance from the nucleus. The $3s$ electron of sodium is consequently much more effectively shielded than the $2p$ of neon. Even though Z for sodium is one unit larger than Z for neon, Z' for the sodium $3s$ electron is actually less than Z' for the neon $2p$ electron. Coupled with the fact that n for the electron removed from sodium is larger than that for the electron from neon, this results in a larger value of Z'/n for neon and a larger ionization potential.
3) The electron removed from both sodium and magnesium is a $3s$ electron. In addition to the electrons which shield the sodium $3s$ electron, the magnesium electron is shielded by another $3s$ electron. This additional shielding by an electron at the same distance from the nucleus ("in the same shell") is not great enough to offset the larger Z for magnesium; Z'/n is consequently larger for magnesium, and so also is the ionization potential.

Similar reasoning may be applied to the other ionization potentials.

Problems

14.1 Lithium, sodium, and potassium have similar chemical properties. Why?

14.2 How does a knowledge of atomic structure help us understand the differences in the properties of metals and nonmetals?

14.3 Explain the origin of the light in a Howard Johnson's neon sign.

14.4 (a) Calculate the ionization potential of the hydrogen atom in its ground state. (b) Calculate the ionization potential of the hydrogen atom with its electron in an excited $2s$ state.

14.5 Instead of having a positive proton and an electron as a hydrogen atom does, positronium is composed of an electron and a positive particle with the mass of an electron. Comment intelligently on the quantum mechanical treatment of positronium with emphasis on how such a treatment would be similar to and how it would differ from that of the hydrogen atom.

14.6 *Electron affinity* is the energy given off when a neutral atom acquires an electron, for instance

$$\mathrm{Li}(1s^2 2s) + e^- \longrightarrow \mathrm{Li}^{-1}(1s^2 2s^2).$$

Use your knowledge of atomic structure to explain the following sequence of electron affinity values:

Atom	Li	Be	B	C	N	O	F
Electron affinity, eV	0.5	−0.6	0.2	1.3	−0.1	1.5	3.5

14.7 The ground state configuration of sodium is $1s^2 2s^2 2p^6 3s^1$. The first ionization potential, discussed in Example 84, is the energy for the process

$$\mathrm{Na}(1s^2 2s^2 2p^6 3s^1) \longrightarrow \mathrm{Na}^+(1s^2 2s^2 2p^6) + e^-.$$

Similarly, the second ionization potential is the energy required for

$$\mathrm{Na}^{+}(1s^2 2s^2 2p^6) \longrightarrow \mathrm{Na}^{+2}(1s^2 2s^2 2p^5) + \mathrm{e}^-.$$

The third and fourth ionization potentials correspond to

$$\mathrm{Na}^{+2}(1s^2 2s^2 2p^5) \longrightarrow \mathrm{Na}^{+3}(1s^2 2s^2 2p^4) + \mathrm{e}^-$$

and

$$\mathrm{Na}^{+3}(1s^2 2s^2 2p^4) \longrightarrow \mathrm{Na}^{+4}(1s^2 2s^2 2p^3) + \mathrm{e}^-$$

respectively.
(a) The first, second, third, and fourth ionization potentials for sodium are respectively 119, 1090, 1650, and 2280 kcal mole^{-1}. Explain the relative magnitudes of these values. (b) Calculate Z' for the $3s$ electron of Na, for the $2p$ electrons of Na^+, for the $2p$ electrons of Na^{+2}, and for the $2p$ electrons of Na^{+3}. (c) A $2p$ electron is shielded from the nuclear charge by the $1s$ and $2s$ electrons and by the other $2p$ electrons. Calculate the amount of shielding of a sodium $2p$ electron resulting from the presence of *each* of the other $2p$ electrons.

14.8 (a) Determine the value of r at which the radial distribution function for the $1s$ wave function of the hydrogen atom has its maximum value. (b) Repeat part (a) for the $2s$ and $2p$ radial probability functions. (c) Calculate the probability of an electron in a $1s$ orbital being located at a distance from the nucleus less than that of the maximum value determined in part (a). (d) Calculate the probability of an electron in a $1s$ orbital being located at a distance from the nucleus greater than that of the maximum value of part (a).

15 / Molecular Structure

15.1. Introduction

We have now reached a most important stage in our efforts to obtain a better understanding of what molecules are and how they react.

A molecule is understood to be a stable grouping of atoms, these atoms having a definite[1] arrangement in space relative to each other. A *chemical* or *molecular bond* is the combination of forces that holds two atoms in fixed juxtaposition within a molecule. Chemical reactions involve the formation and the breaking of bonds. Chemical reactivity depends on the strength of bonds and on their spatial arrangement. Indeed, these bonds are the ultimate concern of virtually all work in chemistry.

Any significantly useful theory must explain why bonds form, why some bonds are stronger and some more reactive than others, and why a particular spatial arrangement is the most stable. We propose to explore these and related questions with the aid of the insights and methods of quantum chemistry.

At two extremes of molecular bonding are *ionic bonds* and *covalent bonds*. The stability of the former is a result of electrostatic attraction between ions. That of the latter is a result of the equal sharing of electrons in common *molecular orbitals*.

15.2. Ionic Bonds

An ion with n more electrons than protons has a charge of $-ne$, where e is the charge on a proton. An ion with m fewer electrons than protons possesses a charge of $+me$. When oppositely charged ions approach, there is a force of attraction which pulls them closer together. The attraction potential energy of the two ions as a function of the distance between their nuclei is

$$V_A = -mne^2/r. \tag{289}$$

At large distances of separation, V_A is essentially zero; as the ions approach (as r decreases), V_A becomes increasingly negative. A more negative energy therefore corresponds to a more stable arrangement of the ions. In order to minimize their energy the ions might be expected to move together until $r = 0$. This does not occur, however. As the ions approach more closely, the electrons of one ion begin to repel those of the other. Also the nuclei of the two ions repel. There is therefore a repulsion potential energy term. M. Born and J. E. Mayer have observed that this repulsion is exponential in form

$$V_R = be^{-r/c}, \tag{290}$$

[1] The effects of deviations from constant equilibrium relative positions will be explored in Chapter 16.

where b and c are constants which must be determined experimentally for each ionic bond. The total potential function is thus

$$V = V_A + V_R = -mne^2/r + be^{-r/c}. \tag{291}$$

As r decreases, V_A decreases but V_R increases. Figure 15.1 is a diagram of the net result in a typical case (Na^+Cl^-). Note that the repulsion term becomes important only at low r, where it quickly overcomes the attraction term. At the equilibrium distance r_e the potential energy has its minimum value. The stability of the ionic molecule is greatest at this internuclear separation.

While this purely ionic approach clearly accounts for the formation of a stable molecular bond, it has the major drawback of requiring the ionization of the atoms involved in the bond formation. Before two neutral atoms can form an ionic bond, one of them must donate one or more electrons to the

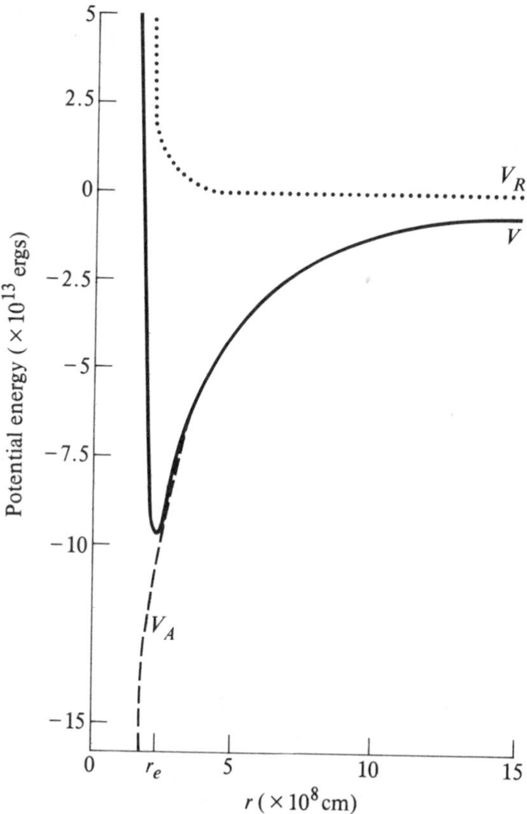

Figure 15.1

other. For substances such as crystalline NaCl the ionic bond description nevertheless forms the basis for a molecular picture which is quite adequate (see Section 17.7). Sodium has a low ionization potential and a low *electron affinity* (a value which quantifies the ability of a neutral atom to attract and hold an extra electron). Chlorine, on the other hand, has a high ionization potential and a high electron affinity. Consequently sodium readily donates an electron to chlorine. The resulting ions maintain their ionic identities in close proximity, the opposite charges attract, and ionic bonds are formed.

For molecular bonds between atoms whose ionization potentials and electron affinities are more nearly equal, the ionic model is not adequate. Two atoms must be considerably unlike if one is to completely take over an electron from the other. Some ionic character is found to persist in all bonds, even those between like atoms such as in the molecules H_2 and Cl_2. But the more dissimilar the ionization potentials and electron affinities, the more ionic will be the nature of the resulting bond. At the extreme, the bond may indeed be essentially totally ionic, as is the case with crystalline NaCl.

A useful concept in dealing with the ionic character of a bond (and its *polarity*; see Section 15.6) is the *electronegativity* X of the atoms involved. This is a relative measure of the ability of an atom to attract and hold electrons. R. S. Mulliken arrived at values for this parameter by using the following formula:

$$X = (I + A)/5.6, \tag{292}$$

Table 15.1 Electronegativities for Selected Atoms

F	4.0	B	2.0
O	3.5	Si	1.8
N	3.0	Al	1.5
Cl	3.0	Be	1.5
Br	2.8	Mg	1.2
I	2.5	Ca	1.0
S	2.5	Li	1.0
C	2.5	Na	0.9
H	2.2	K	0.8
P	2.1		

Linus Pauling, *The Nature of the Chemical Bond*, 3rd ed., Ithaca, N.Y.: Cornell University Press, 1960.

where I is the ionization potential and A the electron affinity. Table 15.1 is a compilation of electronegativity values determined by Linus Pauling using a somewhat different approach. The values determined by the two methods are in good agreement, however. Atoms with widely different values of X may be expected to form bonds with a great deal of ionic character. Those with more similar X values will be more covalent in nature.

Example 85 List the following bonds in order of increasing ionic character:

C—C	S—H	Na—F
C—H	O—H	Na—Cl
C—Cl	N—H	Na—Br

Answer: Using Table 15.1 we may calculate the magnitude of the electronegativity difference for each bond:

C—C	0.0	N—H	0.8
C—H	0.3	Na—F	3.1
C—Cl	0.5	Na—Cl	2.1
S—H	0.3	Na—Br	1.9
O—H	1.3		

Thus in order of increasing ionic character:
C—C < C—H = S—H < C—Cl < N—H < O—H < Na—Br < Na—Cl < Na—F.

15.3. Wave Mechanical Treatment of Molecules: Introduction

In dealing with the ionic model we did not need to use quantum mechanics directly. The development of a more generally applicable theory of chemical bonds will now be undertaken.

As usual let us begin with the most simple case, the hydrogen molecule H_2. This molecule consists of two protons and two electrons. Figure 15.2, a schematic representation of this molecule, defines the various important distances. The Schrödinger equation for this molecule contains four terms for the translational kinetic motion of the four particles. It also has two repulsion terms: one for the repulsion between the nuclei, the other for the repulsion between the two electrons. There are also four attraction terms: between electron 1 and nucleus A, electron 1 and nucleus B, electron 2 and nucleus A, and electron 2 and nucleus B.

This Schrödinger equation has no exact solutions. We must again look for wave functions which are approximate solutions to the wave equation

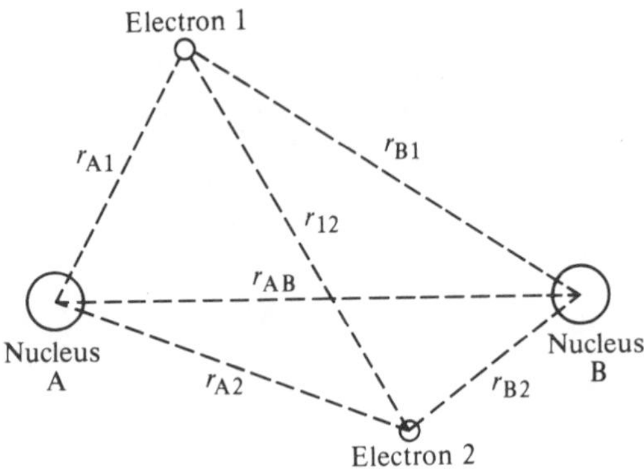

Fig. 15.2. Schematic representation of the hydrogen molecule.

and which are of a form which will be helpful in our quest for a better understanding of molecules.

The first simplification we shall employ is the Born-Oppenheimer approximation, which allows us to treat the electrons separately from the nuclei. This amounts to holding r_{AB} constant in the Hamiltonian operator and looking for wave functions which are associated with the electrons only. The energies associated with these electronic functions will depend on r_{AB} and we must deal with that fact, but the Born-Oppenheimer approximation allows us to tackle one problem at a time. Chapter 16 will be devoted to an examination of other consequences of the motion of the nuclei.

15.4. The LCAO Method; The H_2 Molecule

There are numerous approaches to the problem of developing approximate electronic wave functions for molecules. We shall employ the one most useful for our purposes. Its chief advantage is that it is based on atomic orbitals and therefore provides a pictorially more accessible model. The justification for most of the approximations made in this approach is simple: the results agree reasonably well with physical observables—and they are quite useful.

We first *assume* that the total wave function ψ is a product of one-electron wave functions ψ_j, there being a ψ_j for each of the n electrons of the molecule:

$$\psi = \prod_{j=1}^{n} \psi_j. \tag{293}$$

Each of these one-electron wave functions ψ_j is called a *molecular orbital* (MO).

For the hydrogen molecule H_2 with two electrons,

$$\psi = \psi_1 \psi_2 .$$

In other, more quantum mechanical, words, our assumption is that we can write the total electronic Hamiltonian $\mathscr{H}$ as a sum of one-electron Hamiltonians $\mathscr{H}_j$, one for each electron:

$$\mathscr{H} = \sum_{j=1}^{n} \mathscr{H}_j . \tag{294}$$

For the hydrogen molecule,

$$\mathscr{H} = \mathscr{H}_1 + \mathscr{H}_2 .$$

The Schrödinger equation may then be separated into n one-electron equations:

$$\mathscr{H}_j \psi_j = E_j \psi_j , \tag{295}$$

and the total wave function is given by Eq. (293).

The energy associated with the total wave function is the sum of energies associated with each one-electron function:

$$E = \sum_{j=1}^{n} E_j . \tag{296}$$

As our next approximation we write each molecular orbital ψ_j as a *linear combination of atomic orbitals* (LCAO). This linear combination will include one atomic orbital from each atom.

For H_2 the linear combination for the molecular orbital for electron 1 may be written

$$\psi_1 = C_{A_1} \chi_A + C_{B_1} \chi_B . \tag{297a}$$

The χ functions are 1s wave functions on hydrogen atoms A and B in this case. The values of the coefficients C_A and C_B are determined by the variation method (Section 14.11). In other words we determine what values of these coefficients give the lowest answer for the energy. The resulting function will then correspond to the best wave function *of this type*.

Similarly the linear combination for the molecular orbital for the second electron of H_2 is

$$\psi_2 = C_{A_2} \chi_A + C_{B_2} \chi_B . \tag{297b}$$

Using the variation method and solving for the best values of C_A and C_B yields the following possible functions for Eqs. (297a) and (297b):

$$\psi_+ = (1/\sqrt{2(1+S)})(\chi_A + \chi_B) \tag{298}$$

$$\psi_- = (1/\sqrt{2(1-S)})(\chi_A - \chi_B). \tag{299}$$

S is the *overlap integral* (the integration is carried out over all space):

$$S = \int \chi_A^* \chi_B \, d\tau. \tag{300}$$

The energies which correspond to these two wave functions are respectively

$$E_+ = (H_{AA} + H_{AB})/(1 + S) \tag{301}$$

$$E_- = (H_{AA} - H_{AB})/(1 - S). \tag{302}$$

In these expressions H_{AA} is the *coulomb integral* and H_{AB} the *exchange integral*:

$$H_{AA} = \int \chi_A^* \mathscr{H}_1 \chi_A \, d\tau \tag{303}$$

$$H_{AB} = \int \chi_A^* \mathscr{H}_1 \chi_B \, d\tau. \tag{304}$$

Of course these integrals must be evaluated to obtain a numerical result for a given molecular orbital. Both H_{AA} and H_{AB} will generally be negative, however. This means that ψ_+ will be lower in energy (more stable) than ψ_-. These integrals will also depend on r_{AB}; hence their evaluation at various internuclear distances will give different energies.

For electron 1 there are thus two possible LCAO molecular orbitals

$$\psi_+ = C_+(\chi_A + \chi_B) \tag{298b}$$

$$\psi_- = C_-(\chi_A - \chi_B), \tag{299b}$$

where C_+ and C_- are constants at a given internuclear separation. When the energies of these two wave functions are calculated and plotted as a function of the internuclear separation r_{AB}, the curves illustrated in Fig. 15.3 are obtained. $E = 0$ is assigned to the state in which there are two hydrogen atoms at such a distance apart that there is no interaction between them, that is at large r_{AB} values. A minimum in the E_+ curve occurs at the equilibrium internuclear distance r_e. Thus the molecular orbital ψ_+ corresponds to a stable or bonding situation. At r_e the atoms involved in this MO are at a lower energy than they would be as separated atoms. For the orbital ψ_-, on the other hand, there is no minimum in the energy; E_- is larger at all values of r than the energy

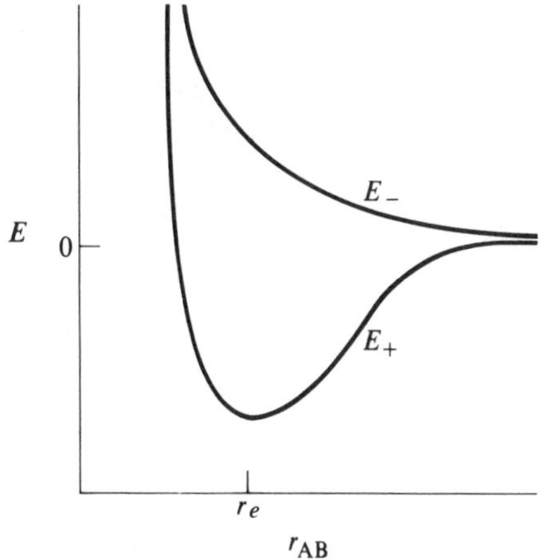

Figure 15.3

of the two separate 1*s* orbitals on two separated hydrogen atoms (χ_A and χ_B). Here ψ_+ is called a *bonding orbital*, whereas ψ_- is an *antibonding orbital.*

For the second electron of H_2, the bonding and antibonding molecular orbitals involving 1*s* atomic orbitals are identical to those for the first electron. In the ground state of the hydrogen molecule the wave function for each electron is thus ψ_+, and the total wave function of H_2 is

$$\psi = \psi_{+_1} \psi_{+_2} = \psi_+^2 .$$

How do we visualize these molecular orbitals? Figures 15.4 and 15.5 represent respectively the formation of the ψ_+ and ψ_- molecular orbitals from the atomic orbitals on two separate hydrogen atoms. For the bonding orbital ψ_+, the signs of the two atomic orbitals are both positive (Eq. 298). As the two atoms approach each other their electron wave functions begin to *overlap*, i.e. there is an area in space (the cross-hatched area of Fig. 15.4b) in which both functions have significant values. Here the functions are both positive. Hence the resulting LCAO molecular orbital is enhanced in this region; it may be pictured as in Fig. 15.4(c). This MO is cylindrically symmetric about a line between the two nuclei.

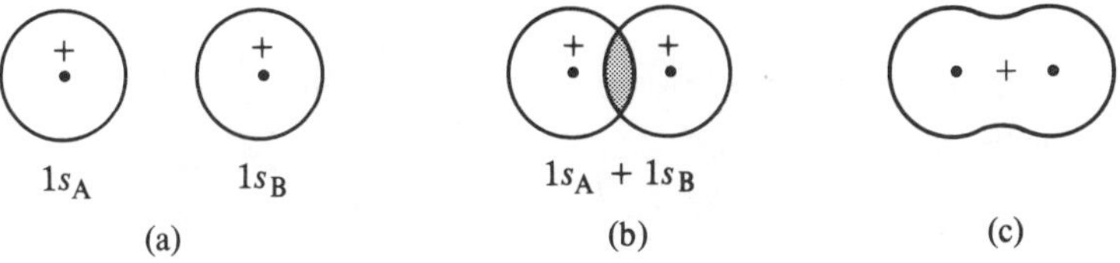

Fig. 15.4. The formation of the bonding ψ_+ molecular orbital.

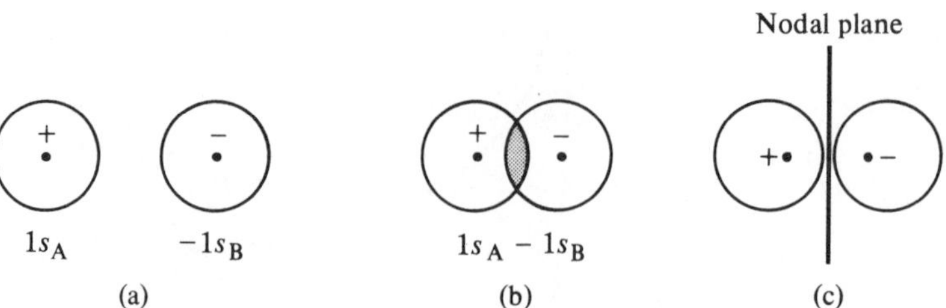

Fig. 15.5. The formation of the antibonding ψ_- molecular orbital.

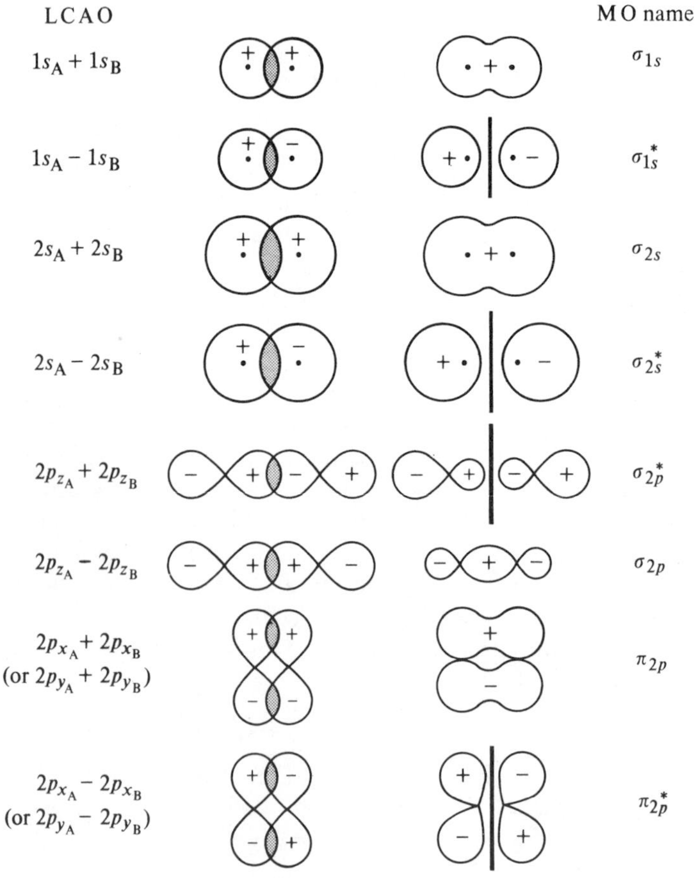

Figure 15.6

The formation of the antibonding orbital is illustrated in Fig. 15.5. In this case the $1s$ orbital on atom A is positive but $1s_B$ is negative (Eq. 299). In the cross-hatched region of Fig. 15.5(b) both functions have significantly large magnitudes but are of opposite signs. Rather than being enhanced in the region between the nuclei, the antibonding orbital has a *nodal plane*, since the two atomic orbitals subtract rather than add.

The functions obtained by squaring Eqs. (298) and (299) may be similarly represented. Recall that these squared functions represent the probability of finding an electron in a given direction relative to the nuclei.

Figure 15.6 summarizes the various molecular orbitals obtained from linear combinations of s and p atomic orbitals of the form of Eq. (297). A molecular orbital resulting from the overlap of two atomic s orbitals or the overlap of two atomic p orbitals each with a lobe extending in space directly toward the other is called a *sigma orbital* σ. A sigma orbital may also be formed between an s orbital and a p orbital when the latter has a lobe pointing directly toward the s orbital. In general, a sigma orbital is one having no nodal plane which contains the line between the two bonded atoms.

When the overlap is between parallel lobes of p orbitals the function is called a *pi orbital* π. In general, a pi orbital is one which has one nodal plane containing the internuclear axis. These may also be formed by the overlap of atomic orbitals other than p orbitals.

In Fig. 15.6 the subscript to σ and π indicates which atomic orbitals are used in the linear combination. The superscript * denotes an antibonding orbital.

As with atomic orbitals, we may place two electrons in each molecular orbital if their spin wave functions are different. Figure 15.7 illustrates one

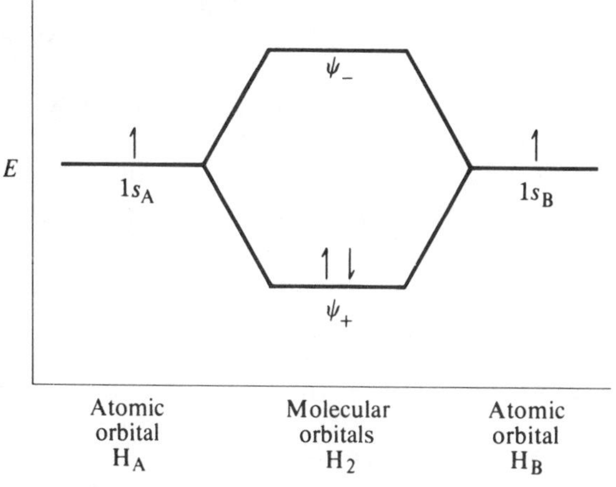

Figure 15.7

method of schematically representing the lowest energy state of the hydrogen molecule. The total spatial wave function is the product of the two one-electron molecular orbitals (Eq. 297):

$$\begin{aligned}
\psi_1 &= \sigma_{1s}(1) = C_+[1s_A(1) + 1s_B(1)] \\
\psi_2 &= \sigma_{1s}(2) = C_+[1s_A(2) + 1s_B(2)] \\
\psi_{H_2} &= \psi_1\psi_2 = \sigma_{1s}(1)\sigma_{1s}(2) = \sigma_{1s}^2 \\
&= C_+^2[1s_A(1)1s_A(2) + 1s_A(1)1s_B(2) + 1s_A(2)1s_B(1) + 1s_B(1)1s_B(2)] \qquad (305)
\end{aligned}$$

The total energy of this wave function is the sum of the energies of each occupied molecular orbital (Eqs. 296 and 301):

$$E = E_1 + E_2 = 2E_+ . \qquad (306)$$

The total spatial wave function (Eq. 305) represents a chemical bond: a pair of electrons in a single molecular orbital which involves both atoms. The energy of the bond (Eq. 306) is lower than the energy of the isolated atoms; the molecule is stable relative to the separated atoms.

We should now ask how accurate is this view of the hydrogen molecule. We can compare the energy and the equilibrium internuclear distance which we calculate for H_2 based on the model developed above with the experimental values. Determining the energy as a function of internuclear distance by evaluating the integrals of Eqs. (300), (303), and (304), we obtain a minimum energy of -2.68 eV (-4.30×10^{-12} erg) at a distance of 0.850×10^{-8} cm (0.850 Angstrom). The experimental values are -4.75 eV (-7.60×10^{-12} erg) and 0.741 Å respectively. While our model predicts a minimum in the electronic energy at a bond distance only 0.1 Å from the experimental value, the actual bond is almost twice as strong as predicted (7.6×10^{-12} erg is required to separate the two hydrogen atoms, that is to break the bond, rather than 4.3×10^{-12} erg). Can we improve the approximate wave functions while retaining the model of molecular bonds based on atomic orbitals?

Equation (305) may be thought of as a sum of the possible *resonance* forms for H_2. The first term, $1s_A(1)1s_A(2)$, places both electrons on nucleus A. This is the wave function for a hydride ion H^- and would correspond to the resonance form H^-H^+. Similarly the fourth term, $1s_B(1)1s_B(2)$, corresponds to H^+H^-. These two terms are therefore *ionic*. The second and third terms are called *covalent* terms: in each there is one electron associated with each nucleus. One interpretation of the total wave function as a sum of these resonance terms is that the stability of the bond results from the electrons being under the influence of both nuclei. The two electrons move about separately within this field of influence and are shared by the two nuclei involved in the bond, rather than belonging totally to either nucleus. It is also implicit in this formulation, however, that the electrons are just as likely to find themselves both in the same region immediately adjacent to one nucleus (the ionic terms), as more evenly shared between the two nuclei (the covalent terms). As we have discussed, for atoms of identical electronegativity (as is necessarily the case for *homonuclear diatomic* molecules like H_2) we would not expect ionic forms to be of very great importance. Therefore, we could possibly improve our model by deemphasizing the ionic terms of the wave function, writing instead

$$\psi_{H_2} = N\{[1s_A(1)1s_B(2) + 1s_A(2)1s_B(1)] + C[1s_A(1)1s_A(2) + 1s_B(1)1s_B(2)]\}, \tag{307}$$

where N is simply the normalizing constant but C is a weighting constant. If C is less than 1, it will have the effect of decreasing the contribution of the ionic terms relative to that of the covalent terms, and vice versa. Applying the variation principle to find that value of C which gives the minimum in energy, we obtain $C = \frac{1}{4}$, $E = -3.23$ eV (-5.17×10^{-12} erg), and $r_e = 0.88$ Å. This produces a significant improvement in E, but the bond distance is caused to deviate even further from the experimental value.

Another improvement may be accomplished by using a variable atomic number Z' rather than Z in the atomic wave function. This helps to account for the electronic repulsion term in the correct Hamiltonian which we completely ignored in Eq. (293). Minimizing the energy with respect to Z' we obtain $E = -4.02$ eV (-6.4×10^{-12} erg) at an equilibrium bond distance $r_e = 0.77$ Å with $C = 0.256$ (Eq. 307).

Further modifications in the wave functions for hydrogen can be made. Wave functions have been proposed which result in calculated values for E and r_e which agree almost perfectly with the experimental value. Although they are better mathematically, these modified wave functions are unfortunately extremely difficult, if not impossible, to visualize. For our purposes molecular orbitals obtained by the LCAO approach will be quite sufficient.

15.5. LCAO-MO Method: Homonuclear Diatomic Molecules

We now have a quantum mechanical model or picture of the bond in the hydrogen molecule. The same principles may now be applied to other homonuclear diatomic molecules.

The He_2 molecule has four electrons. Its LCAO-MO's are made up of two atomic $1s$ orbitals, one from each helium atom. Figure 15.8(a)

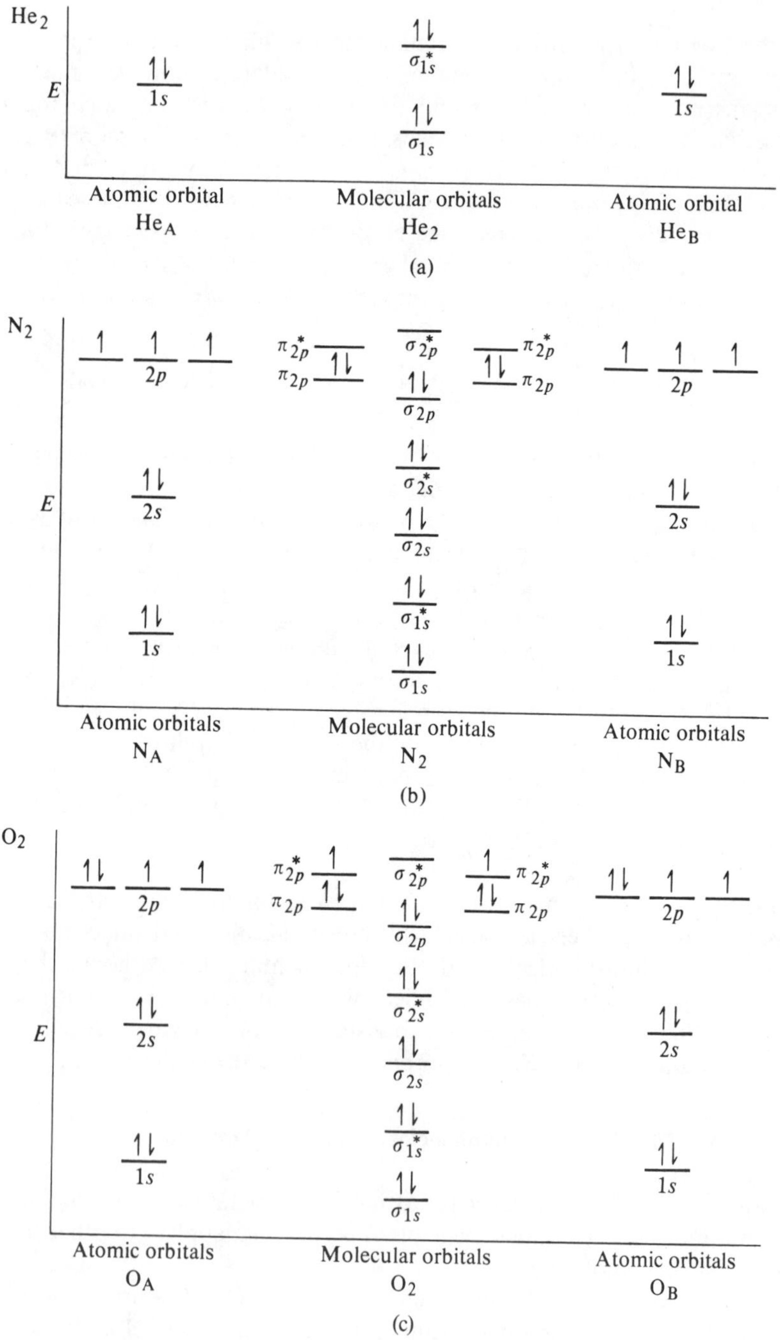

He_2

E

1↓ $1s$ | 1↓ σ_{1s}^* | 1↓ σ_{1s} | 1↓ $1s$

Atomic orbital He_A — Molecular orbitals He_2 — Atomic orbital He_B

(a)

N_2

E

1 1 1 $2p$ | π_{2p}^* π_{2p} 1↓ | σ_{2p}^* | 1↓ π_{2p}^* π_{2p} | 1 1 1 $2p$

1↓ σ_{2p}

1↓ $2s$ | 1↓ σ_{2s}^* | 1↓ σ_{2s} | 1↓ $2s$

1↓ $1s$ | 1↓ σ_{1s}^* | 1↓ σ_{1s} | 1↓ $1s$

Atomic orbitals N_A — Molecular orbitals N_2 — Atomic orbitals N_B

(b)

O_2

E

1↓ 1 1 $2p$ | π_{2p}^* 1 π_{2p} 1↓ | σ_{2p}^* | 1 π_{2p}^* 1↓ π_{2p} | 1↓ 1 1 $2p$

1↓ σ_{2p}

1↓ $2s$ | 1↓ σ_{2s}^* | 1↓ σ_{2s} | 1↓ $2s$

1↓ $1s$ | 1↓ σ_{1s}^* | 1↓ σ_{1s} | 1↓ $1s$

Atomic orbitals O_A — Molecular orbitals O_2 — Atomic orbitals O_B

(c)

Figure 15.8

schematically illustrates the bonding situation. Two of the electrons go into the bonding σ_{1s} orbital, but the other two must be in antibonding σ_{1s}^* orbital. The total spatial wave function is

$$\psi_{He_2} = N\sigma_{1s}^2\sigma_{1s}^{*2},$$

and the total bonding energy is essentially zero: the stabilization gained by having two electrons in a bonding orbital is lost by having two in the antibonding orbital. In fact the total wave function is for all practical purposes the same as a separate 1*s* atomic wave function containing two electrons on each of the two helium atoms. A stable He_2 molecule does not exist.

A nitrogen atom has seven electrons; its configuration is $1s^22s^22p^3$. Figure 15.8(b) illustrates the formation of LCAO-MO's for N_2. The atomic 1*s* orbitals combine to give σ_{1s} and σ_{1s}^* molecular orbitals; the atomic 2*s* functions give σ_{2s} and σ_{2s}^* orbitals; the two $2p_z$ orbitals yield a σ_{2p} and a σ_{2p}^*, while the $2p_x$ and $2p_y$ orbitals give a pair of degenerate bonding π_{2p} molecular orbitals and a degenerate pair of antibonding π_{2p}^* orbitals. The relative energies are represented schematically in the figure. As in the *aufbau* method of filling atomic orbitals, we start with the molecular orbital of lowest energy, placing two electrons in each orbital. The resulting configuration for N_2 is $\sigma_{1s}^2\sigma_{1s}^{*2}\sigma_{2s}^2\sigma_{2s}^{*2}\sigma_{2p}^2\pi_{2p}^4$. The stabilization due to both the σ_{1s} and the σ_{2s} bonds is nullified because the corresponding antibonding orbitals are also occupied. N_2 does, however, have one sigma bond (σ_{2p_z}) and two pi bonds (π_{2p_x} and π_{2p_y}). The nitrogen molecule therefore has three bonds or is said to have a *triple bond.*

Example 86 Describe the bonding in the oxygen molecule.

Answer: Figure 15.8(c) diagrams the occupation of the molecular orbitals of the O_2 molecule. There are a total of sixteen electrons in the O_2 molecule. The occupation of the molecular orbitals is the same as for N_2 except for the last two oxygen electrons. These two go into the antibonding π_{2p}^* orbitals, one in each of the two degenerate functions. Both do not go into the same orbital because of electron repulsion (see Section 14.4). The two electrons in antibonding orbitals cancel the energy gained by a pair of electrons in a bonding π_{2p} orbital. O_2 is therefore said to have a *double bond.* Electron spin resonance measurements confirm the presence of two unpaired electrons (see Section 18.9).

15.6. Heteronuclear Diatomic Molecules

The preceding development may be modified to cover bonds between unlike atoms. In contrast to the homonuclear case, the atomic orbitals which make up the linear combinations in a heteronuclear bond usually do not have the same energy. The case of carbon monoxide (CO) is a good illustration.

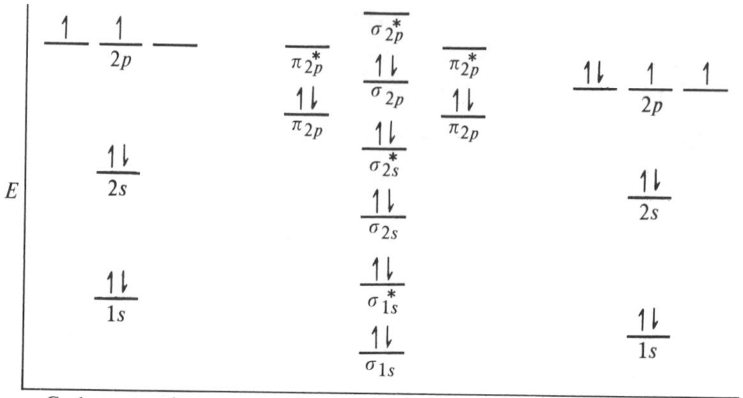

E

Carbon atomic orbitals: $2p$ (1, 1, –); $2s$ (1↓); $1s$ (1↓)

Carbon monoxide molecular orbitals: σ_{2p}^*; π_{2p}^*, π_{2p}^*; σ_{2p} (1↓); π_{2p} (1↓), π_{2p} (1↓); σ_{2s}^* (1↓); σ_{2s} (1↓); σ_{1s}^* (1↓); σ_{1s} (1↓)

Oxygen atomic orbitals: $2p$ (1↓, 1, 1); $2s$ (1↓); $1s$ (1↓)

Figure 15.9

Figure 15.9 is a schematic diagram of the formation of molecular orbitals in CO. Due to the larger nuclear charge of oxygen, the oxygen atomic orbitals are of lower energy than the corresponding carbon orbitals. Consequently the orbital contribution made by one atom to the LCAO-MO's will differ from that made by the other:

$$\psi_{\text{MO}} = N(\chi_{\text{C}} + \lambda\chi_{\text{O}}). \tag{308}$$

Here the χ functions are atomic orbitals and λ is a scale factor. When $\lambda > 1$, the oxygen orbital will contribute more; in molecular orbitals in which $\lambda < 1$, the carbon will contribute more. In the bonding CO orbitals, $\lambda > 1$. Figure 15.10 illustrates this effect for the σ_{1s} orbital of carbon monoxide. The wave function has a greater value near oxygen than near carbon. Recalling that the square of this function gives the probability of finding an electron, this means that the electrons in this molecular orbital are more likely to be found near the oxygen nucleus than near the carbon nucleus. The oribtal is said to be *polarized*: the center of negative charge is nearer the oxygen since the negatively charged electrons are more likely to be found near the oxygen, and the center of positive charge is nearer the carbon atom.

Since $\lambda < 1$ for the antibonding molecular orbitals of CO, both the polarity and the net bonding resulting from σ_{1s} and σ_{2s} are cancelled because σ_{1s}^* and σ_{2s}^* are each occupied by a pair of electrons. However, since σ_{2p} and both π_{2p} orbitals are fully occupied and the corresponding antibonding orbitals are empty, CO has a triple bond and is polar, all three bonds being polar with the oxygen atom more negative. This is schematically represented by

$$\overset{+\!\!\longrightarrow}{\text{C}\equiv\text{O}}.$$

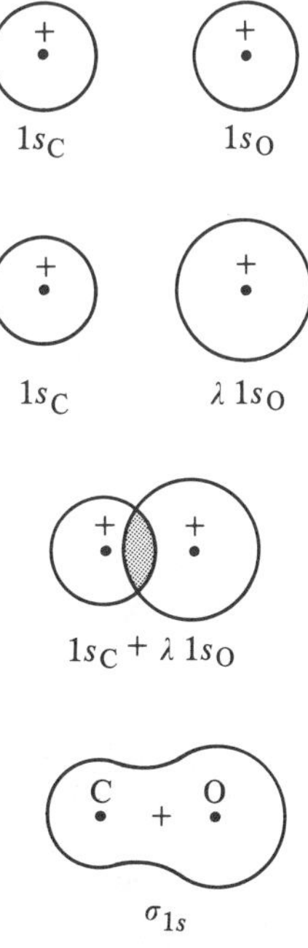

Figure 15.10

The greater the energy difference of the *valence* orbitals (atomic orbitals which contribute to the bonds of the molecule), the more polar the bond becomes. Said another way, the more the two atoms of a bond differ in electronegativity, the more polar the resulting bond. At one extreme we find the homonuclear cases in which $\lambda = 1$ and the *dipole moment* μ (an experimental measure of polarity) is zero. At the other extreme are the ionic cases in which $\lambda = 0$.

Let us explore the latter case and see how ionic bonds may be treated as a special case of molecular orbital theory.

Consider KCl. Figure 15.11 illustrates the bonding situation for the valence orbitals. Notice that when the thirty-six electrons of KCl are placed in the molecular orbitals all the MO's resulting from the atomic $3p$ orbitals are filled. There are an equal number of electrons in bonding and antibonding orbitals. Our first thought might be that there is therefore no bonding. But let us look at the situation more closely.

Recall (Section 15.5) that when both the σ_{1s} and the σ^*_{1s} orbitals are filled, there is no net bonding and the electrons may be assumed to be located in atomic $1s$ orbitals, two electrons on one atom and two on the

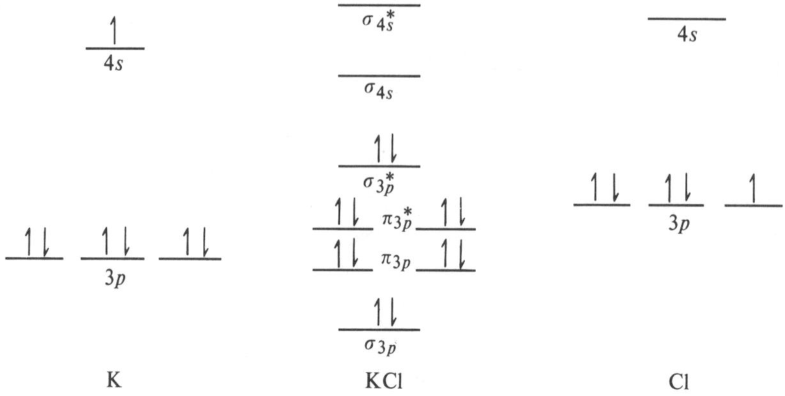

Figure 15.11

other. Therefore in KCl all bonding orbitals are cancelled and there are no covalent bonds; the electrons may be considered to be located on the atoms, eighteen on potassium and eighteen on chlorine. This means, of course, that potassium is one electron short and is K^+, while chlorine now has an extra electron and is Cl^-. Potassium has donated an electron to chlorine; the bond is ionic.

In some cases it may not be clear which atomic orbitals to use in a linear combination. One rule worth remembering is that the atomic orbitals which may be included in a LCAO-MO should have similar energies. The HCl molecule provides a good example of this and of other considerations which help in the choice of atomic orbitals. In hydrogen the 1*s* orbital is of lowest energy. The chlorine 3*p* orbitals are of roughly the same energy as the hydrogen 1*s* orbital because of the much larger charge of the chlorine nucleus. The bond will be formed by the overlap of a hydrogen 1*s* orbital and a chlorine 3*p* orbital:

$$\psi_{\mathrm{HCl}} = N[1s_{\mathrm{H}} + \lambda(3p_{\mathrm{Cl}})].$$

But is this the 3_{p_x}, the 3_{p_y}, or the 3_{p_z} orbital? Following the convention which places the *z*-axis along the molecular bond, Fig. 15.12(a) illustrates what happens when a p_x or p_y orbital overlaps with an *s* orbital. The overlap of the positive *p* lobe is exactly cancelled by that of the negative lobe. Thus a linear combination of $1s_{\mathrm{H}}$ with $3p_{x_{\mathrm{Cl}}}$ or $3p_{y_{\mathrm{Cl}}}$ does not lead to bond formation. Figure 15.12(b) illustrates how the overlap of $1s_{\mathrm{H}}$ and $3p_{z_{\mathrm{Cl}}}$ does lead to a bonding orbital. For this molecule $\lambda > 1$ in the bonding orbital and there is a dipole moment.

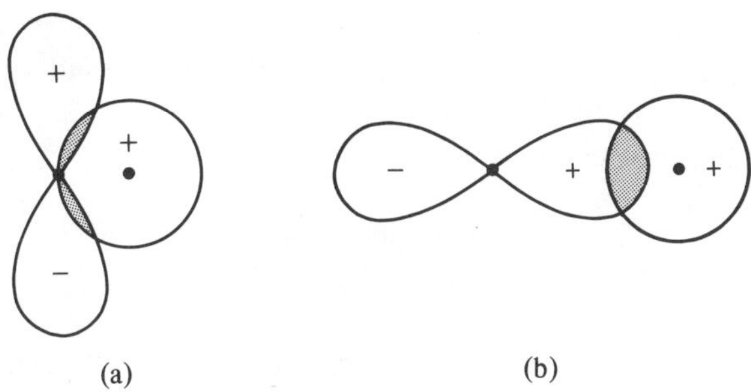

Figure 15.12

15.7. Polyatomic Molecules: Localized Bonds

Now let us apply these bonding ideas to molecules with more than two atoms. These are the molecules whose structures most often concern us. We shall assume initially that all bonds are *localized*. This means that the bond between two adjacent atoms in a polyatomic molecule will be considered to be made up of only the atomic orbitals of those two atoms. We further assume that we can treat each bond separately. The total electronic wave function of the molecule will be the product of localized bond wave functions, one function for each bond of the molecule, and the total electronic energy will be the sum of the energies of the separate bonds. These assumptions generally work well for σ bonds and for isolated π bonds, but a different approach must be taken for *conjugated* systems—molecules in which the π bonds are not isolated. This situation will be treated in Section 15.9.

Molecules with three or more atoms differ from diatomic molecules in two further ways. With only two atoms, a diatomic molecule must be linear. A molecule with more than two atoms may be nonlinear: the bonds may make angles other than 180°. The problem of the two- or three-dimensional geometry of the molecule is introduced. Also an atom now may be bonded to two or more other atoms. In H_2O, for instance, it becomes necessary to select which atomic orbital of the oxygen will be used to form a linear combination with one hydrogen and which with the other hydrogen.

The differences and similarities between the diatomic approach and that used for polyatomic molecules will be clearer if we contrast two actual molecules, O_2 and H_2O for example. The details of the O_2 treatment were considered in Section 15.5. Let us now examine the water molecule.

There are two bonds in H_2O, each between a hydrogen and the central oxygen. The atomic configuration of oxygen is $1s^2 2s^2 2p^4$. Placing two electrons in $2p_z$ results in one unpaired electron in the $2p_x$ orbital and one in $2p_y$. The first bond may then be made up of a linear combination of the 1s orbital of one hydrogen (A) and the 2p_x of the oxygen. The second will be a linear combination of the 1s of the other hydrogen (B) and the 2p_y of the oxygen. The two bonding molecular orbitals are then

$$\psi_1 = N(1s_{\mathrm{H_A}} + \lambda_1 2p_{x_\mathrm{O}})$$

$$\psi_2 = N(1s_{\mathrm{H_B}} + \lambda_2\, 2p_{y_\mathrm{O}}).$$

Each of these MO's contains two electrons, one originally from oxygen and one from hydrogen. Figure 15.13(a) illustrates the fact that since p_x and p_y are perpendicular, the angle between the two O—H bonds in water would be predicted to be 90°. Experimental measurement of this angle (see Chapter 18) gives a value of 104.5°. Our simple approach has successfully predicted a nonlinear bonding geometry, but the angle is off by 14.5°. Can we explain why?

When the best wave functions ψ_1 and ψ_2 are determined using the variation principle, it is found that $\lambda > 1$. This should be expected, since oxygen is more electronegative than hydrogen. The two bonds are therefore polar, with the hydrogen ends somewhat positive. The repulsion of the

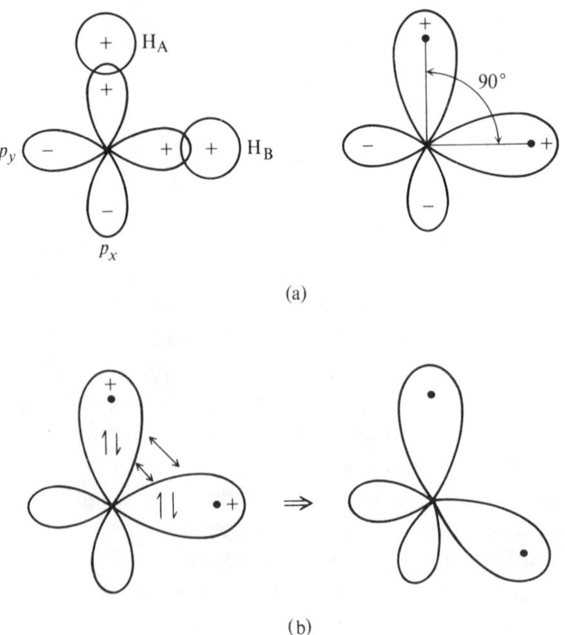

Figure 15.13

positive ends of these two dipoles and the repulsion of the negatively charged electrons in the bonds leads to an increase in bond angle (see Fig. 15.13b). There are other effects which may contribute to this increase in bond angle. One, to be discussed in the next section, is partial *hybridization* of the oxygen atomic orbitals.

Example 87 Should the bond angle in H_2S be expected to be larger or smaller than that in H_2O?

Answer: The two S—H bonds in H_2S will each result from the linear combination of the 1*s* orbital of a hydrogen and a sulfur $3p_x$ or $3p_y$ atomic orbital. The resulting molecule should be similar to H_2O with the two S—H bonds forming an angle of 90°. Since a 3*p* orbital extends further into space than a 2*p* orbital, the S—H bond distance is longer than O—H (1.35 Å for S—H, 0.96 Å for O—H). Consequently the two hydrogen atoms in H_2S are further from each other than those in H_2O.

Sulfur is also less electronegative than oxygen. The S—H bond is thus less polar than the O—H bond, and the hydrogen end of the bond dipole is less positive in H_2S than in H_2O.

The net result of these two effects is less repulsion between the two hydrogen atoms in H_2S than in H_2O; the angle in H_2S should be closer to 90° than in H_2O. This angle is found by experiment to be 93°.

The bonds in other polyatomic molecules may be treated in a similar manner. Of particular interest and importance are those of organic molecules. When the bonds formed with carbon in these molecules are considered using the simple approach outlined in this chapter, severe problems arise.

15.8. Hybrid Orbitals

The simplest stable molecule of carbon with hydrogen is methane (CH_4). All the C—H bonds of methane are of equivalent length and strength, and all the angles between the C—H bonds are found experimentally to be 109°28′, that is, the molecule is tetrahedral.

The ground state electronic configuration of atomic carbon is $1s^22s^22p^2$. Based on our previous reasoning, carbon should be expected to be *divalent* and the simplest hydrocarbon should be CH_2. The two bonds in that molecule should each consist of a shared pair of electrons: a 2*p* electron of carbon in conjunction with a 1*s* electron of a hydrogen atom. How can we rationalize four equivalent bonds with mutual bond angles of 109°28′?

First, in order to form four bonds with hydrogens the carbon must have available four atomic orbitals with a single electron in each orbital. This would be possible if one of the $2s$ electrons were placed in the third $2p$ orbital, resulting in a configuration of $1s^2 2s^1 2p^3$. The $2s$ orbital is of lower energy, however, and it is consequently more energetically favorable for an electron to be in a $2s$ than a $2p$ orbital; energy is required to "excite" the $2s$ electron into a $2p$ orbital. This can be accomplished only if the energy released upon formation of four bonds with hydrogen atoms is sufficiently large to make up for this required energy expenditure. In this excited configuration carbon can form four bonds, but they will not be equivalent since three will be linear combinations of a hydrogen $1s$ and a carbon $2p$, while the fourth will be made up of a hydrogen $1s$ and a carbon $2s$.

The next step is to *hybridize* the carbon valence orbitals. To do this we make linear combinations of the $2s$, $2p_x$, $2p_y$, and $2p_z$ atomic orbitals to form four new hybrid atomic orbitals. Before accomplishing this, let us first examine the general procedure of combining atomic orbitals on a single atom.

Figure 15.14(a) illustrates schematically the results of adding an s and a p_x function *on the same atom.* The positive lobe of the p_x is augmented, whereas the negative lobe is decreased in magnitude. Figure 15.14(b) illustrates the results of adding an s, a p_x, and a p_y function on the same atom. The sum function has lobes directed equidistantly between the original p_x and p_y lobes, the positive lobe being increased while the negative lobe is decreased in magnitude. Figure 15.14(c) shows the results when the p_x orbital is subtracted from the sum of the s and p_y orbitals, while Fig. 15.14(d) pictures the sum function when p_y is subtracted from the sum of s and p_x, and Fig. 15.14(e) the sum function when both p_x and p_y are subtracted from s. Clearly the sign of an orbital in the linear combination determines the orientation of the positive lobe of the hybrid orbital.

Now what does a hybrid orbital look like which is made up of the sum of an s and all three p atomic orbitals? It will have an augmented positive lobe extended in a direction exactly between the positive lobes of p_x, p_y, and p_z (orbital t_1 in Fig. 15.15). Similarly the hybrid orbital

$$\psi = N(s - p_x + p_y - p_z)$$

is oriented so that its augmented positive lobe extends between the $-x$, $+y$ and $-z$ axes (orbital t_2 in Fig. 15.15). It is not difficult to confirm that varying the signs of the p orbitals in the linear combination will result in eight possible hybrid orbitals, each directed into a different octant of a cartesian coordinate system (that is, toward the corners of a circumscribing cube).

Recall that we need four bonds with mutual angles of $109°28'$. The four hybrid atomic orbitals of Fig. 15.15 have exactly this arrangement, each positive lobe pointing toward one of the four non-adjacent corners of a cube with the carbon nucleus in the center of the cube (or equivalently, each positive lobe pointing toward one of the apices of a tetrahedron, the nucleus

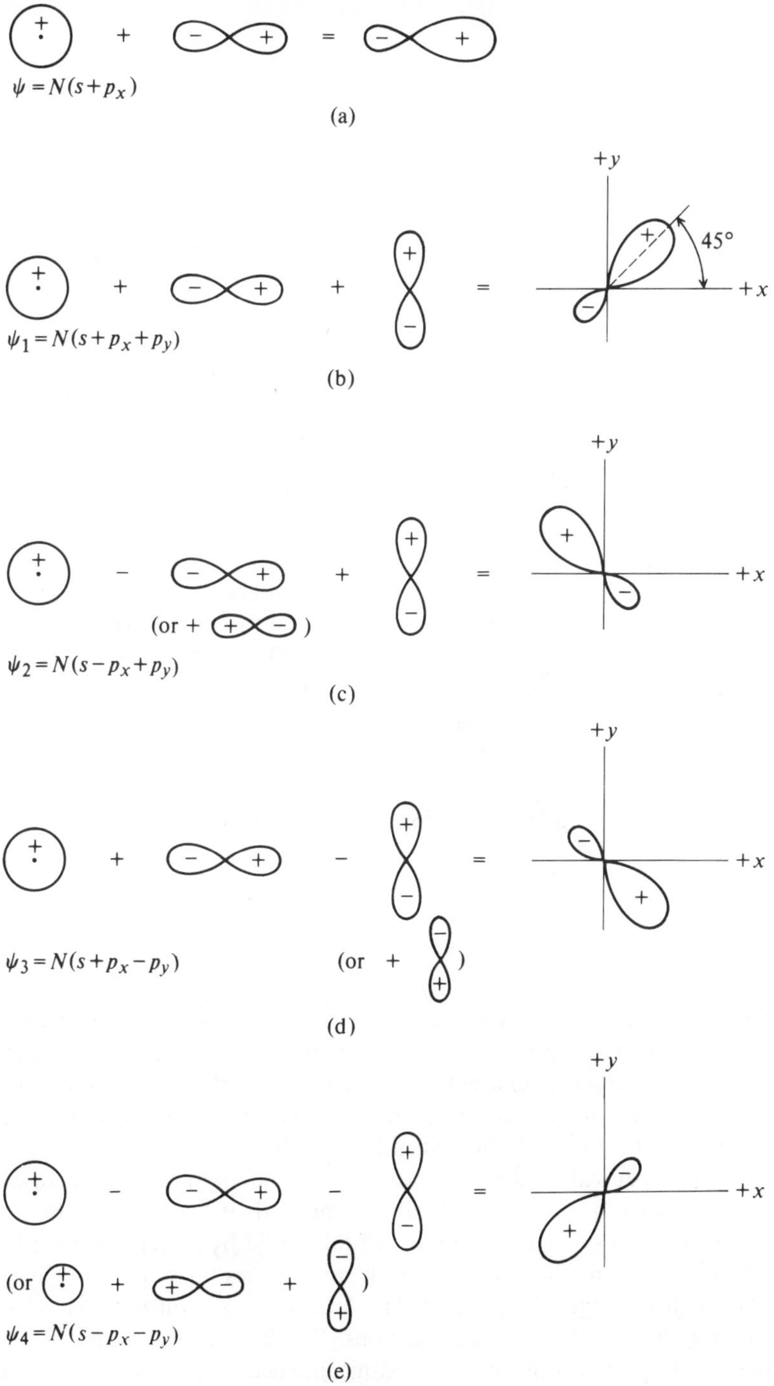

$\psi = N(s + p_x)$

(a)

$\psi_1 = N(s + p_x + p_y)$

(b)

$\psi_2 = N(s - p_x + p_y)$

(c)

$\psi_3 = N(s + p_x - p_y)$

(d)

$\psi_4 = N(s - p_x - p_y)$

(e)

Figure 15.14

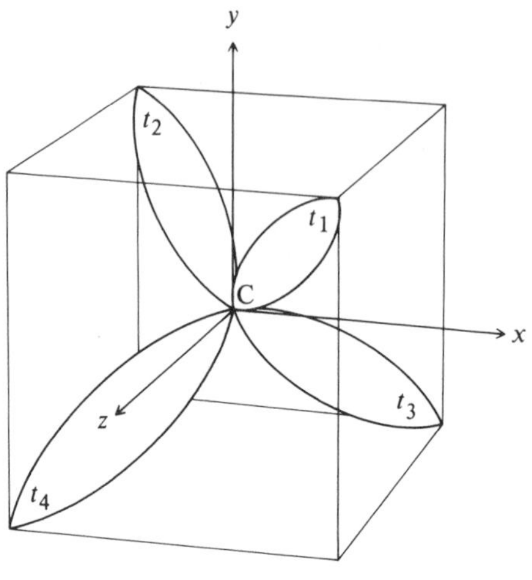

$$t_1 = 0.5\,(s + p_x + p_y + p_z)$$
$$t_2 = 0.5\,(s - p_x + p_y - p_z)$$
$$t_3 = 0.5\,(s + p_x - p_y - p_z)$$
$$t_4 = 0.5\,(s - p_x - p_y + p_z)$$

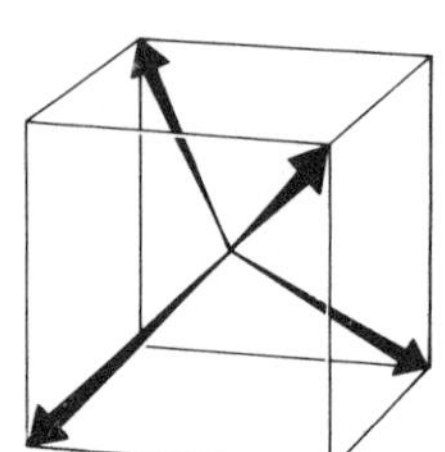

Figure 15.15

being at the tetrahedron's center.) These are the four tetrahedral hybrid orbitals of carbon. They are called sp^3 hybrids because each is made up of three times as much p character as s character; three p orbitals are used while only one s orbital appears in the linear combination. There is one electron in each of these hybrid atomic orbitals.

The four equivalent directed molecular bonds which carbon forms with hydrogen atoms in methane are each composed of a linear combination of one of the carbon sp^3 hybrid atomic orbitals and a hydrogen $1s$ orbital. These are sigma bonds. In ethane (C_2H_6) each carbon uses three of its sp^3 hybrids to overlap with hydrogen $1s$ orbitals. The fourth sp^3 orbital on each is used to form a sigma bond between the carbons. Similar reasoning may be applied to the bonding of carbon in any organic molecule where carbon is bonded to four other atoms.

Example 88 Nitrogen is another important atom whose molecular bonds may be explained by hybrid atomic orbitals. All three H—N—H angles in ammonia (NH_3) are found by experiment to be close to the tetrahedral value. Explain the bonding in ammonia.

Answer: The atomic configuration of nitrogen is $1s^2 2s^2 2p^3$. When four hybrid sp^3 atomic orbitals are formed using $2s$ and $2p$ there are five valence electrons (there were only four in carbon) to be placed in these four orbitals. A single electron may be placed in each of three sp^3 orbitals, but the fourth orbital must contain a pair of electrons. Each of the N—H bonds is thus a linear combination of a $1s$ orbital on hydrogen and one of the singly occupied sp^3 hybrid orbitals of nitrogen.

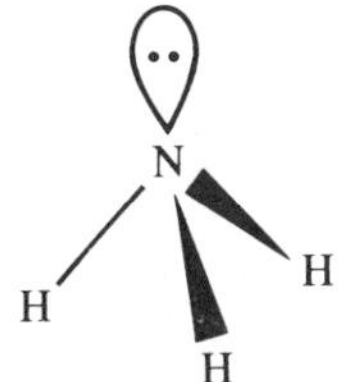

There are many cases when carbon is not bonded tetrahedrally. In ethylene (C_2H_4) and formaldehyde (CH_2O), for instance, carbon is bonded to only three other atoms, the molecules are planar, and the bond angles are all approximately 120°. In acetylene (C_2H_2) each carbon is bonded to only two other atoms and the molecule is linear. How can we account for these situations?

If carbon is in its excited configuration, $1s^2 2s^1 2p^3$, but only the $2s$, $2p_x$, and $2p_y$ orbitals are used in the linear combinations for hybrid atomic orbitals, the three orbitals shown in Fig. 15.16 may be obtained. These orbitals are equivalent; they are located 120° from each other and are symmetric with respect to the xy plane. Notice also that in addition to these three sp^2 hybrids (or *trigonal hybrids*), there is an atomic p_z orbital (perpendicular to the xy plane in Fig. 15.16). Each of these four atomic orbitals contains one electron.

In forming the ethylene molecule each carbon is sp^2 hybridized, two of the hybrid orbitals of each carbon being used to form sigma bonds with hydrogens. The third sp^2 of each carbon is used in the formation of an sp^2-sp^2 sigma bond between the two carbons. The remaining p_z orbital on each of the carbons may then form a pi bond if they are parallel. In order that they may have this orientation the molecule must be planar. Ethylene is said to have a *double bond*; a sigma and a pi bond are formed between the carbons. A similar rationale accounts for the geometry of formaldehyde and for similar bonding situations in other organic molecules.

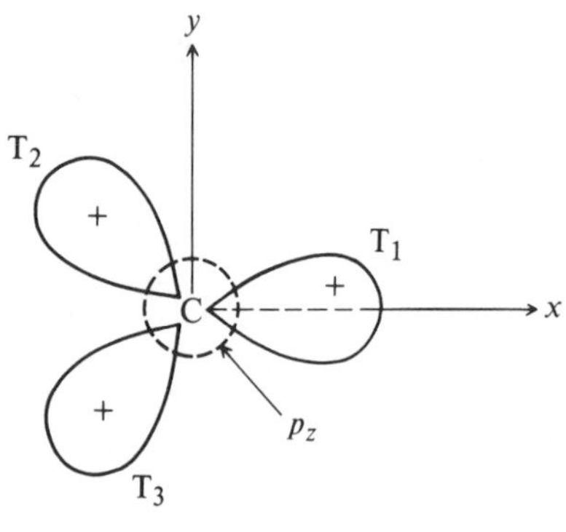

$$T_1 = \frac{1}{\sqrt{3}}\,(2s) + \sqrt{\frac{2}{3}}\,(2p_x)$$

$$T_2 = \frac{1}{\sqrt{3}}\,(2s) - \frac{1}{\sqrt{6}}\,(2p_x) + \frac{1}{\sqrt{2}}\,(2p_y)$$

$$T_3 = \frac{1}{\sqrt{3}}\,(2s) - \frac{1}{\sqrt{6}}\,(2p_x) - \frac{1}{\sqrt{2}}\,(2p_y)$$

Fig. 15.16.
The sp^2 or trigonal hybrid orbitals.

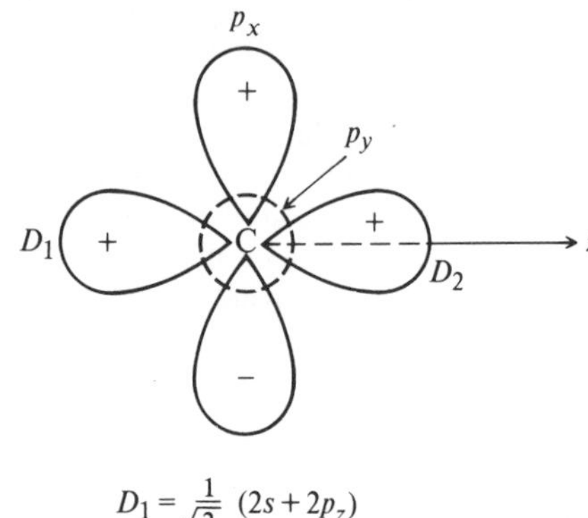

$$D_1 = \frac{1}{\sqrt{2}}\,(2s + 2p_z)$$

$$D_2 = \frac{1}{\sqrt{2}}\,(2s - 2p_z)$$

Fig. 15.17.
The *sp* or digonal hybrid orbitals.

Acetylene is explained by the *sp* hybrids of Fig. 15.17. The p_x and p_y orbitals, each containing a single electron, are not used in the hybrid orbitals. In acetylene one of the *sp* hybrids on each carbon is involved in bonding with a hydrogen; the other participates in an *sp-sp* sigma bond between the two carbons. The p_x orbitals on the two carbons form a pi bond, as do the p_y atomic orbitals. There are then three bonds between the two carbons, two pi bonds and one sigma bond. Acetylene therefore has a *triple bond.*

Example 89 Describe the geometry of the allene molecule $CH_2{=}C{=}CH_2$.

Answer: Each of the carbons at the ends of the molecule is sp^2 hybridized. Two of the sp^2 orbitals of each end carbon are used to form sigma bonds with 1*s* hydrogen atomic orbitals. The third forms a sigma bond by combining with one of the *sp* hybrid orbitals of the central atom.

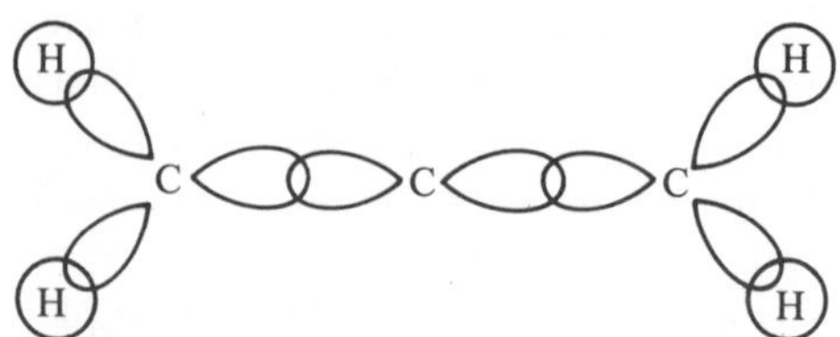

The remaining unhybridized *p* orbitals of the three carbons are free to overlap to form pi bonds. The *p* orbitals of adjacent carbons must, however, be oriented parallel to form a pi bond. Therefore the p orbitals of the two end carbons must be oriented perpendicular to each other in order to both be parallel to one of the p orbitals of the central carbon:

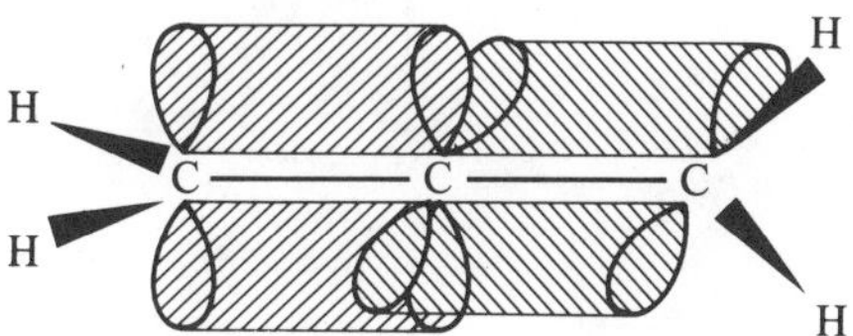

The $\mathrm{C}\!<^{\mathrm{H}}_{\mathrm{H}}$ plane at one end of the allene molecule must therefore be perpendicular to that at the other end.

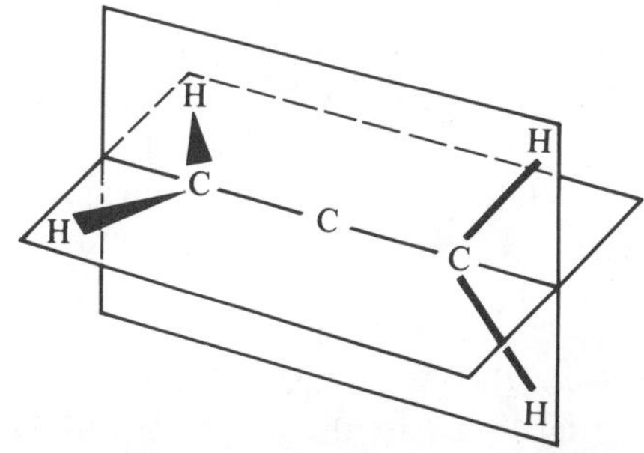

In describing molecules containing transition metals it is often advantageous to use hybrid atomic orbitals involving d orbitals. Some of these molecules, such as hemoglobin and the cytochromes, are of great importance in physiological reactions. Two of the more important types of such hybrids are the *square planar* dsp^2 and the *octahedral* d^2sp^2 hybrid orbitals.

In the square planar case four equivalent atomic orbitals are needed to form bonds between the metal and four other atoms. All the bonded atoms including the metal lie in a plane, and all bond angles between adjacent bonds are 90°. The dsp^2 orbital is made up of a linear combination of one d orbital ($d_{x^2-y^2}$), one s orbital and two p orbitals (p_x and p_y). Since $d_{x^2-y^2}$ has positive lobes along $\pm x$ and negative lobes along $\pm y$, a hybrid orbital with a positive lobe extended along the x-axis would be obtained by adding s, $d_{x^2-y^2}$, and p_x. The hybrid orbital composed of the sum of s and *negative* $d_{x^2-y^2}$ and p_y orbitals would have a positive lobe extended along the negative y-axis. Similar reasoning applies to the other two dsp^2 orbitals.

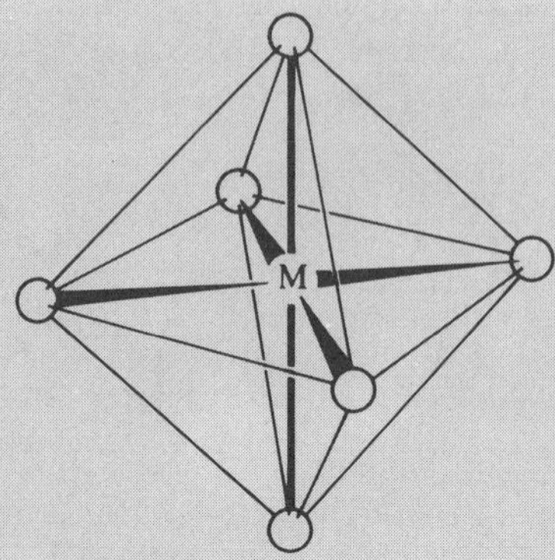

Figure 15.18

In the octahedral case six atomic orbitals are needed to form bonds between the metal and six atoms located at the apices of an octahedron with the metal at the center (see Fig. 15.18). Each of the six equivalent d^2sp^3 hybrid atomic orbitals is made up of a linear combination of two d orbitals (the d_{z^2} and $d_{x^2-y^2}$), one s orbital and all three p orbitals (p_x, p_y, and p_z).

In many cases the covalent bonds formed involving transition metals are *coordinate covalent*: the atom to which the metal is bonded (the *ligand*) supplies both electrons to the bond formed by the overlap of a filled atomic orbital of the ligand with an empty hybrid orbital belonging to the metal. Some of the molecules (or *complexes*) formed by transition metals with other atoms are ionic.

Example 90 Hemoglobin is an octahedral coordinate covalent complex of ferrous ion (Fe^{+2}), a *porphyrin* ring system, an imidazole group, and either water or oxygen. In the blood, hemoglobin acts as a carrier of oxygen. It replaces its oxygen with a water molecule in the cells and returns to the lungs where it swaps the water for another oxygen, thence returning to the cells in various parts of the body where oxygen is needed for biological oxidation. Figure 15.19 gives a simplified picture of the hemoglobin molecule. Explain the bonding in this molecule when oxygen is attached to the iron. In this case electron spin resonance measurements indicate that there are no unpaired electrons.

Answer: If the bonding between Fe^{+2} and the rest of the molecule were purely ionic, no hybrid orbitals would be expected to form, and the five d orbitals would be effectively degenerate (assuming their interactions with the other groups are weak). The configuration of Fe^{+2} is $1s^22s^22p^63s^23p^63d^6$. The six d electrons would distribute themselves among the five essentially degenerate d orbitals in such a way that as many electrons as possible would be unpaired. Thus there would be a pair of electrons in one of the d orbitals. The other four d orbitals

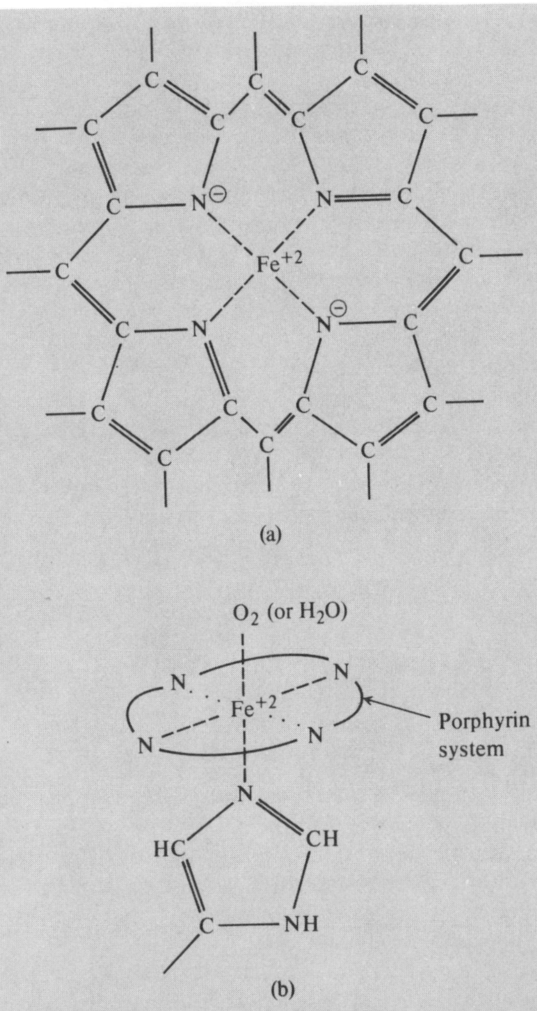

Fig. 15.19. (a) Hemoglobin bonding to the porphyrin (one of the resonance forms). (b) Hemoglobin bonding to imidazole and O_2 (or H_2O).

would contain one electron each, with a total of four unpaired electrons. Since this is *not* what is observed for this molecule when O_2 is bonded, we must look further.

If hybrids are formed they will be made up of the $3d_{x^2-y^2}$, $3d_{z^2}$, $4s$, and $4p$ atomic orbitals of Fe^{+2}. The $3d_{xy}$, $3d_{yz}$, and $3d_{xz}$ atomic orbitals remain undisturbed and will be lower in energy than the d^2sp^3 hybrid orbitals. The six d electrons therefore go into these three d orbitals, a pair into each orbital. The hybrid orbitals are empty and there are no unpaired electrons. If bonds are to be formed using three hybrid orbitals, they must be coordinate covalent: the ligand must donate both electrons.

Four of the six hybrid orbitals point toward the nitrogens of the porphyrin ring. Figure 15.19(a) depicts only one of the resonance forms. Actually the system is completely conjugated: the four nitrogens are all equivalent (Section 15.9). These nitrogens are sp^2 hybridized in a manner similar to carbon, but nitrogen has one more electron. A single electron is located in two of the sp^2 hybrid orbitals, which may then form σ bonds by overlapping with sp^2 hybrid orbitals on neighboring carbons. Another electron goes into the nitrogen p_z orbital, which may then participate in the conjugated π system that stretches over the entire ring system (Section 15.9). This leaves a pair of electrons in the third sp^2 hybrid orbital, which interestingly enough points directly toward the Fe^{+2} ion and overlaps with an empty d^2sp^3 iron hybrid orbital, forming a coordinate covalent bond. Thus there are four of these bonds in the plane of the ring system, one bond between each of the four nitrogens and the Fe^{+2}.

The coordinate covalent bond with the imidazole nitrogen also involves a filled sp^2 hybrid overlap with another d^2sp^3 hybrid of the Fe^{+2}.

The question of the bonding of O_2 has not at this time been settled. Two of the possibilities are

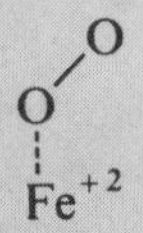

```
      O
     /
    O             O=O
    ¦     and      ¦
  Fe+2           Fe+2
```

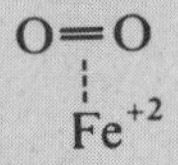

In the first case the oxygen atoms are sp^2 hybridized with a single electron in one of the hybrid orbitals, a single electron in the p_z orbital, and a pair of electrons in each of the other two sp^2 hybrid orbitals. A σ bond and a π bond may be formed between the two oxygen atoms by overlap of their sp^2 and p orbitals, each containing a single electron. One of the filled sp^2 orbitals of one of the oxygen atoms may then overlap with an empty d^2sp^3 orbital of Fe^{+2}, forming a coordinate covalent bond.

In the second case a filled π molecular orbital of O_2 may overlap with an empty d^2sp^3 orbital of Fe^{+2}, resulting in a coordinate covalent bond.

In any case the bond with oxygen is rather weak. O_2 is readily released in the cells. O_2 is also readily replaced by poisons such as carbon monoxide.

When the hemoglobin molecule contains H_2O instead of O_2 there are four unpaired electrons. In this case the iron ion behaves essentially as a free ion; the stability of hemoglobin is due to the attraction between the Fe^{+2} ion and the dipoles of the surrounding groups. An alternative explanation of bonding in such transition metal complexes, called crystal field theory, accounts for both the O_2 and the H_2O cases without using hybrids.[2] The model we use here is quite sufficient for our purposes, however.

[2] See for instance Walter J. Moore, *Physical Chemistry* 4th ed., Englewood Cliffs, N.J.: Prentice-Hall (1972).

15.9. Delocalized Molecular Orbitals: Conjugated Pi Bonds

Although in most cases molecular structure may be discussed in terms of a collection of atoms held together by localized σ or π bonds, there are some important exceptions. One of the most interesting of these occurs when there is a series of π bonds located between alternate pairs of bonded atoms, that is when each atom of a series of adjacent atoms has a p_z atomic orbital containing a single electron. The porphyrin ring system of hemoglobin discussed in the previous section is one example. Other examples include butadiene

$$CH_2{=}CH{-}CH{=}CH_2$$

and benzene

```
        CH
     //    \
  HC        CH
   |        ||
  HC        CH
     \\    /
        CH
```

Localized bonds fail to explain some very important experimental observations for these and related molecules. Butadiene is a planar molecule. The bonds shown above as double bonds are actually both 1.35 Å long, slightly longer than the normal 1.34-Å double bond between two carbon atoms, while the central bond written as a single σ bond is found to be 1.46 Å, significantly shorter than the usual carbon-carbon single bond length of 1.54 Å. In benzene all six carbon-carbon bonds are found to be equivalent with all bond lengths equal to 1.40 Å, intermediate between a normal single bond and a normal isolated double bond.

These observations may be explained by assuming that the π bonds are *delocalized.* In butadiene, we begin by setting up the localized sigma bonds. These will be composed of sp^2 hybrid orbitals on the carbons and $1s$ orbitals on the hydrogens. There will be remaining on each of the four carbons a p_z orbital containing one electron (Fig. 15.20a). If π bonding occurred only by the overlap of two adjacent p_z orbitals, the two resonance extremes of localized π bonds could be obtained (Fig. 15.20b). If all four p_z orbitals are allowed to overlap simultaneously, however, a molecular orbital is obtained which extends over all four carbon atoms. It is delocalized (Fig. 15.20c). In other words, there will be molecular orbitals made up of the linear combinations of four p_z atomic orbitals, one from each of the four carbon atoms. These linear combinations will be of the form

$$\psi_i = a_i(2p_{z_1}) + b_i(2p_{z_2}) + c_i(2p_{z_3}) + d_i(2p_{z_4}), \qquad (309)$$

where a_i, b_i, c_i, and d_i are constant coefficients for a given LCAO-MO, ψ_i. The values of these coefficients may be found by minimizing the energy

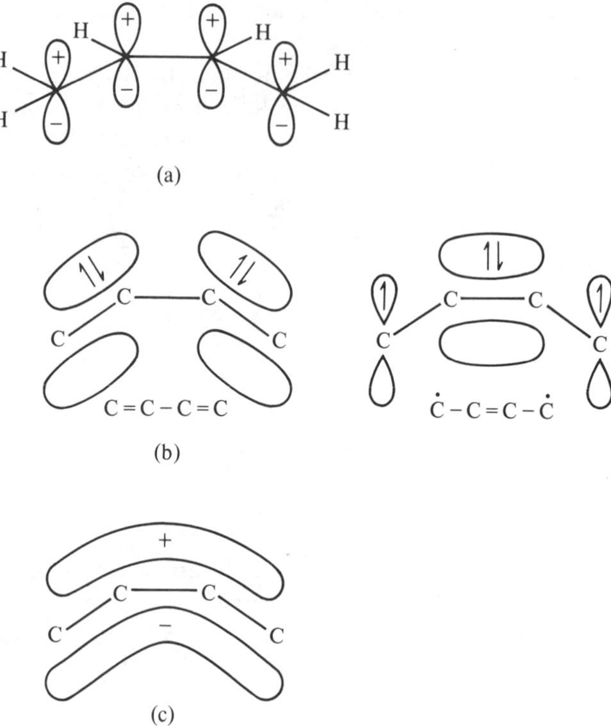

Figure 15.20

(variation principle, Section 14.4).[3] The four wave functions obtained are given in Fig. 15.21. The first two are bonding orbitals; the last two are antibonding. Notice the number of nodes. There are four electrons (one from each atom's p_z orbital); two will be in ψ_1 and two in ψ_2 since these two orbitals are lower in energy than ψ_3 and ψ_4. The total π bonding wave function will be the product of the occupied molecular orbitals. There is π character between atoms 1 and 2 and between atoms 3 and 4 in both occupied molecular orbitals. Thus these pairs of carbons may be considered to be linked by a double bond (a π bond and a σ bond). There will be only a partial double bond between carbons 2 and 3; while there is π character between these two atoms in ψ_1, there is none in ψ_2.

The π bonds in butadiene are said to be *conjugated*. Rather than being localized in a bonding orbital between two atoms, the electrons belong to orbitals which extend over the entire conjugated system.

---

[3] A complete discussion of this procedure is given in M. W. Hanna, *Quantum Mechanics in Chemistry*, 2nd ed., Menlo Park, Calif.: W. A. Benjamin (1969) Chapter 8.

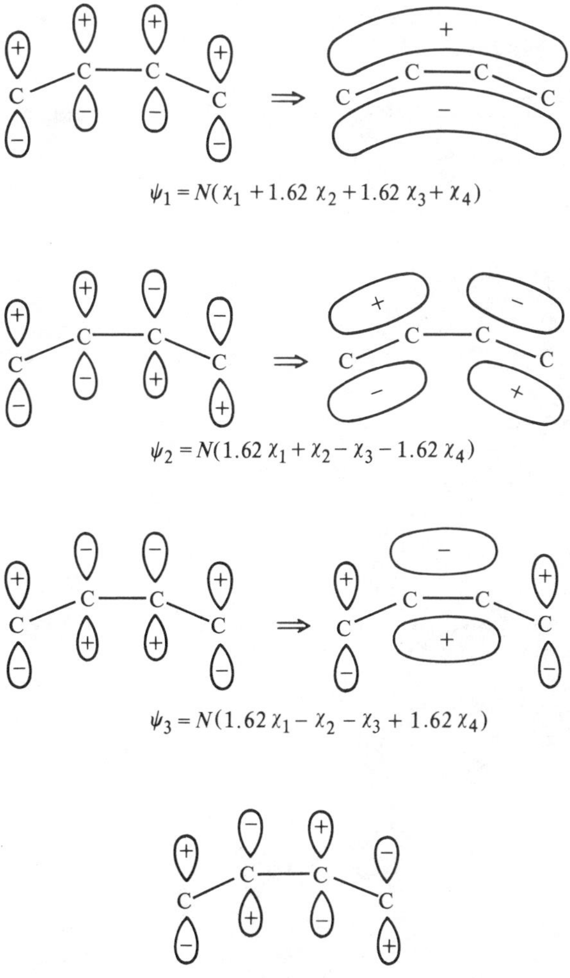

$$\psi_1 = N(\chi_1 + 1.62\,\chi_2 + 1.62\,\chi_3 + \chi_4)$$

$$\psi_2 = N(1.62\,\chi_1 + \chi_2 - \chi_3 - 1.62\,\chi_4)$$

$$\psi_3 = N(1.62\,\chi_1 - \chi_2 - \chi_3 + 1.62\,\chi_4)$$

$$\psi_4 = N(\chi_1 - 1.62\,\chi_2 + 1.62\,\chi_3 - \chi_4)$$

Figure 15.21

The π bonds of benzene are also conjugated and the development of a complete molecular orbital model for this molecule is similar to that for butadiene.[3] The six π electrons are paired in three bonding molecular orbitals which are linear combinations of the six atomic p_z orbitals. Again these π electrons are not localized but are free to move over the entire conjugated system.

15.10. Applications

Many molecular properties may be rationalized or predicted by appealing to the model that quantum theory provides. Some of these properties and their relationship to molecular structure have been discussed or alluded to in the foregoing sections. In the examples that follow and in the problems at the end of the chapter some of these will be amplified and other properties introduced.

Example 91 Predict the relative order of bond length and bond strength in the following molecules: (a) O_2^{-2}, (b) O_2^-, (c) O_2, and (d) O_2^+.

Answer: The molecular orbital diagram of O_2 was given in Section 15.5. The occupation of the valence orbitals is

	___		σ_{2p}^*
↑		↑	π_{2p}^*
↑↓		↑↓	π_{2p}
	↑↓		σ_{2p}

The *bond order* is equal to half the number of electrons in bonding orbitals minus half the number in antibonding orbitals. The filled orbitals with energies below σ_{2p} obviously contribute nothing to the bonding order since there are exactly the same number of bonding and antibonding electrons. Considering the valence orbitals, the bond order is $(\frac{1}{2})(6-2)=2$. The O_2 molecule has a double bond. Now consider the ions:

a) O_2^{-2} Bond order $= \frac{1}{2}(6-4) = 1$
18 electrons

	___	
↑↓		↑↓
↑↓		↑↓
	↑↓	

b) O_2^- Bond order $= \frac{1}{2}(6-3) = 1.5$
17 electrons

	___	
↑↓		↑
↑↓		↑↓
	↑↓	

c) O_2, 16 electrons, bond order = 2 (see above)

d) O_2^+ Bond order $= \frac{1}{2}(6-1) = 2.5$
15 electrons

	___	
↑		___
↑↓		↑↓
	↑↓	

The value of the equilibrium internuclear distance, the minimum in the E versus r_{AB} curve (see Fig. 15.1), is determined by two opposing effects. As the nuclei move closer, their atomic orbitals overlap more and the stability of the bond increases. Opposing this is the repulsion of the nuclei and the nonbonding electrons. The greater the number of pairs of orbitals which overlap and form bonds, the greater the energy of stabilization and the closer the nuclei will move before the repulsion term balances that energy. Thus the bond length will be shorter when the number of bonds (the bond order) is larger. We would predict, then, that the bond lengths should be:

$$O_2^+ < O_2 < O_2^- < O_2^{-2}.$$

The experimental bond lengths are:

$$O_2^+: 1.12 \text{ Å} \qquad O_2^-: 1.28 \text{ Å}$$
$$O_2: 1.21 \text{ Å} \qquad O_2^{-2}: 1.49 \text{ Å}$$

The bond strength is the total energy necessary to break all the bonds between the two bonded atoms, that is the *dissociation energy*. All of the above molecular species have a bonding sigma orbital. The difference in dissociation energy will be a result of the difference in bonding and antibonding π electrons. The bond strength should therefore be directly related to the bond order and should be in the order

$$O_2^+ > O_2 > O_2^- > O_2^{-2}.$$

This is observed experimentally. For example, the dissociation energy for O_2^+ is 6.48 eV, while for O_2 it is 5.08 eV.

Differences in the strengths of single bonds are due to the amount of orbital overlap. Thus a π bond resulting from the overlap of parallel orbitals is weaker than a σ bond made up of orbitals directed toward each other, allowing for greater overlap.

Example 92 The *resonance energy* or *delocalization energy* is defined to be the extra energy of stabilization a conjugated system possesses. It is calculated by subtracting the energy of formation of the nonconjugated molecule from that of the conjugated molecule. The latter can be determined thermodynamically; the former cannot be measured directly because the nonconjugated molecule does not actually exist. We calculate this quantity from *bond energies*, the energies of formation of particular bonds, determined by thermodynamic measurements of bond dissociation energies. Using the experimental enthalpy of combustion of benzene (Table 2.2) and the bond combustion enthalpies given in Table 15.2, calculate the delocalization enthalpy of benzene.

Table 15.2. Bond Combustion Enthalpies

Bond	ΔH_C, kcal mole^{-1}	Reaction
C—H	−54.0	$C{-}H + \frac{5}{4}O_2 \rightarrow CO_2 + \frac{1}{2}H_2O$
C—C	−49.3	$C{-}C + 2O_2 \rightarrow 2CO_2$
C═C	−117.5	$C{=}C + 2O_2 \rightarrow 2CO_2$

Answer: The calculated molar enthalpy of combustion of a mole of one of the resonance forms of benzene

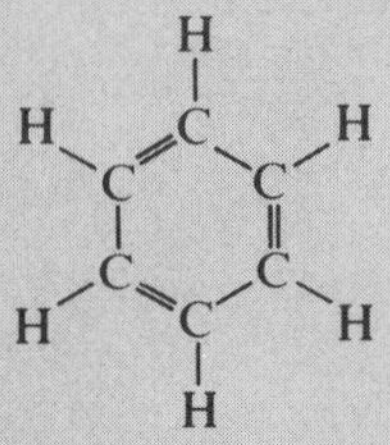

is the sum of the bond combustion enthalpies: $6[\Delta H_C(\text{C—H})] + 3[\Delta H_C(\text{C—C})] + 3[\Delta H_C(\text{C=C})] = 6(-54.0) + 3(-49.3) + 3(-117.5) = -824.4$ kcal mole^{-1}. The experimental enthalpy of combustion value from Table 2.2 is -789.1 kcal mole^{-1}. The difference gives us the additional stabilization due to the delocalization of the π electrons: 35.3 kcal mole^{-1}.

Example 93 Arrange the following in the order of increasing acidity: (a) Cl_3CCOOH, (b) $Cl_2HCCOOH$, (c) ClH_2CCOOH, and (d) H_3CCOOH.

Answer: The ability of these molecules to act as acids will be determined to a great extent by the stability of the resulting ion $RCOO^-$. Since electronegativity (see Section 15.2) is a relative measure of the ability of an atom to attract and hold an electron and since oxygen has a relatively high electronegativity (3.5), this ion can be expected to be relatively stable. But what effect does a change in R have on the stability of the ion?

If the oxygen bearing the "extra" electron could be made more electronegative, the stability of the ion would be increased. Since the attraction of an atom for an electron is determined by the effective nuclear charge Z' (Section 14.4), any change which increases Z' (or anything which decreases the shielding σ) increases the electronegativity. The shielding is determined by the electrons associated with the atom which are in a position between the electron in question and the nucleus so that they effectively decrease the nuclear charge felt by the outer electron. Altering the $C—O^-$ bond so that the electrons of the bond spend more of their time closer to the carbon (increasing λ in $\psi_{MO} = \chi_O + \lambda\chi_C$) would decrease the contribution of these electrons to the shielding and increase the stability of O^-.

This latter effect may be accomplished by making the carbon atom of the C—O bond more electronegative. And by similar reasoning this may be done by making the carbon to which it is attached more electronegative. In other words, if atom A in a molecule such as A—B—C—D is made more electronegative, then the effect is passed down the chain, resulting in a general shift of the electrons in all three bonds toward the A-end of the molecule.

Chlorine is more electronegative than hydrogen. Thus when a chlorine atom is substituted for one of the hydrogens in H_3CCOOH to give ClH_2CCOOH, the ion ClH_2CCOO^- is more stable than H_3CCOO^-. Consequently ClH_2CCOOH is more acidic than H_3CCOOH. Replacing a second and then a third hydrogen with chlorines increases the electronegativity further, resulting in even greater stability for the ion. Therefore the acidities of the molecules are in the order

$$H_3CCOOH < ClH_2CCOOH < Cl_2HCCOOH < Cl_3CCOOH.$$

Example 94 1,4-dioxane may exist in either the "chair" form or the "boat" form (see Fig. 15.22a). How may we use a measurement of the dipole moment to distinguish between the two forms?

Answer: The dipole moment μ is defined as

$$\mu = qr, \tag{310}$$

where r is the distance between two equal and opposite electrical charges of magnitude q. As mentioned before, if the effective center of positive and negative charge is not at the same point, the bond is polar and a dipole moment results.

Since oxygen is more electronegative than carbon, C—O bonds are generally polar. Figure 15.22(b) illustrates that even though there are bond dipoles, the center

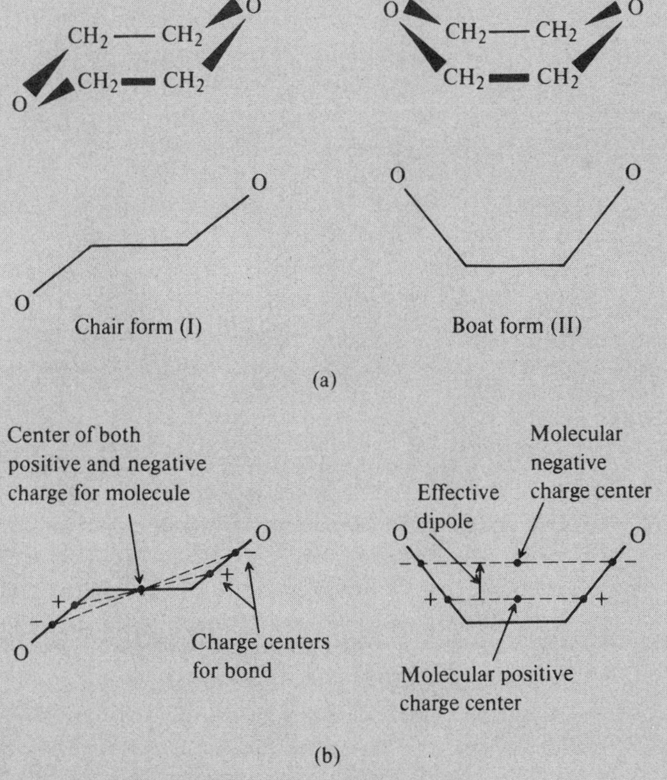

Fig. 15.22. Dioxane

of positive and negative charges for the entire molecule may still be located at the same point, so that the entire molecule has no dipole. The "chair" form of dioxane has no separation of charge centers, while the "boat" form does. The "chair" form has no dipole moment, but the "boat" form does.

Problems

15.1 Discuss the molecular orbital treatment of the molecular ion HeH^+ relative to that for H_2.

15.2 Describe the sigma bond between nitrogen and boron in $H_3N{-}BCl_3$.

15.3 Everything else being equal, sigma bonds formed using sp^3 hybrid atomic orbitals are stronger than those formed using p atomic orbitals, which are in turn stronger than those formed from s atomic orbitals. Why do you suppose this is true?

15.4 Explain the following trends in molecular parameters of the hydrogen halides in terms of quantum chemistry:

Molecule	Bond length ($\times 10^8$ cm)	Bond dissociation energy, kcal/mole^{-1}	Ionic character, %	Dipole moment, Debyes
HF	0.92	135	40	1.91
HCl	1.27	102	17	1.03
HBr	1.41	87	11	0.78
HI	1.61	71	5	—

15.5 NH_3 has a dipole moment but BF_3 does not. Discuss the molecular structures of these compounds.

15.6 Discuss the bonds and the molecular structure of the planar molecule phosgene $COCl_2$.

15.7 The polyunsaturated fatty acid linoleic has several types of C—C bonds.

$$\mathrm{CH_3{-}\overset{2}{C}H_2{-}\overset{3}{C}H_2{-}CH_2{-}CH_2{-}\overset{6}{C}H{=}\overset{7}{C}H{-}\overset{8}{C}H_2{-}CH{=}CH{-}CH_2{-}CH{=}CH{-}CH_2{-}CH_2{-}CH_2{-}CH_2{-}C({=}O)OH}$$

linoleic acid

(a) Discuss the bonds between carbons 2 and 3, between carbons 6 and 7, and between carbons 7 and 8. (b) Arrange these bonds in order of increasing length, explaining why this relative order should be expected.

15.8 In which of the following oximes will the C(1)—C(2) bond distance be shorter? Explain.

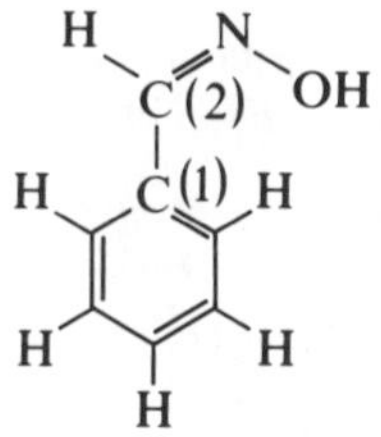

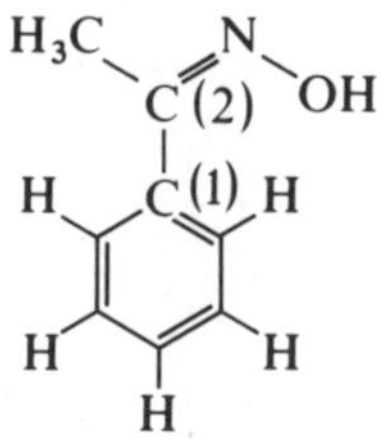

15.9 Discuss the bonding and the details of the structure of LSD.

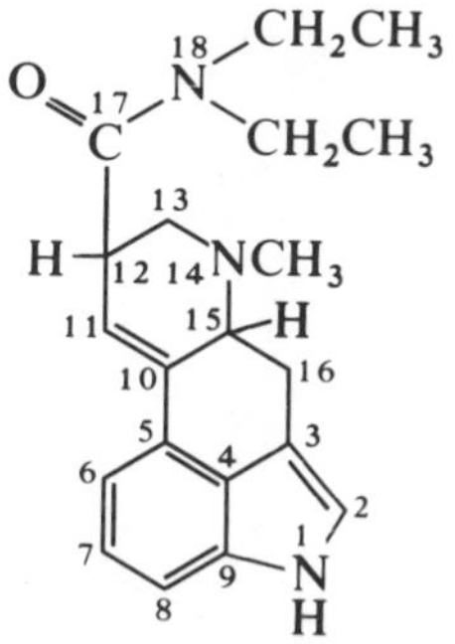

15.10 Is pyridine *aromatic* (that is, do its π electrons behave like those of benzene)? Explain.

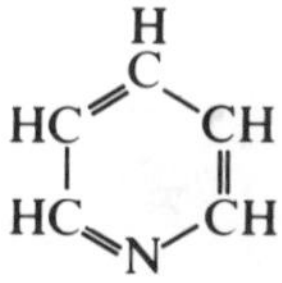

15.11 Discuss the structure of and the bonding in the nucleic acid component adenine.

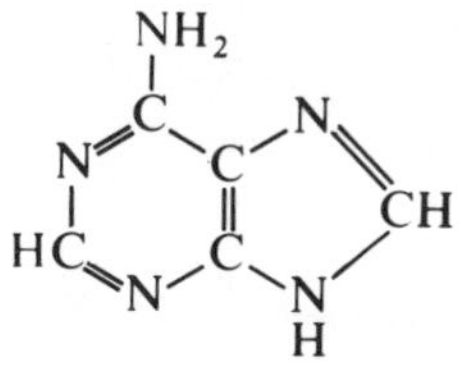

15.12 The structure of diborane is

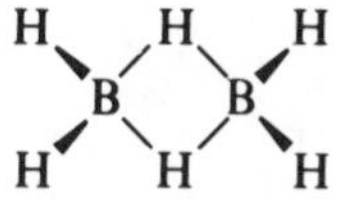

The two central hydrogen atoms are each bonded to *both* boron atoms. Show how these "three-centered" bonds (B—H—B) may be explained by the LCAO-MO approach using linear combinations of atomic orbitals on three atoms to form a molecular orbital.

15.13 Discuss the molecular orbital treatment of the vitamin A_1 structure. Contrast this approach with the particle-in-the-box treatment (see Example 73).

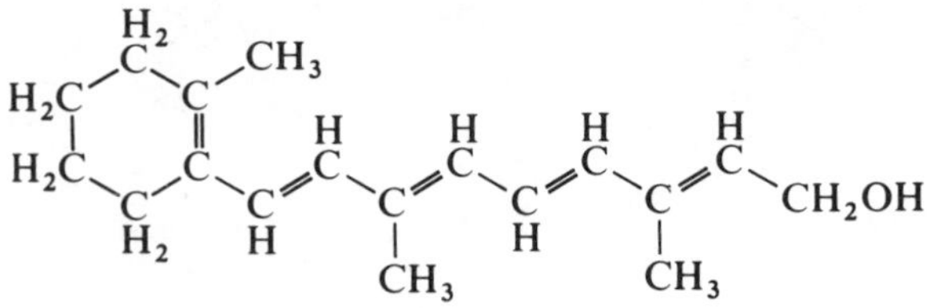

16 / Molecular Rotation and Vibration

16.1. Introduction

In the preceding chapter we concerned ourselves with the electronic wave functions and energies of molecules. Implicit in this treatment was a picture of a stationary molecule whose bond lengths are fixed at their equilibrium values. The Born-Oppenheimer approximation (Section 15.3) allowed us to focus our attention on the electronic functions and to postpone consideration of the effects which depend on the location of the nuclei.

Actually a molecule is far from stationary. It moves *translationally* through the space of its container. It changes its relative orientation in space by *rotating*. Molecular *vibration* results in the oscillation of bond lengths and bond angles about equilibrium values. Since each of these phenomena depends on the positions and the motions of the nuclei which make up the molecule, a quantum mechanical treatment of the nuclear positions and motions should provide valuable additional insight into the behavior of molecules.

16.2. The Translation and Internal Wave Equations

We wish to examine the wave mechanical consequence of the motion of the atoms which comprise a molecule. In our treatment we will consider each atom to be a single particle with its position equal to the location of the atomic nucleus. The wave mechanical results for the motion of the electrons relative to the atomic or nuclear positions are given in Chapter 15. The total behavior of the molecule will be the result of a combination of the nuclear and electronic effects.

Let us initially confine our attention to a diatomic molecule AB, such as HCl, H_2 and NO.

The Hamiltonian operator for the system composed of two particles A and B held together by a mutual potential energy V is

$$\mathscr{H} = -(h^2/8m_A\pi^2)\nabla_A^2 - (h^2/8m_B\pi^2)\nabla_B^2 + V(r_{AB}). \tag{311}$$

The potential energy term is a function of the internuclear separation r_{AB}. This term contains the repulsion between nuclei and between electrons as well as the attraction of nuclei for electrons. The potential energy as a function of r_{AB} is illustrated in Fig. 15.1.

Since the potential energy is a function of the relative distance between the nuclei rather than the position of the molecule in the container, the Schrödinger equation for the diatomic molecule may be simplified somewhat. The translation of the molecule may be treated by considering the translational

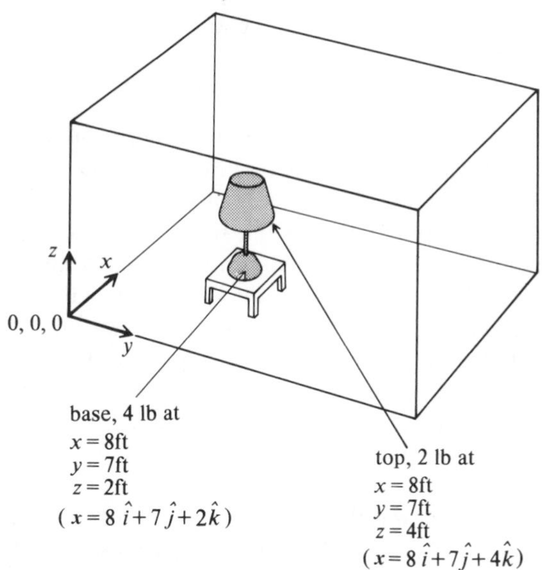

Figure 16.1

motion of the molecule's *center of mass* rather than the translation of each individual atom. The remaining motion of the molecule is *internal*: it does not depend on the location of the molecule in the container but only on the relative positions of the atoms.

The center of mass $\boldsymbol{R}$ of a diatomic molecule is given by

$$\boldsymbol{R} = (m_{\mathrm{A}}\,\boldsymbol{x}_{\mathrm{A}} + m_{\mathrm{B}}\,\boldsymbol{x}_{\mathrm{B}})/(m_{\mathrm{A}} + m_{\mathrm{B}}), \tag{312}$$

where m is the mass and $\boldsymbol{x}$ the vector position of the atoms (see Section 13.3).

Example 95 Determine the center of mass of a lamp whose location in the room is given in Fig. 16.1.

Answer: $R_x = [(2)(8) + (4)(8)]/(2 + 4) = 8$ ft

$R_y = [(2)(7) + (4)(7)]/(2 + 4) = 7$ ft

$R_z = [(2)(4) + (4)(2)]/(2 + 4) = 2\frac{2}{3}$ ft

$\boldsymbol{R} = (8\hat{\imath} + 7\hat{\jmath} + 2\frac{2}{3}\hat{k})$ ft

The Schrödinger equation, rewritten in terms of the coordinates of the center of mass and the internal coordinates, may be separated into two parts, an equation in terms of the center of mass:

$$-[h^2/8(m_{\mathrm{A}} + m_{\mathrm{B}})\pi^2]\nabla_R^2\,\psi_{\mathrm{trans}} = E_{\mathrm{trans}}\,\psi_{\mathrm{trans}} \tag{313}$$

and an equation in terms of the internal coordinates:

$$[-(h^2/8\mu\pi^2)\nabla_r^2 + V(r_{AB})]\psi_{int} = E_{int}\psi_{int}. \tag{314}$$

The total energy is the sum of the internal energy and the translational energy, and the total wave function is the product of the two wave functions.

Equation (313) is identical with the Schrödinger equation for a particle of mass $m_A + m_B$ (the total mass of the molecule) possessing only kinetic energy and moving in a three-dimensional box (Section 13.5). The wave functions for this situation are given by Eq. (250) and the corresponding energies by Eq. (251). In other words, the translational motion of a molecule is the same as that of a particle of mass equal to the total mass of the molecule located at the center of mass of the molecule. As pointed out in Chapter 13, if the container has macroscopic dimensions the translational energy levels are so close together that they may be treated equally well by classical mechanics.

Of much greater interest is the wave equation in terms of internal variables (Eq. 314). Here μ is the reduced mass, $m_A m_B/(m_A + m_B)$. The coordinates involved in this equation do not change with the molecule's position in the container but only when the positions of the atoms change relative to the center of mass. The energy E_{int} will be a function only of the relative positions of the atoms.

Equation (314) is not easy to solve. The process involves first changing the coordinates from cartesian to spherical (positions given in terms of the distance from the origin r and the angles θ and ϕ; see Fig. 14.1). It may then be separated into two equations:

$$\left[-(h^2/8\mu\pi^2r^2)\frac{\partial}{\partial r}(r^2)\frac{\partial}{\partial r} + V(r)\right]\psi_{vib} = E_{vib}\psi_{vib} \tag{315}$$

and

$$-(h^2/8\mu\pi^2r^2)\left[(1/\sin\theta)\frac{\partial}{\partial\theta}\sin\theta\frac{\partial\psi_{rot}}{\partial\theta} + (1/\sin^2\theta)\frac{\partial^2\psi_{rot}}{\partial\phi}\right] = E_{rot}\psi_{rot}. \tag{316}$$

The total internal energy and the total internal wave function are then

$$E_{int} = E_{vib} + E_{rot} \tag{317}$$

$$\psi_{int} = \psi_{vib}\psi_{rot}. \tag{318}$$

The use of the subscripts $_{vib}$ and $_{rot}$ anticipates the results of the next few sections.

16.3. The Rigid Rotor

If r is held constant (the bond length of the molecule does not change), Eq. (316) has only the angular variables θ and ϕ. This is an approximation since the atoms of the molecule do oscillate about the equilibrium internuclear separation r_e, but the error introduced is not great. This assumption is known as the *rigid rotor* approximation. We shall discuss a more accurate approach in Section 16.5.

Equation (316) may now be solved (with a reasonable amount of mathematical maneuvering). The wave functions are made up of two parts, one dependent only upon θ and one only upon ϕ. These solutions are exactly the same as the $\psi(\theta)$ and $\psi(\phi)$ functions of Section 14.5, except that we now call the quantum number J instead of ℓ. The functions $\psi_{\text{rot}}(J, m)$ may be obtained from Table 14.1.

The allowed *rotational* energies are

$$E_{\text{rot}} = J(J + 1)(h^2/8\pi^2 I), \tag{319}$$

where I is the *moment of inertia* and is defined as

$$I = \sum_j m_j r_j^2, \tag{320}$$

where m_j is the mass of the jth atom of the molecule and r_j is the perpendicular distance between the jth atom and the *axis* about which the molecule is rotating. The sum is over all the atoms of the molecule. Rotation axes always pass through the molecular center of mass. For a diatomic molecule AB there is but one unique rotation axis (see Section 16.6) and

$$I = m_{\text{A}} r_{\text{A}}^2 + m_{\text{B}} r_{\text{B}}^2,$$

which reduces to

$$I = \mu r_{\text{AB}}^2 \tag{321}$$

since the internuclear distance $r_{\text{AB}} = r_{\text{A}} + r_{\text{B}}$.

Note that E_{rot} is a function of the quantum number J but not of m. Since the wave function is determined by the value of m as well as J and since m may have the values $-J, -J + 1, \ldots, 0, \ldots, J - 1, J$, there will be $2J + 1$ wave functions with the same energy. The degeneracy of each rotational energy level is $2J + 1$.

A rotating diatomic molecule, therefore, has associated with it a rotational wave function ψ_{rot} which is a function of the angular coordinates θ and ϕ and quantum numbers J and m. Its rotational energy E_{rot} is determined by its rotational wave function and depends on the moment of inertia, the value being given by Eq. (319). It may only possess rotational energy of this magnitude and may change its wave function and energy only by absorbing or emitting exactly the right amount of energy that will leave it in another quantized state with energy given by Eq. (319).

Example 96 Calculate the amount of energy which a H_2 molecule must absorb in order to change its rotational state from $J = 1$, $m = 0$ to $J = 4$, $m = -3$. The bond distance of H_2 is 0.74 Å.

Answer: The moment of inertia of the H_2 molecule is

$$I = \mu r^2 = \{(1/N)(1/N)/[(1/N) + (1/N)]\}(0.74 \times 10^{-8})^2$$

$$= \frac{1}{2N}(0.74 \times 10^{-8})^2 = \left[\frac{1}{2(6.02 \times 10^{23})}\right](0.74 \times 10^{-8})^2$$

$$= 4.5 \times 10^{-41} \text{ g-cm}^2 \text{ molecule}^{-1}$$

The energy of the initial ($J = 1$, $m = 0$) rotational state is

$$E_1 = 1(1 + 1)(6.6 \times 10^{-27} \text{ erg-sec})^2/8\pi^2(4.5 \times 10^{-41} \text{ g-cm}^2)$$
$$= 2.4 \times 10^{-14} \text{ erg molecule}^{-1} \text{ or } 1.46 \times 10^3 \text{ joules mole}^{-1},$$

and that of the final state is

$$E_2 = 4(4 + 1)(6.6 \times 10^{-27} \text{ erg-sec})^2/8\pi^2(4.5 \times 10^{-41} \text{ g-cm}^2)$$
$$= 2.4 \times 10^{-13} \text{ erg molecule}^{-1} \text{ or } 1.46 \times 10^4 \text{ joules mole}^{-1}.$$

The energy needed for the transition is

$$\Delta E = E_2 - E_1 = 2.2 \times 10^{-13} \text{ erg molecule}^{-1} \text{ or } 1.3 \times 10^4 \text{ joules mole}^{-1}.$$

16.4. The Harmonic Oscillator

Equation (315) is a function only of r, the distance between the two nuclei. It also contains $V(r_{AB})$, and before this equation can be solved we must substitute an algebraic function for the potential energy. Plots such as Fig. 15.1 show the exact experimental dependence of the energy on r_{AB}; we must find an algebraic function which gives this same dependence.

Perhaps the simplest approach to this problem is to consider the molecular bond to be like a spring whose potential energy is given by *Hooke's law*. Such a spring is called a *harmonic oscillator* and

$$V = \tfrac{1}{2}\kappa(r - r_e)^2, \tag{322}$$

where r is the instantaneous internuclear separation, r_e is the equilibrium separation, and κ is the *force constant* of the spring, a measure of the strength of the spring. In this case it is related to the strength of the molecular bond.

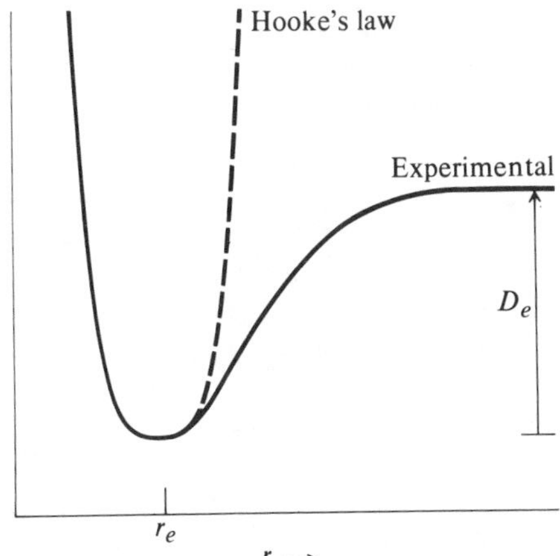

Figure 16.2

A comparison of this function and the experimental function appears in Fig. 16.2. The curves agree well at r values less than r_e but not when r is much larger than r_e. Nevertheless the agreement with experiment is surprisingly good. We shall discuss a more accurate approach in Section 16.5.

When this V is substituted into Eq. (315), solutions may be obtained. The resulting wave functions contain the *Hermite polynomials* and depend on a quantum number v which may be equal to zero or to any positive integer. Table 16.1 contains these functions for $v = 0$ through $v = 3$. The *vibrational energy* is given by

$$E_{\text{vib}} = (h/2\pi)(v + \tfrac{1}{2})\sqrt{\kappa/\mu}. \tag{323}$$

Table 16.1 Hermite Polynomial Solutions for the Harmonic Oscillator (Not Normalized). In these expressions a_0 and a_1 are constants, and $f = (r - r_e)/[(h^2/4\pi^2)/\mu\kappa]^{1/4}$.

v	ψ_{vib}
0	$(e^{-f^2/2})a_0$
1	$(e^{-f^2/2})a_1 f$
2	$(e^{-f^2/2})a_0(2f^2 - 1)$
3	$(e^{-f^2/2})a_1(\frac{2}{3}f^3 - f)$

The vibrational energy is quantized, depending on the quantum number v. It is nondegenerate and depends on the strength of the bond. The energy of vibration may also be written as

$$E_{\text{vib}} = h\nu_0(v + \tfrac{1}{2}), \tag{324}$$

where the constants $(\frac{1}{2}\pi)\sqrt{\kappa/\mu}$ have been grouped together and called ν_0, the "*natural frequency*" of the harmonic oscillator.

Notice that when $v = 0$, $E_{\text{vib}} = \frac{1}{2}h\nu \neq 0$. This is called the *zero point energy*. There will *always* be at least this much vibrational energy present.

Example 97 The difference in the energy of the $v = 0$ and $v = 4$ vibrational states of carbon monoxide is 1.72×10^{-12} erg molecule^{-1}. Calculate the force constant for CO.

Answer: The difference in energy between the $v = 0$ and $v = 4$ levels is

$$\begin{aligned}\Delta E = E_4 - E_0 &= \frac{h}{2\pi}(4 + \tfrac{1}{2})\sqrt{\kappa/\mu} - \frac{h}{2\pi}(0 + \tfrac{1}{2})\sqrt{\kappa/\mu}\\ &= 4(h/2\pi)\sqrt{\kappa/\mu}.\end{aligned}$$

The force constant is therefore

$$\kappa = (2\pi\Delta E/4h)^2\mu.$$

The reduced mass of CO is

$$\begin{aligned}\mu &= (12/N)(16/N)/[(12/N) + (16/N)] = 6.86/(6.02 \times 10^{23})\\ &= 1.14 \times 10^{-23}\text{ g},\end{aligned}$$

and κ is

$$\begin{aligned}\kappa &= [2\pi(1.72 \times 10^{-12}\text{ erg})/4(6.6 \times 10^{-27}\text{ erg-sec})]^2(1.14 \times 10^{-23}\text{ g})\\ &= 1.91 \times 10^6\text{ g sec}^{-2}.\end{aligned}$$

This is equal to 1910×10^3 dynes cm^{-1} or 1910 newtons meter^{-1}.

A vibrating diatomic molecule therefore has associated with it a wave function ψ_{vib} which is a function of the instantaneous displacement from the equilibrium bond distance and of a quantum number v. The vibrational energy is determined by the vibrational wave function and depends on the strength of the bond, the value being given by Eq. (323). The vibrational energy may only change by exactly the difference between two allowed energy levels.

Figure 16.3 gives the energy levels of a typical harmonic oscillator. Notice that zero vibrational energy is conventionally chosen as the lowest point of the potential energy curve. Zero vibrational energy is hypothetical, of course, since the vibrational energy is never less than the zero point ($v = 0$) energy $\frac{1}{2}h\nu_0$. The molecule vibrates about the equilibrium bond length with its minimum

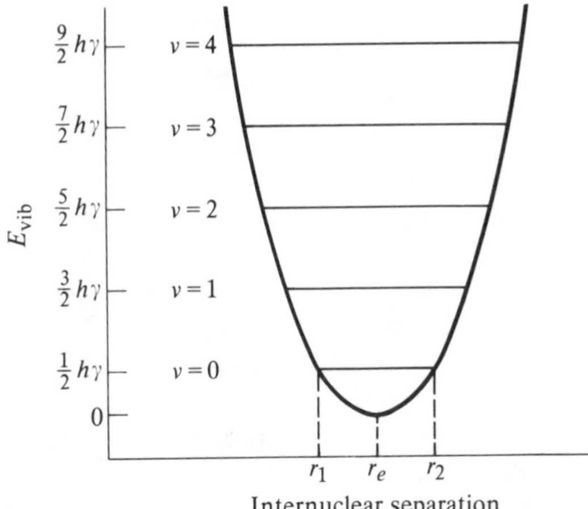

Fig. 16.3. Energy-level diagram for a harmonic oscillator.

and maximum values (for example, r_1 and r_2) determined approximately by the intersection of the energy level line and the potential energy curve. Note also that the energy levels of a harmonic oscillator are evenly spaced, the adjacent levels being separated by an energy $h\nu_0$.

16.5. The Vibration and Rotation of Real Diatomic Molecules

Real molecules are not rigid rotors, nor is their vibration harmonic. Nevertheless the energies of rotation and vibration derived assuming these approximations are accurate enough for many applications. If more accurate results are needed corrections may be applied.

Instead of assuming harmonic motion, we could use a function for the potential energy which fits the observed curve (Fig. 16.2) better. One such function is the *Morse potential*:

$$V = D_e(1 - e^{-b(r-r_e)})^2, \tag{325}$$

where D_e is the *dissociation energy* (see Fig. 16.2) and b is a constant. When this function is used for V in Eq. (315) the vibrational energy is

$$E_{\text{vib}} = \frac{h\omega}{2\pi}(v + \tfrac{1}{2}) - X\omega(v + \tfrac{1}{2})^2, \tag{326}$$

where $\omega = b\sqrt{2D_e/\mu}$ and $X = h^2\omega/16\pi^2 D_e$. Actually the correction is not so great as the complexity of this relationship might imply. The major change results from the second term, which becomes more important as v increases. (Figure 16.4 illustrates the net effect.) Whereas the vibrational energy levels for a harmonic oscillator are evenly spaced (Fig. 16.3), the levels for a real diatomic molecule are spaced closer together as v increases. Notice in Fig. 16.4 that as E_{vib} approaches the dissociation energy D_e, the separation of the energy levels approaches zero and the maximum departure from the equilibrium bond length r_e becomes progressively larger. When the vibrational energy of the bond exceeds the dissociation energy the molecule breaks apart; it dissociates into the constituent atoms or ions.

The use of the rigid rotor approximation involves the assumption that the bond length remains constant at the equilibrium value r_e while the molecule rotates. When this approximation is not used, the quantized energy levels are $E_{\text{rot}} = hB[J(J+1)] - hD[J(J+1)]^2 + hH[J(J+1)]^3 + hK[J(J+1)]^4 \cdots$. In most cases the correction terms beyond the D term are negligibly small and are ignored so that

$$E_{\text{rot}} = hB[J(J+1)] - hD[J(J+1)]^2. \tag{327}$$

B is the *rotational constant* and is equal to $h/8\pi^2 I$, while D is called the *centrifugal distortion constant* and is equal to $h^3/32\pi^4 I^2 r^2 \kappa$. This correction term causes the separation of rotational energy levels to be smaller than

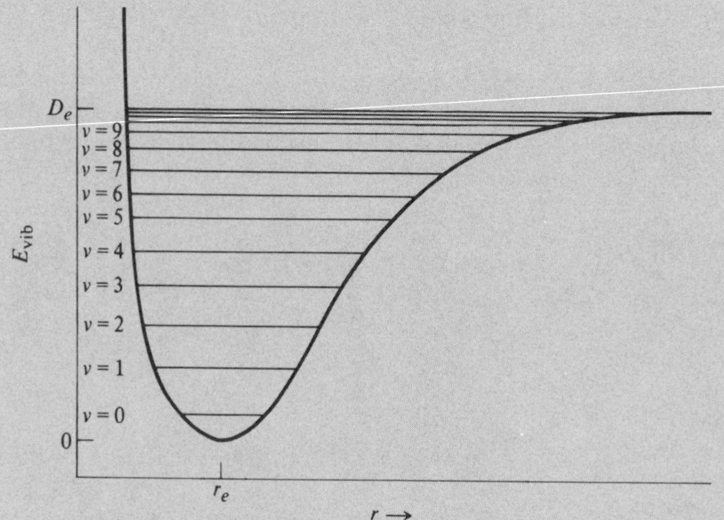

Fig. 16.4. Energy-level diagram for the vibration of a real diatomic molecule (Morse potential).

for the rigid rotor. The effect increases as the rotational quantum number increases. The change is a small one, the centrifugal distortion constant D usually being about 10^{-4} as large as the rotational constant B.

Although these corrections are important in some situations, especially when molecules are in states with large v and/or J values, for most cases they can be safely overlooked.

16.6. The Rotation of Polyatomic Molecules

The translation of an object may be described in terms of three mutually perpendicular components. We speak of the total velocity being made up of an x-component, a y-component and a z-component, the total velocity being expressed as

$$\mathbf{v} = \hat{\imath}v_x + \hat{\jmath}v_y + \hat{k}v_z. \tag{328}$$

In a similar manner, the rotation of any three-dimensional object may be expressed in terms of three components, one for the rotation of the object about each of three mutually perpendicular rotation axes. The total rotational motion is the resultant of these three components. The three rotation axes intersect at the center of mass of the object.

For a diatomic molecule these three axes are pictured in Fig. 16.5. For rotation about the axis which is along the bond, the moment of inertia I_a is extremely small and consequently the spacing of the energy levels (Eq. 319) is very large. Except at very high temperatures the molecules may be

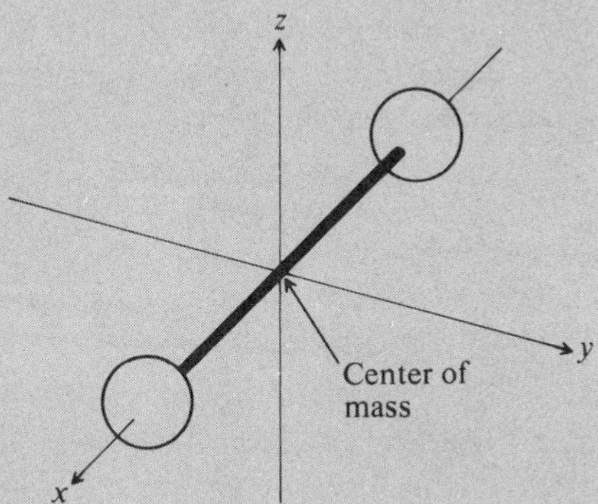

Fig. 16.5. The three mutually perpendicular rotation axes x, y, and z for a diatomic molecule.

considered to be in the lowest quantum state relative to rotation about this axis. Thus $J_a = 0$ and $E_{\text{rot}_a} = 0$; the molecule does not rotate about this axis.

Rotation of the diatomic molecule about either the y-axis or the z-axis, both of which are perpendicular to the bond, is clearly identical to rotation about the other. Rotation about either axis gives the same energy relationship (Eq. 319),

$$E_{\text{rot}_b} = J_b(J_b + 1)(h^2/8\pi^2 I_b)$$
$$E_{\text{rot}_c} = J_c(J_c + 1)(h^2/8\pi^2 I_c),$$

so that in addition to the m degeneracy discussed in Section 16.3 there is a two-fold degeneracy which results from the fact that $I_b = I_c$. The equation for the energy of a diatomic rigid rotor is usually written without the b and c subscripts, but we must remember that $E_{\text{rot}} = E_{\text{rot}_b} + E_{\text{rot}_c}$.

For linear polyatomic molecules such as CO_2 and C_2H_2 (acetylene), we again have one axis with $I_a \simeq 0$ and the other two axes having equal moments of inertia. The energy of rotation for a linear polyatomic molecule is therefore given by Eq. (319) also. Remember, however, that the equality $I = \sum_j m_j r_j^2$ now has more than two terms.

The situation becomes somewhat more complicated when nonlinear molecules are considered. There are several possible cases. When all three moments of inertia are equal, the molecule is called a *spherical top*. Examples include CH_4, SiH_4, and SF_6. As we shall see in Chapter 18, little use can be made of a knowledge of the energy levels of this class of molecules, and we shall not discuss them further.

When all three of the moments of inertia are different, the molecule is called an *asymmetric top*. Most molecules fall into this class. The treatment of asymmetric tops is extremely difficult and no general expression for the energy exists.

The third type of nonlinear polyatomic molecule occurs when two of the moments of inertia are equal but different from the third: $I_a \neq I_b = I_c$. These molecules are called *symmetric tops*. Examples include CH_3Cl and benzene.

Example 98 Show that benzene is a symmetric top.

Figure 16.6(a) shows the most convenient arrangement of the rotation axes for benzene. The x-axis is perpendicular to the plane of the molecule, while the y- and z-axes lie in the plane.

Answer: Consider first the x-axis. Figure 16.6(b) is the yz plane of the molecule. The x-axis is perpendicular to the page and passes through the center of the molecule. The perpendicular distance from all the carbon atoms to this axis is the same, r_{Cx}. This distance for all the hydrogens is also the same r_{Hx}. The moment of inertia is then $I_a = 6m_C r_{Cx}^2 + 6m_H r_{Hx}^2$.

The calculation of the moment of inertia about the y-axis is more complicated. From Fig. 16.6(c) and an elementary knowledge of trigonometry, the following relationships can be obtained:

$$r_{C_1y} = r_{C_4y} = r_{CC}$$
$$r_{C_2y} = r_{C_3y} = r_{C_5y} = r_{C_6y} = \tfrac{1}{2}r_{CC}$$
$$r_{H_1y} = r_{H_4y} = r_{C_1y} + r_{CH} = r_{CC} + r_{CH}$$
$$r_{H_2y} = r_{H_3y} = r_{H_5y} = r_{H_6y} = (r_{C_1y} + r_{CH}) \sin 30^\circ = 0.5(r_{CC} + r_{CH}).$$

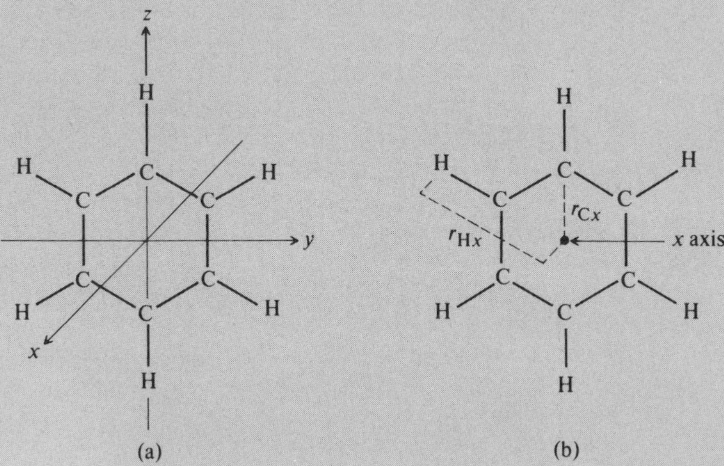

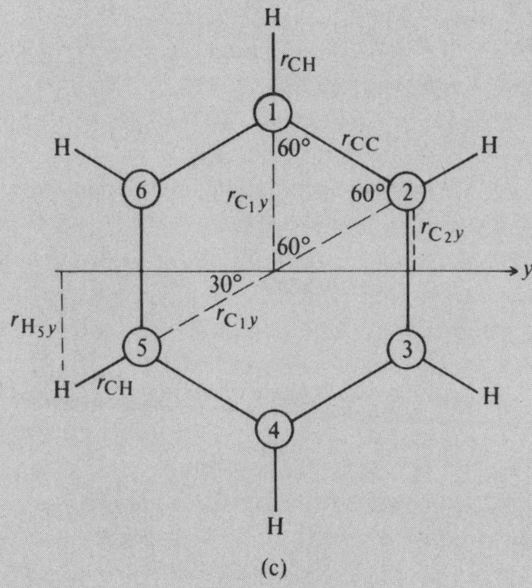

Figure 16.6

Figure 16.6 (*continued*)

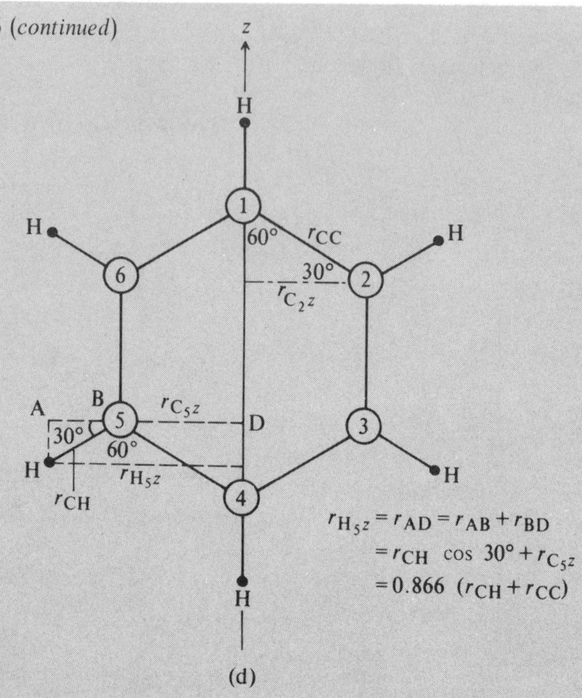

(d)

The moment of inertia about the y axis is then

$$I_b = m_C[2r_{CC}^2 + 4(\tfrac{1}{2}r_{CC})^2] + m_H\{2(r_{CC} + r_{CH})^2 + 4[0.5(r_{CC} + r_{CH})]^2\}$$
$$= m_C(2r_{CC}^2 + r_{CC}^2) + m_H[2(r_{CC} + r_{CH})^2 + (r_{CC} + r_{CH})^2]$$
$$= 3m_C r_{CC}^2 + 3m_H(r_{CC} + r_{CH})^2.$$

Similarly, Fig. 16.6(d) gives the appropriate information for determining the following relationships for the z-axis of rotation:

$$r_{C_1z} = r_{C_4z} = 0$$
$$r_{C_2z} = r_{C_3z} = r_{C_5z} = r_{C_6z} = r_{CC} \sin 60° = 0.866 r_{CC}$$
$$r_{H_1z} = r_{H_4z} = 0$$
$$r_{H_2z} = r_{H_3z} = r_{H_5z} = r_{H_6z} = 0.866(r_{CH} + r_{CC}),$$

and the moment of inertia about the z-axis is

$$I_c = m_C[2(0)^2 + 4(0.866 r_{CC})^2] + m_H[2(0)^2 + 4(0.866)(r_{CH} + r_{CC})^2]$$
$$= 3m_C r_{CC}^2 + 3m_H(r_{CH} + r_{CC})^2.$$

Thus $I_b = I_c \neq I_a$, as required for a symmetric top.

The quantum mechanical treatment of the symmetric top does yield a general equation for the energy of rotation:

$$E_{\text{rot}} = (h^2/8\pi^2 I_b)J(J+1) + (h^2/8\pi^2)\left(\frac{1}{I_a} - \frac{1}{I_b}\right)K^2. \tag{329}$$

There are two quantum numbers required. As before, J may have any positive integer value including zero, whereas K may be any positive or negative integer (including zero) for which the magnitude of K is less than or equal to J, that is, $|K| \leq J$.

16.7. The Vibration of Polyatomic Molecules

The motion of a molecule with n atoms may be completely characterized by specifying the x-, y-, and z-components of the velocity of each of the atoms of the molecule. There are thus $3n$ variables to be specified. These variables are known as *degrees of freedom.*

An alternative approach treats the translational motion of the molecule in terms of the three velocity components of the molecular center of mass, the rotational motion in terms of the three axes of rotation, and the vibrational motion in terms of changes in the bond distances and angles. For a molecule with n atoms we must still account for $3n$ degrees of freedom. For all molecules three variables are accounted for by the translation of the center of mass. There are thus $3n - 3$ internal degrees of freedom. For linear molecules (including diatomics) there is rotation about only two of the three axes of rotation. Rotation therefore accounts for two of the internal degrees of freedom, and there must be $3n - 5$ vibrational degrees of freedom. For H_2 there are $3n = 6$ total degrees of freedom, three translational, two rotational, and one vibrational. This one vibrational degree of freedom is the stretching of the bond.

For nonlinear molecules there is rotation about all three axes. Thus for these molecules there are $3n - 6$ vibrational degrees of freedom.

Unfortunately the simplicity of vibration in a diatomic molecule which involves only the stretching of a single bond is not found for polyatomic molecules. The vibrations of these molecules are complicated functions of changes in the lengths and angles of all the bonds of the molecules. Each of these $3n - 6$ (or $3n - 5$) motions (degrees of freedom) is called a *normal mode*. Each normal mode has its own set of energy levels given by Eq. (324) with its own vibrational quantum number and natural frequency, so that for a polyatomic molecule the vibrational energy is given by

$$E_{\text{vib}} = h \sum_j \nu_{0j}(v_j + \tfrac{1}{2}). \tag{330}$$

The sum is over the $3n - 6$ (nonlinear) or $3n - 5$ (linear) normal modes. Each natural frequency ν_0 is a function of the force constants of the molecular bonds, a measure of the ease of changing the length of the bonds and the size of the angles of the molecule.

Picturing the actual vibrations occurring in a particular normal mode is rather difficult. Considering several examples may prove instructive.

First consider the linear triatomic molecule CO_2. There are $3n - 5 = 4$ vibrational modes. We shall not carry out the details of determining these modes, but the results are given in Fig. 16.7(a). The symmetric stretching mode involves the simultaneous stretching and then contracting

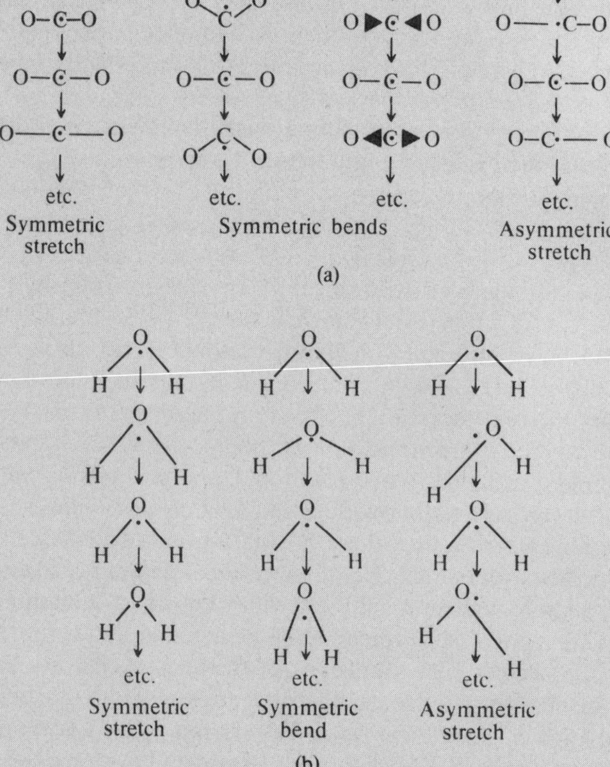

Fig. 16.7. (a) Vibrational modes for linear CO_2. (b) Vibrational modes for nonlinear H_2O.

of both carbon-oxygen bonds. The carbon position remains fixed at the center of gravity of the molecule. The two symmetric bending modes are degenerate. One involves the changing of the bond angle, the motion taking place in the plane of the page, the other involves the same motion but perpendicular to the page. Note that all the atoms move, the carbon in the opposite direction from the oxygens, but the center of mass of the molecule remains fixed. If the center of mass moved, this motion would involve translation as well as vibration.

Now consider the nonlinear H_2O molecule. Its $3n - 6 = 3$ vibrational modes are shown in Fig. 16.7(b). Notice that there is only one symmetric bend in this case. The out-of-plane bend of the linear molecule would result in rotational motion in this nonlinear molecule.

Example 99 Two of the natural frequencies ν_0 of the normal vibration modes for H_2O are approximately equal and are about twice the value of the third. Assign these three frequencies by referring to Fig. 16.7(b). What does this say about the relative ease of bending and stretching?

Answer: The symmetric and asymmetric stretching modes both involve the lengthening and shortening of bonds, whereas the bending mode involves only a change in angle at the oxygen atom. The two stretching modes may thus be expected to display similar characteristics and may be assigned the larger natural frequencies, while the smaller frequency is assigned to the bending mode.

Since $\nu_0 = (1/2\pi)\sqrt{\kappa/\mu}$, a larger value of ν_0 means a larger force constant κ. There is then more resistance to bond deformation than to changes in angle.

For larger molecules sorting out the normal modes can be quite a chore, but it can be done in many cases. A simplification sometimes occurs when the normal mode is primarily made up of the stretching of only one bond or only one type of bond (for example the C—H or C=O stretching modes in organic molecules) or when it is primarily made up of only one bond angle bend (for instance the H—C—H bending mode.)

Problems

16.1 Calculate the moment(s) of inertia for each of the following:

a) CO_2: $\quad r_{C-O} = 1.13$ Å

b) H_2O: $\quad r_{O-H} = 0.96$ Å, H—O—H = 104.5°

c) octahedral $PtCl_6^{-2}$: $\quad r_{Pt-Cl} = 2.30$ Å

d) phosgene, $COCl_2$: O=C 1.17 Å; C—Cl 1.74 Å; angle O—C—Cl 124°; angle Cl—C—Cl 112°

16.2 The internuclear separation in the carbon monoxide molecule is 1.13 Å. Calculate the energies of the $J = 1$ through $J = 5$ rotational states. Give the degeneracy of each energy level.

16.3 Calculate the energies of the three lowest rotational energy states for CO_2 (see Problem 16.1).

16.4 Determine the values of the energy for the five lowest rotational energy states of benzene. The C—C and C—H bond distances are 1.40 Å and 1.08 Å, respectively. (See Example 98).

16.5 The vibrational force constant is related to the strength of the bond in a diatomic molecule. Explain the following in terms of their molecular bonds:

a) O_2 $\kappa = 11.4 \times 10^5$ dynes cm^{-1}
   N_2 $\kappa = 22.6 \times 10^5$ dynes cm^{-1}

b) F_2 $\kappa = 4.5 \times 10^5$ dynes cm^{-1}
   Cl_2 $\kappa = 3.2 \times 10^5$ dynes cm^{-1}
   Br_2 $\kappa = 2.4 \times 10^5$ dynes cm^{-1}
   I_2 $\kappa = 1.7 \times 10^5$ dynes cm^{-1}

16.6 The natural vibrational frequencies of H_2, HD and D_2 are 1.26×10^{14} sec^{-1}, 1.08×10^{14} sec^{-1}, and 8.9×10^{13} sec^{-1} respectively. Calculate the force constant for each. Is this result surprising?

16.7 For carbon monoxide the internuclear separation is 1.13 Å and the dissociation energy is 11.1 eV. Sketch the Morse potential curve for CO (see Fig. 16.4). What difference does the value of the constant b make in the appearance of the curve?

16.8 If we could construct a spring of length 10 cm with a force constant equal to that of carbon monoxide (see Example 97), how far would it stretch when we attached a 100-g weight?

16.9 One of the normal modes of the following molecules is primarily the stretching of the C—O bond. Arrange these molecules in order of increasing ν_0 for this mode.

a) R—C(=O)—Cl

b) R—C(=O)—O$^-$

c) R—C(=O)—H

16.10 How many translational, rotational, and vibrational degrees of freedom does each of the molecules listed below have? How many normal modes?

a) carbon dioxide CO_2

b) ethanol C_2H_5OH

c) glucose $C_6H_{12}O_6$

16.11 In which of the linoleic acid C—C bonds specifically mentioned in Problem 15.7 will the normal mode corresponding primarily to the C—C stretch have the largest ν_0?

17 / Intermolecular Forces: Liquids and Solids

17.1. Introduction

In Chapter 15 we dealt in considerable detail with the formation of stable molecular bonds between atoms. The two types of bonds we considered were ionic and covalent. The forces holding the atoms together in an ionic bond involve only the attraction between oppositely charged ions. A covalent bond involves attractive forces between the nuclei and the electrons located in associated molecular orbitals. In both cases the atoms or ions approach each other closely and become firmly linked. Considerable energy is required to break them apart. The molecules resulting from two or more atoms held together by such bonds form an identifiable entity which acts as a unit in chemical reactions and physical processes.

We recognize the three states of matter from simple experience: gases flow and are compressible; liquids flow but are much less compressible; solids have rigidity. In Chapter 4 we considered some of the thermodynamic consequences of the existence of, and the transitions between, the various *phases* of pure substances. Let us now use our quantum chemical ideas to explore the states in which matter may exist. In our treatment we shall examine other forces whose action at the molecular level helps explain such behavior as the nonideality of gases, the existence of solids and liquids, differences in solubility among solvents and solutes, and the structure of macromolecules such as proteins and DNA.

17.2. Gases

In Section 1.12 the kinetic theory treatment of gases was presented in terms of a hypothetical ideal gas. An ideal gas is by definition composed of point particles which have no volume themselves and do not interact with each other. These particles are pictured as moving randomly and independently throughout the container, colliding elastically with its walls. The average velocity with which the particles are moving is a function of the temperature (Eq. 23). The equation of state for such a gas is $PV = nRT$ (Eq. 21), the ideal gas equation.

Many (but not all) gases composed of real atoms or molecules behave as the ideal gas equation predicts if the temperature is relatively high and the pressure relatively low. Still, all real gases deviate significantly from ideal behavior at low temperatures and high pressures. Real atoms and molecules are not point particles but occupy a significant volume; moreover real atoms and molecules do exert forces on each other. Modifications of the ideal gas equation to take these factors into account yield equations of state which are in much closer agreement with reality. One of the most widely used modified equations is that of van der Waals:

$$\left(P + \frac{an^2}{V^2}\right)(V - nb) = nRT. \tag{3}$$

In this equation the constant b is related to the volume occupied by a mole of particles, and a is related to the interactive forces among the particles. We shall discuss these forces in the next section.

A gas is thus pictured as a large number of randomly moving atoms or molecules evenly distributed throughout the container. The forces among these atoms or molecules only become significant when the density (n/V) is large, that is, at high pressure and low temperature. These forces may then become large enough to cause the liquefaction of the gas.

The translation of a gaseous molecule is treated quantum mechanically as the translation of a single particle of mass equal to the molecular mass located at the molecule's center of mass (Section 16.2). In containers of macroscopic dimensions, translational motion may be treated equally well by classical mechanics (Section 13.5).

Since each molecule of a gas is essentially independent, its vibrational and rotational motions may be treated by the methods of the previous chapter.

17.3. van der Waals Forces

The forces of interaction between the neutral molecules of gases and liquids are of several types. Collectively they are called van der Waals forces. The three main types are *dipole-dipole* forces, *dipole-induced dipole* forces, and *London dispersion* forces.

If a molecule possesses a permanent dipole, the positive end of the dipole of one molecule can attract the negative end of the dipole of another molecule. The potential energy of such an attraction between molecules of the same species is found to be

$$V_{\mathrm{dd}} = -2\mu^4/3kTr^6, \tag{331}$$

where μ is the dipole moment (Eq. 310) of the molecule, k is Boltzmann's constant, r is the distance between the two molecules, and T is the absolute temperature.

The presence of an electrical charge or permanent dipole on one molecule can cause a shift in the most probable positions of the electrons in a nearby molecule. The resulting charge separation is evidenced as an *induced dipole moment* in the second molecule. This induced dipole interacts with the permanent dipole, the potential energy being

$$V_{\mathrm{di}} = -2\alpha\mu^2/r^6, \tag{332}$$

where μ is the value of the permanent dipole moment and α is the *polarizability*, a measure of the ease with which the electrons of the second molecule can be shifted.

The interaction of atoms and molecules which do not possess permanent dipoles (such as neon, CO_2, and CH_4) cannot be accounted for by either of the above. This effect was explained by F. London in 1930. We have regarded the 1*s* orbital as spherically symmetric since an electron may be found with equal probability at any angle relative to the origin. Averaged over even a short period of time, such an orbital will have no dipole moment: the center of the positive and negative charges will be at the same point, the nucleus. *Instantaneously*, however, the electrons will be at particular positions in the orbital and unless they are exactly opposite each other, the center of negative charge will not be situated at the nucleus and there will be an *instantaneous orbital dipole*. This instantaneous dipole can induce a dipole in an orbital of another molecule. Even though a single interaction of this type lasts only an instant, the combined effect of many of these instantaneous interactions results in a net attractive force, called the *London dispersion* force. The resulting potential energy was calculated by London to be

$$V_L = -3h\nu'\alpha^2/4r^6, \tag{333}$$

where h is Planck's constant, α is the polarizability, and ν' is the frequency with which the orbital dipole changes.

Notice that in all three cases the attraction is a function of r^{-6}. The closer the molecules are to each other the greater the attraction, the attraction falling off rapidly as the two move apart. When the molecules move very close together, however, their filled molecular orbitals begin to overlap and the electrons in these orbitals repel each other. In addition, the nuclei of the two molecules repel when they approach closely. It is found experimentally that this repulsion potential energy is approximately of the form A/r^{12} where A is a constant which depends on the molecules involved.

The total potential energy of interaction is a combination of the attraction terms and the repulsion terms proportional to r^{-6} and r^{-12} respectively. This total potential is often written

$$V_{LJ} = Ar^{-12} - Br^{-6} \tag{334}$$

and is called the *Lennard-Jones potential.*

These, then, are the forces primarily responsible for the deviations from ideal behavior in gases and for the liquefaction of gases.

Example 100 One mole of a typical pure material occupies 20.0 ml when it is a liquid and 22.4 liters as a gas. Calculate the ratio of the dipole-dipole interaction energies of the liquid and gas at the boiling point of the material.

Answer: First calculate the volume occupied by a single molecule in each phase:

$$\text{Liquid:}\quad \frac{20.0\ \text{cm}^3\ \text{mole}^{-1}}{6.02\times10^{23}\ \text{molecules mole}^{-1}} = 3.32\times10^{-23}\ \text{cm}^3\ \text{molecule}^{-1}$$

$$\text{Gas:}\quad \frac{22400\ \text{cm}^3\ \text{mole}^{-1}}{6.02\times10^{23}\ \text{molecules mole}^{-1}} = 3.72\times10^{-20}\ \text{cm}^3\ \text{molecule}^{-1}$$

The average distance of molecular separation is equal to the length of a side of the cubic volume occupied by a single molecule:

$$\text{Liquid:}\quad r_{\text{liq}} = (3.32\times10^{-23}\ \text{cm}^3)^{1/3} = 3.21\times10^{-8}\ \text{cm}$$

$$\text{Gas:}\quad r_{\text{gas}} = (3.72\times10^{-20}\ \text{cm}^3)^{1/3} = 3.33\times10^{-7}\ \text{cm.}$$

The ratio of the dipole-dipole interaction (Eq. 331) in the two phases is

$$\frac{(V_{\text{di}})_{\text{liq}}}{(V_{\text{di}})_{\text{gas}}} = \frac{(-2\mu^4/3kTr^6)_{\text{liq}}}{(-2\mu^4/3kTr^6)_{\text{gas}}} = \frac{r_{\text{gas}}^6}{r_{\text{liq}}^6}$$

$$= \left(\frac{33.3\times10^{-8}}{3.21\times10^{-8}}\right)^6 = 1.25\times10^6.$$

17.4. Liquids

When the temperature is lowered and/or the pressure is increased sufficiently, all gases will become liquids. When the motion of the gas molecules is slowed sufficiently and the molecules are brought closer together (in other words the density is increased) the attractive forces of interaction are able to overcome at least a part of the random chaotic motion of the molecules.

The molecules of a liquid are much closer together than those of a gas. Even though they still move about translationally, the motion is not free and random as in the gas phase. At any instant every liquid molecule has roughly the same semiordered relation to its nearest neighbors. This is not a static situation, however; the molecules ceaselessly jostle each other and change their positions. The arrangement of a molecule relative to the molecules beyond its nearest neighbors is also not ordered but relatively random.

The development of a quantitative microscopic theory for liquids similar to the kinetic theory for gases has not been successful. The qualitative model of a liquid as intermediate between the total order of a solid and the chaos of a gas has proved helpful in the understanding of physical and chemical phenomena, however. Intermolecular forces may be used to explain such phenomena as *viscosity* (the resistance to flow) and *surface tension* (the net force on a liquid surface resulting from unbalanced intermolecular forces). The energy and enthalpy of vaporization would be expected to be positive, since energy must be supplied to overcome the intermolecular forces before a molecule can escape into the gas phase. The entropy of vaporization should likewise be positive since the gas phase is less ordered than the liquid.

The vibration of a molecule is virtually identical in the gas and liquid phases. The rotational motion, however, changes significantly when a gas liquefies. Though the liquid molecules still rotate, the rotation is no longer free. Interactions with adjacent molecules cause significant changes in the rotational energy levels; consequently a given level is represented by slightly differing energies among the various molecules of the system.

17.5. The Hydrogen Bond

One effect which can lead to seemingly anomalous values for such observations as the boiling point, the viscosity, the infrared spectrum (see Chapter 18), and the enthalpy of vaporization is the *hydrogen bond.*

A hydrogen bond results from an interaction between two electronegative atoms (usually oxygen, nitrogen, or fluorine), one of which is covalently bonded to a hydrogen atom. The electrons of a covalent bond between a highly electronegative atom and hydrogen are located much closer to the electronegative atom. This results in a polar bond with the hydrogen at the positive end of the dipole. The second electronegative atom (the *acceptor*) is the negative end of its dipole. The two dipoles interact, the net result being a relatively stable arrangement with the hydrogen located between the two electronegative atoms and attracted to both (see Fig. 17.1).

The attractive force which results in a hydrogen bond is stronger than an ordinary dipole-dipole interaction. This results from the fact that the hydrogen atom is so small it can come very close to the electronegative acceptor atom before repulsion becomes important. Since r is smaller than usual,

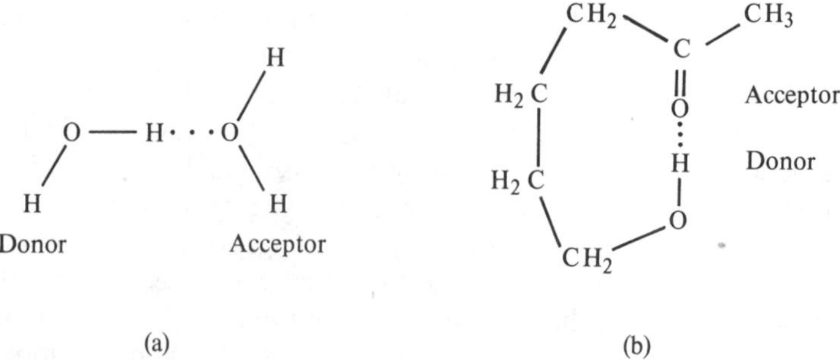

Fig. 17.1. (a) Intermolecular hydrogen bond between two water molecules. (b) Intramolecular hydrogen bond.

the interaction is correspondingly stronger (Eq. 331). Also electronegative atoms which act as acceptors all have *lone pairs* of electrons; these are exposed atomic valence orbitals which contain two unshared electrons. The overlap of this filled atomic orbital with a hydrogen 1*s* orbital which is "almost empty" (due to the shift of the electrons toward the electronegative atom to which the hydrogen is bonded) could result in a filled molecular orbital. Although the hydrogen bond is primarily a result of the electrostatic dipole-dipole interaction, some added stability is gained through this orbital overlap.

Hydrogen bonds may be formed between two different molecules (*intermolecular*) or between two electronegative atoms of the same molecule (*intramolecular*). These two types are illustrated in Fig. 17.1. One of the most interesting examples of intramolecular hydrogen bonding occurs in the deoxyribonucleic acid (DNA) molecule. The double helix formed by this molecule is a result of hydrogen bonds established between the two chains (Fig. 17.2).

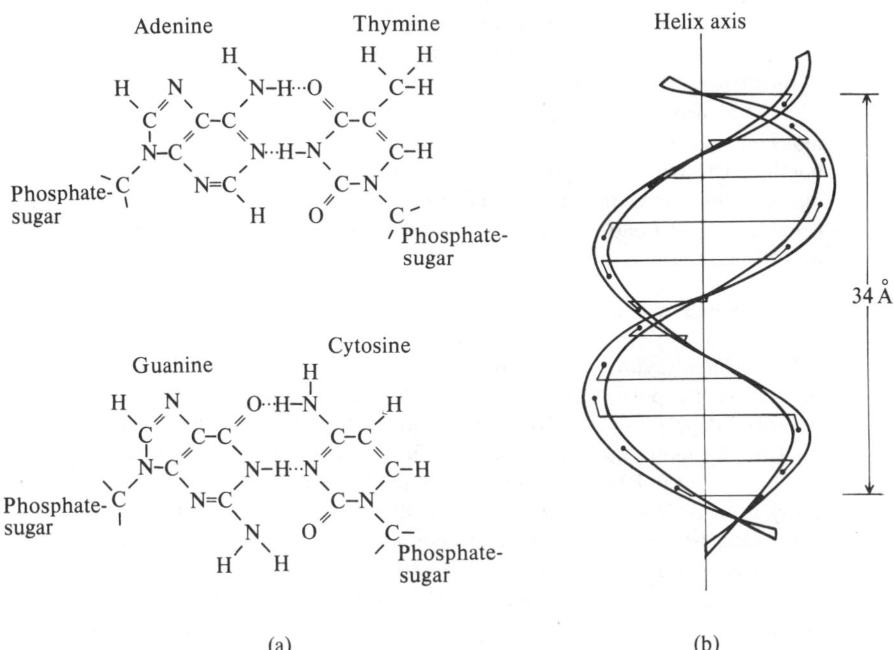

Fig. 17.2. (a) The specific pairing of hydrogen bonds (dotted lines) of the bases thymine and adenine and the bases cytosine and guanine as they occur in DNA. (b) Schematic diagram of DNA. The two spirals represent the phosphate-sugar backbone; the rods represent hydrogen-bonded base pairs. Adapted from R. B. Setlow and E. C. Pollard, *Molecular Biophysics* (1962), Addison-Wesley, Reading, Mass.

The fact that these bonds are strong enough to provide stability, but weak enough so that the strands of the DNA molecule can separate relatively easily, is essential for the activity of this most important genetic agent. The conformation of proteins and various fibers is also at least partially the result of intramolecular hydrogen bonding. The denaturation of proteins (for instance cooking an egg) involves the breaking of hydrogen bonds.

With a dissociation energy of about 5 kcal mole^{-1}, the hydrogen bond is significantly weaker than a covalent bond (approximately 100 kcal mole^{-1}). Liquids with hydrogen bonds have higher boiling points, larger vaporization energies, and higher viscosities than would otherwise be expected.

Example 101 Arrange the following in order of their expected boiling points: CH_3Cl, CH_2Cl_2, $CHCl_3$, H_2O, ethanol, argon, helium, and neon.

Answer: The relative magnitudes of boiling points are directly related to the relative intermolecular forces: substances with small intermolecular forces may be expected to have low boiling points.

Hydrogen bonding is generally stronger than van der Waals forces; dipole-dipole interactions are stronger than dipole-induced dipole interactions, which are in turn stronger than London dispersion forces.

Let us first collect the molecules listed into groups on the basis of the intermolecular forces which they may exhibit. Water and ethanol are capable of hydrogen bonding. CH_3Cl, CH_2Cl_2 and $CHCl_3$ cannot form hydrogen bonds but do have dipole moments. Argon, helium, and neon do not have dipole moments and are capable only of London dispersion forces. Tentatively we may arrange these in order of increasing boiling points:

$$\text{argon, helium, neon} < CH_3Cl,\ CH_2Cl_2,\ CHCl_3 < H_2O,\ \text{ethanol}.$$

For liquids with only London dispersion forces the relative size of the forces depends on the polarizability (Eq. 333). The ease with which a molecule may be polarized depends on the "tightness" with which the electrons are held by the nucleus. The looser the grip, the more diffuse the electron cloud about the nucleus, the higher the polarizability. In general the further the electron cloud extends away from the nucleus, the weaker the attraction between the nucleus and the outer electrons. Argon, helium, and neon are spherical atoms with argon larger than neon which is larger than helium. The polarizabilities, and therefore the London dispersion forces, would be expected to be in the order $He < Ne < Ar$. Since the forces holding the atoms in the liquid state are greatest for argon, its boiling point should be the highest of the three. The boiling points should be in the order

$$He < Ne < Ar.$$

For liquids whose intermolecular forces result primarily from dipoles, the larger the dipole the greater the intermolecular forces and thus the higher the boiling point. The dipole moments of the chloromethanes are in the order

$$CH_3Cl < CH_2Cl_2 < CHCl_3,$$

and the boiling points should be in the same order.

Both hydrogens of water are capable of forming hydrogen bonds with other water molecules. This, coupled with the relative simplicity of the H_2O molecules, results in liquid water being strongly hydrogen bonded throughout. The ethanol molecule forms fewer hydrogen bonds and should have a lower boiling point:

$$\text{ethanol} < H_2O.$$

The resulting order of boiling points is therefore:

$$\text{helium} < \text{neon} < \text{argon} < CH_3Cl < CH_2Cl_2 < CHCl_3 < \text{ethanol} < H_2O.$$

The experimentally determined boiling points are

He	−268.6°C	CH_2Cl_2	40°C
Ne	−245.9°C	$CHCl_3$	61.7°C
Ar	−185.7°C	ethanol	78.5°C
CH_3Cl	−23.8°C	H_2O	100°C

17.6. Liquid Solutions

The vapor pressure of a volatile solvent is related by Raoult's law (Eq. 97) to its mole fraction in dilute solution. An ideal binary solution is defined as one in which both components obey Raoult's law over the entire concentration range. The majority of solutions exhibit either positive or negative deviations: the vapor pressure over much of the concentration range is respectively either higher or lower than Raoult's law would predict.

The vapor pressure is a measure of the *escaping tendency* of molecules from the liquid solution. The more inclined are the molecules of a particular species to leave the solution and enter the gas phase, the higher will be the vapor pressure of that species. If the vapor pressure of a substance is to follow Raoult's law over the entire concentration range, the forces attempting to hold the molecules within the solution must be the same at all concentrations. For a given species, the variation of its vapor pressure with concentration will then be determined by the concentration of molecules of that species in the liquid surface. The vapor pressure will thus be proportional to the concentration.

The forces resisting the vaporization of an A molecule in the pure liquid are forces between A molecules, F_{AA}. As more and more of component B is introduced, forces between A and B molecules, F_{AB}, become important. At a mole fraction of 0.5, half the forces holding an average A molecule back will be AB forces and half will be AA forces. As the solution becomes very concentrated in B, the escape of A will be almost totally determined by F_{AB} since almost all of A's nearest neighbors will be B molecules. The only way for the escaping tendency of A (and therefore the vapor pressure) to be determined solely by the concentration is for the forces retarding the escape of A to be the same at all concentrations; the forces between A and B molecules must therefore be the same as those between two A molecules, $F_{AB} = F_{AA}$. Unless the two molecular species are very similar in structure, this is not very likely. If the van der Waals and/or hydrogen bonding forces between A molecules are significantly different from those between B molecules, we should expect a departure from Raoult's law. When $F_{AA} > F_{AB}$, the forces retarding the escape of an A molecule into the vapor phase decrease as the concentration of B increases: the vapor pressure of A will be larger than that predicted by Raoult's law. When $F_{AA} < F_{AB}$, the reverse is true and the vapor pressure of A is less than the Raoult's law prediction: a negative deviation is observed.

Example 102 Which of the following binary solutions would be expected to follow Raoult's law over the entire concentration range: H_2O-pyridine, benzene-toluene, methanol-CCl_4 or $SiCl_4$-CCl_4?

Answer: The types of intermolecular attractions present between water molecules and between pyridine molecules are certainly different. The same can be said for CCl_4 and methanol. These solutions should depart significantly from ideality.

On the other hand, both benzene and toluene ($C_6H_5CH_3$) have cyclic conjugated π systems. The forces between benzene molecules and between toluene molecules are quite similar. The same may be said for the very similar molecules $SiCl_4$ and CCl_4. Consequently these solutions follow Raoult's law reasonably well.

Other solution phenomena may be understood in terms of intermolecular forces. In Section 7.10 we discussed the effect of molecular association in solution on colligative properties. A most important type of association is intermolecular hydrogen bonding. For example, van't Hoff *i*-factor values less than unity are found for weak carboxylic acids dissolved in nonpolar solvents (benzoic acid in benzene, for example). This results from the formation of dimers or even higher polymers by hydrogen bonding (Fig. 17.3).

The solubility of a solid substance in various solvents depends on the relative magnitudes of the intermolecular forces in the solid and those between solvent and solute. The weaker the former and the stronger the latter, the higher the solubility.

The stability resulting from the potential energy of interaction between ions and dipolar solvents allows ions to exist in solution. This is a familiar and important phenomenon.

The effects of the intermolecular forces operating in solution are thus manifested in a variety of ways. There are such thermodynamic effects as

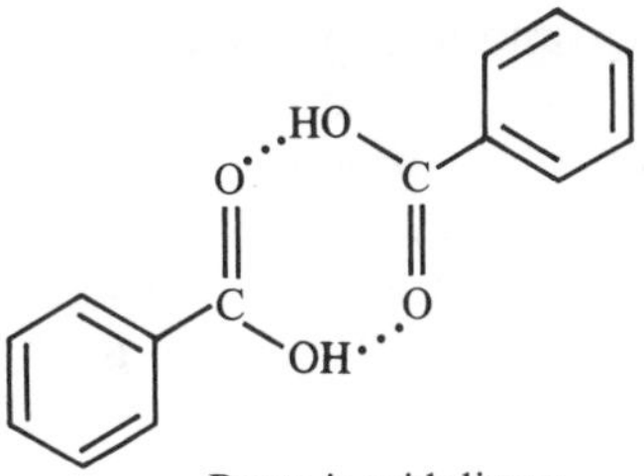

Benzoic acid dimer

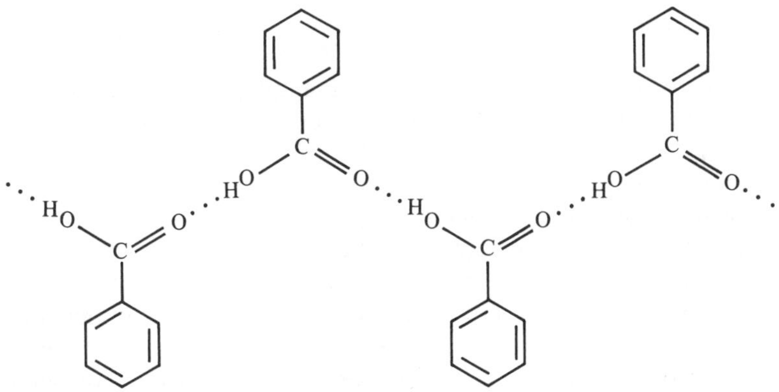

Benzoic acid polymer

Figure 17.3

mentioned above; chemical kinetics also provides numerous examples; shifts in absorption frequencies ranging from subtle to dramatic are caused by solvent interactions (see Chapter 18). The study of intermolecular forces in solutions is a most interesting and important area of physical chemistry.

17.7. Solids

In analogy with the results of the kinetic theory of gases, the motion of the molecules of a liquid is related to the temperature; as the temperature of a liquid decreases, the molecular motion slows. In the liquid state, the intermolecular forces tending to hold the molecules together are not able to overcome the kinetic energy that causes the molecules to continue to jostle and move about. As the temperature decreases, however, a point is reached where the kinetic energy is not sufficiently large to overpower the intermolecular forces, and the liquid solidifies. All translational and rotational motion ceases; the molecules are frozen into definite, stable positions.

The molecules of the resulting solid may bear the same relationship to each other as those of a liquid: a short-range semiordered relationship with their nearest neighbors, with no long-range order relative to molecules further away. In this case the solid is called a *glass*. Most solids, however, are totally ordered, possessing both short-range and long-range order (Fig. 17.4). Such solids are called *crystals*. The development of a general treatment of glasses shares many of the difficulties of attempts to develop such a treatment for liquids. On the other hand, as a result of their fully-ordered arrangement, a highly successful general treatment of the crystalline state has been possible.

Crystals may be classified in various ways. One criterion is based on the nature of the forces which hold the units of the solid together.

An *ionic crystal* consists of ions held together primarily by purely electrostatic forces. Table salt is a classic example of such a crystal. In the NaCl crystal the nearest neighbors of each sodium ion are six chloride ions, while the nearest neighbors of each Cl^- are six Na^+ ions, this ordered arrangement being repeated throughout the crystal. It is impossible to identify in the crystal a simple NaCl molecule made up of a single Na^+ and its particular Cl^-. The distance between a sodium ion and each of its six neighboring chloride ions is the same. The crystal is thus one huge molecule made up of Na^+ and Cl^- ions in equal numbers, bonded together by ionic bonds.

A *covalent crystal* consists of atoms held together *throughout the crystal* by covalent bonds. The classic example here is diamond. Each carbon atom in a diamond crystal is bound by covalent tetrahedral sp^3 bonds to four other carbons, each of which is bonded to four carbons, and so on throughout the crystal. Like the ionic crystal, each covalent crystal is one giant molecule.

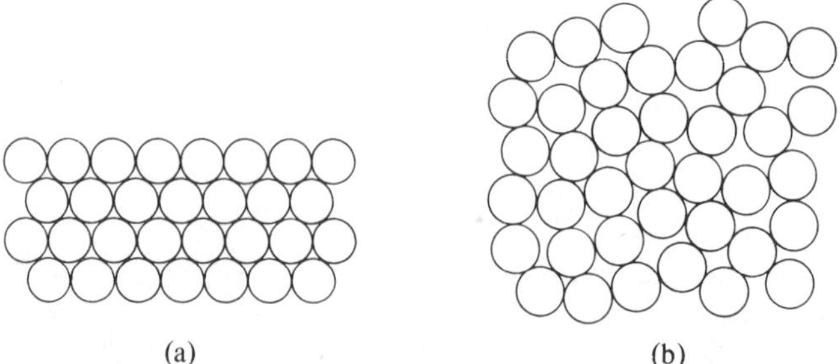

Fig. 17.4. Schematic two-dimensional representation of a crystalline solid (a) and a liquid (b). Each circle represents a molecular unit. The molecular units are arranged in a similar manner in the third dimension as well. Adapted from G. W. Castellan, *Physical Chemistry*, 2nd ed. (1971), Addison-Wesley, Reading, Mass.

In a *molecular crystal* the units are molecules which retain their identity. In the orderly arrangement of the crystal, these individual molecules are held to each other by van der Waals forces and/or hydrogen bonds. The molecules *pack* together, attempting to attain the closest approach to a maximum number of nearest neighbors. The equilibrium situation is reached when the forces of intermolecular attraction are exactly balanced by intermolecular repulsion (see the Lennard-Jones equation, Eq. 334). Most organic molecules form molecular crystals. Figure 17.5 is a drawing of the *unit cell* of such a crystal. The crystal consists of millions of these basic units stacked together three-dimensionally. Since the intermolecular forces in molecular crystals are rather weak, the vibrational motion and energy of such crystals is very similar to that of the same molecules in the liquid and gas phases.

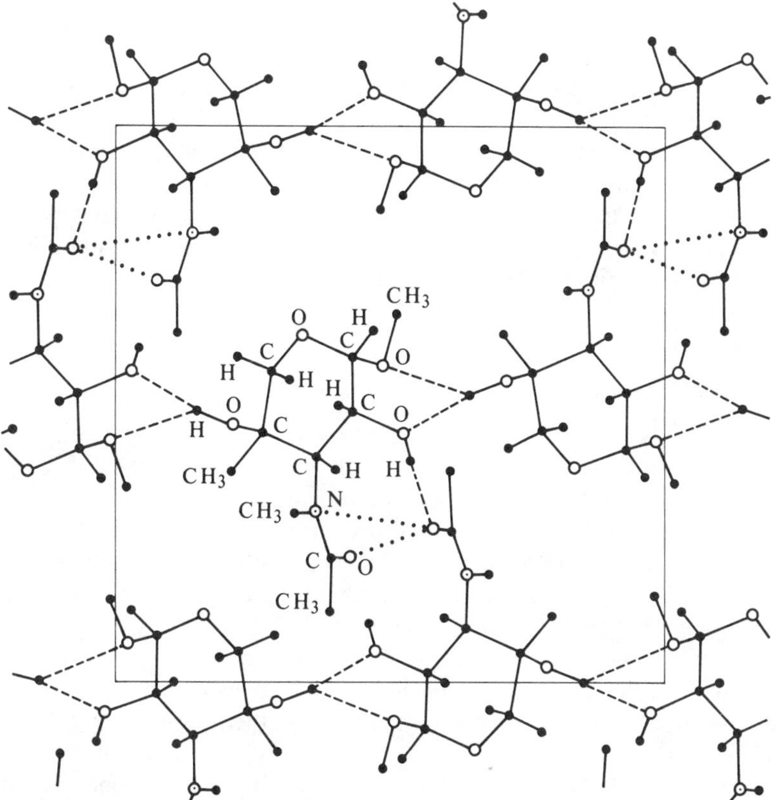

Fig. 17.5. The unit cell of the crystal structure of methyl-N-acetylgarosaminide. Hydrogen bonds are indicated by - - - and dipole-dipole interactions by ·····.

The units of a *metallic crystal* are positive metal ions. The stabilizing forces in such a crystal are obviously not simply ionic since all the ions are positive. Rather, these positive ions are surrounded by a "sea" of electrons which move throughout the crystal, belonging not to a particular metal ion but to *all* the ions of the crystal. The potential energy of interaction between these mobile electrons and the positive ions stabilizes the crystal. Such materials conduct electricity well since the electrons flow easily when an electrical potential is applied.

Some crystals do not fit into a single category. Graphite, for example, consists of planes of covalently bonded carbons with these planes held together by van der Waals forces.

Such properties of crystals as the melting point and the enthalpy of fusion depend on the magnitude of the forces which hold the units of the crystal together. Ionic and covalent bonds are stronger than hydrogen bonds, which are stronger than van der Waals forces. Ionic and covalent crystals generally have higher melting points and larger enthalpies of fusion than molecular and metallic crystals. Molecular crystals with hydrogen bonds have higher melting points and larger ΔH_{fus} values than those with only van der Waals forces.

One of the most important of the phenomena characteristic of the crystalline state is the diffraction of x-rays. This effect has enabled x-ray crystallographers to determine the molecular structures of thousands of substances including a number of proteins. We shall discuss the diffraction of x-rays in some detail in the next chapter.

17.8. Liquid Crystals

Not all the physical states of matter fit neatly into one of the three categories: gas, liquid, or solid. One particular type of substance should be mentioned. A *liquid crystal* possesses some of the properties of a liquid and some of a crystal. It is a highly viscous fluid, but the molecular arrangement possesses some long-range order as in a crystal. This long-range order is not three-dimensional, however. In a *smectic* liquid crystal, the molecules are oriented parallel and are arranged in equidistantly separated planes, but within each particular plane the spacing between the parallel molecules varies. In a *nematic* liquid crystal, the molecular axes are parallel but there is no other order.

Detergent molecules form liquid crystals. Cytoplasm, sperm cells, and some physiological membranes are also thought to have liquid crystalline structures.

The terms *paracrystalline* and *mesophase* are sometimes applied to the liquid crystalline state.

Problems

17.1 In heterogeneous catalysis, reactant molecules may be adsorbed on the surface of a solid. If a chemical bond is formed between the reactant and the surface, *chemical adsorption* is said to occur. If the two are held together by van der Waals forces, it is called *physical adsorption*. For which of these two situations would you expect the desorption to have the higher activation energy E_a?

17.2 Why is the ortho isomer of nitrophenol more volatile than the meta or para isomer?

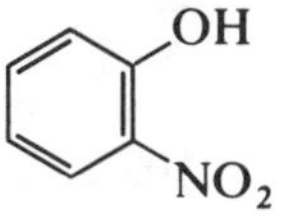

OH

NO_2

17.3 When two molecules are compared which are very similar except that one is capable of *inter*molecular hydrogen bonds, what will be the relative magnitudes of the following properties?

a) boiling point
b) melting point
c) solubility in water
d) density of pure liquid
e) *viscosity* (the resistance to liquid flow)
f) *surface tension* (the force that opposes an increase in the surface area of a liquid)
g) diffusion in pure liquid
h) van't Hoff *i* factor
i) vapor pressure

17.4 Repeat Problem 17.3 for two similar molecules, one of which forms strong *intra*molecular hydrogen bonds.

17.5 Explain why nonpolar molecules are more soluble in nonpolar solvents than in polar solvents.

17.6 Why should the dipole-dipole van der Waals intermolecular interaction depend on the temperature (Eq. 331)?

17.7 Explain the following sequences of polarizabilities:

a)

Atom	Polarizability α, cm^3
He	0.2×10^{-24}
Ar	1.6×10^{-24}
Xe	4.0×10^{-24}

b)

Atom	Polarizability α, cm^3
H_2	0.8×10^{-24}
N_2	1.7×10^{-24}
O_2	1.5×10^{-24}

17.8 In hydrogen bonds between two electronegative atoms in crystals, the hydrogen is usually located on the line between the two (O—H ··· O) rather than at an angle (O⁄H··O). Why do you suppose this is so?

17.9 Table 1.2 contains values of the constants *a* and *b* of the van der Waals gas equation for a number of substances. Recall that the value of *a* is related to the interaction between molecules. Based on your knowledge of intermolecular forces, rationalize the relative sizes of *a* in Table 1.2.

17.10 Use your knowledge of the forces of interaction between molecules in liquid solution to explain the following:

a) Oil (hydrocarbons) and water are immiscible. When a drop of oil is dripped into water it spreads out in a thin layer on the surface.
b) Soap is usually composed of sodium salts of long-chain fatty acids such as lauric acid $CH_3(CH_2)_{10}COOH$, myristic acid $CH_3(CH_2)_{12}COOH$, and palmitic acid $CH_3(CH_2)_{14}COOH$. When placed in water they form a thin layer on the surface of the water, their ionic (hydrophilic) ends extending into the water while their hydrocarbon (hydrophobic) ends extend away from the surface.
c) When both oil and soap are mixed with water they both appear to go into solution, forming microscopic spherical aggregates called *micelles*. In these aggregates the negatively charged ionic ends of the soap molecules are located on the surface of the sphere (see Fig. 17.6) while their hydrocarbon ends extend inward. The oil molecules are located in the interior of the sphere.
d) These micelles remain separated and do not coalesce into larger aggregates.
e) Soaps are rather ineffective in acidic solutions.
f) In "hard water" containing ions such as Ca^{+2} and Mg^{+2}, soaps are not as effective and the bathtub ring is much worse.

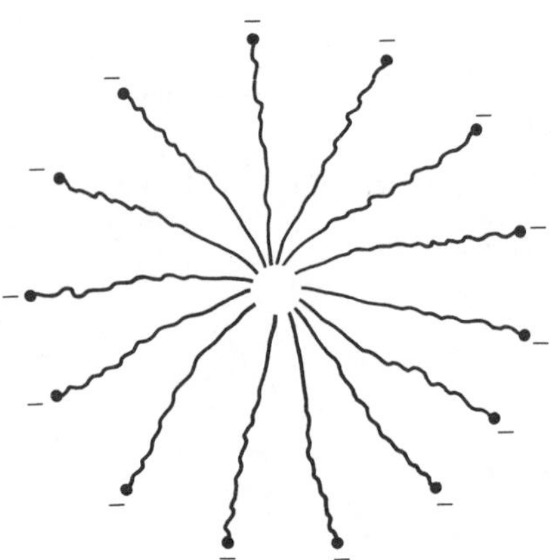

Figure 17.6

17.11 One of the principal functions of bile is to allow the water-insoluble lipids to come into contact with the water-soluble digestive enzymes (lipases). The major components of bile are glycocholic acid and taurocholic acid:

$$C_{28}H_{26}(OH)_3\overset{\overset{\large O}{\|}}{C}NHCH_2COOH$$

glycocholic acid

$$C_{23}H_{26}(OH)_3\overset{\overset{\large O}{\|}}{C}NHCH_2CH_2SO_3H$$

taurocholic acid

(a) How do these bile acids function? (b) What is the role of peristalsis in this process?

17.12 In the hydrogen-bonded system $>C{=}O\cdots H{-}O{-}C<$, what effect will the hydrogen bond have on (a) the O—H bond, (b) the C—O bond, and (c) the C=O bond?

17.13 Calculate the potential energy between an argon atom and a methanol molecule arising from the dipole-induced dipole interaction term (Eq. 332) at a separation of (a) 10 Å and (b) 5 Å. The dipole moment of methanol is 1.7 debyes (1.7×10^{-18} esu-cm or 5.7×10^{-28} coulomb-cm), and the polarizability of argon is 1.63×10^{-24} cm^3.

18 / Molecular Spectroscopy and Diffraction

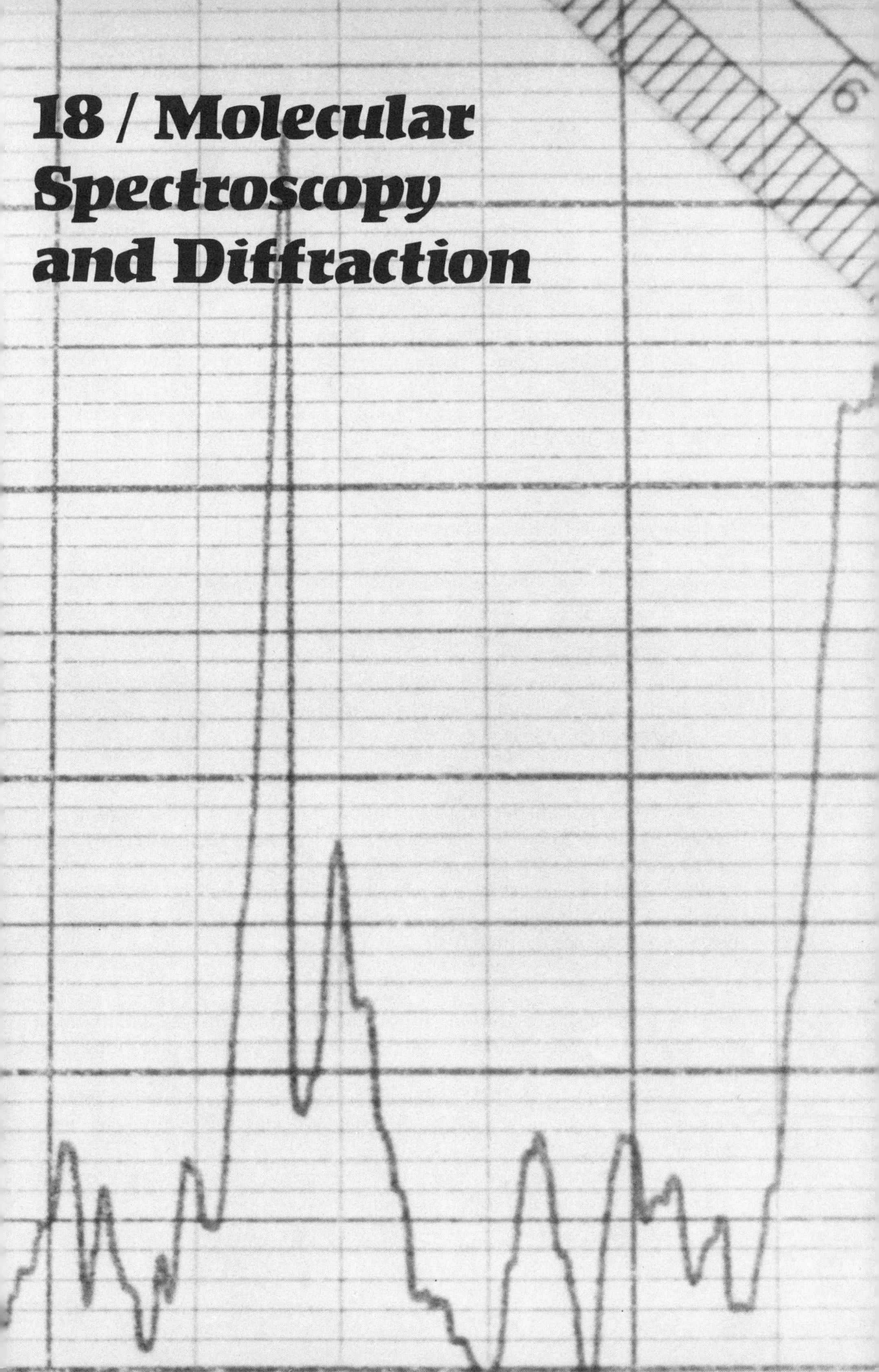

18.1. Introduction

We have been much concerned in previous chapters with the development of expressions for the allowed energies of atoms and molecules. To do this we have used the postulates of quantum mechanics with the understanding that our predictions must agree with experiment. Some of the most important methods of checking out the predictions of wave mechanics are based on the interaction of molecules and atoms with electromagnetic radiation. Of even greater importance, the absorption, emission, and diffraction of electromagnetic radiation also form the bases of highly useful methods for the study of molecular structure. In this chapter we shall examine some of these methods.

18.2 Electromagnetic Radiation

Electromagnetic radiation is composed of periodically varying electric and magnetic fields passing through space as a wave. Consider first only the electric field. If we focused our attention on a single point in space through which electromagnetic radiation from some source is passing, we would discover that the magnitude of the field (the *field strength* or *amplitude*) varies periodically in a sinusoidal manner with time (Fig. 18.1a). This periodic variation may be expressed as

$$\mathscr{E} = \mathscr{E}^{\circ} \sin 2\pi \nu t, \tag{335}$$

where $\mathscr{E}^{\circ}$ is the maximum field strength (amplitude), and ν is the *frequency* of the wave, defined to be the number of times the wave motion repeats itself in one second. The wave motion therefore repeats every $1/\nu$ sec.

If at a particular instant we measured the magnitude of the field at a number of points through which the wave is passing, we would find that its position dependence is given as

$$\mathscr{E} = \mathscr{E}^{\circ} \sin (2\pi x/\lambda), \tag{336}$$

where x is the distance from the source and λ is the *wavelength*, defined as the distance between identical points of the wave. Figure 18.1(b) illustrates this schematically.

The magnitude of the field which an atom or molecule located at point x feels at time t is given by Eqs. (335) and (336). This field will change with time in a periodic manner.

The two quantities λ and ν are most important. They are related to each other by the velocity of the wave motion:

$$(\nu \text{ sec}^{-1})(\lambda \text{ cm}) = \lambda\nu(\text{cm sec}^{-1}) = v \text{ cm sec}^{-1}.$$

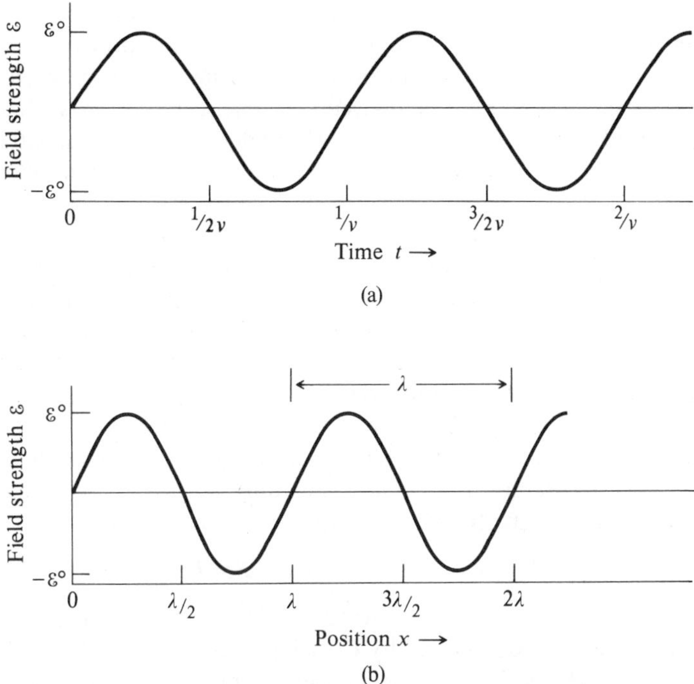

Fig. 18.1. (a) Variation of the amplitude of electromagnetic radiation with time. (b) Variation of the amplitude of electromagnetic radiation with distance.

The velocity of electromagnetic radiation is the speed of light $c = 3.0 \times 10^{10}$ cm sec^{-1}. (Visible light is one kind of electromagnetic radiation.) Therefore

$$\lambda \nu = c. \tag{337}$$

There is also a periodically varying magnetic field associated with electromagnetic radiation. This field is perpendicular to the electric field. It varies with distance and time exactly as the electric field varies.

18.3. The Interaction of Electromagnetic Radiation with Matter

When an atom or molecule is immersed in an electric or magnetic field, its potential energy may be changed significantly if there is a way in which it can interact with that field. (See Example 71b, for instance.) In the case of the electric and magnetic fields of electromagnetic radiation, these fields are changing periodically in time. Consequently their effect on the potential energy of the molecule or atom is a time-dependent one. Our previous assumption of a conservative state which is independent of time cannot be used when we are considering the interaction of electromagnetic radiation with atoms or molecules. Rather, we must use the time-dependent wave mechanical equation. We shall not attempt a detailed derivation and discussion of the time-dependent treatment here[1] but will simply give the results.

As we have pointed out before, a conservative molecular or atomic system may exist in only certain energy states. These states, as predicted by equations we have derived, are dependent upon the values of certain quantum numbers. In order to change states, a system must absorb or emit an amount of energy exactly equal to the difference in energy between the two states. In going from one state with energy E_1 to another with energy E_2, a system must absorb or emit a packet of energy ΔE, such that $\Delta E = E_2 - E_1$.

The results of the time-dependent quantum mechanical treatment confirm the suggestion by Planck that matter can only absorb energy from electromagnetic radiation if the frequency of the radiation is such that

$$\nu = |\Delta E|/h = |E_2 - E_1|/h. \tag{338}$$

Thus the frequency must be equal to the absolute difference in energy of two allowed quantized states, divided by Planck's constant. This same relationship holds for emitted radiation: a molecule can change its energy state to a lower allowed state by emitting radiation of frequency ν given by $\nu = |\Delta E|/h$. This equation is clearly a most important relationship; on it the scientific field of *spectroscopy* is based. Spectroscopy is the study of the absorption and emission of radiation by matter.

The time-dependent treatment yields other limitations on the types of molecules which may absorb or emit radiation, and on the particular energy transitions which may possibly accompany such absorption or emission. We shall discuss these *selection rules* as the occasion arises.

It should be mentioned at this point that there are two major types of interactions which matter may have with electromagnetic radiation. One, *resonance absorption*, is the type we have been discussing: the molecule or atom absorbs energy from the radiation and changes from one allowed conservative state (or *stationary state*) to another. The second type, *nonresonance absorption*, involves the absorbing of an amount of energy that causes the system to change from a stationary state to a *nonstationary state* (a state which is not one of those resulting from the time-independent treatment of the previous chapters). This new state is not stable and the system rapidly emits some or

---

[1] For such a discussion see Walter J. Moore, *Physical Chemistry*, 4th ed., Englewood Cliffs, N.J.: Prentice-Hall (1972), pp. 751–755.

all of the energy it has absorbed, returning either to its original state or to some other conservative state. This nonresonance phenomenon is important in Raman scattering and in x-ray diffraction; it will be considered again when these topics are discussed in Sections 18.8 and 18.11.

18.4. Spectroscopy

In Section 9.7 we discussed some of the aspects of the absorption of radiation by molecules. Particular reference should be made to that section in regard to Beer's law, which relates the intensity of the radiation being absorbed (actually the absorbance A) to the length b of the radiation path through the sample and to the concentration c of the sample:

$$A = \varepsilon bc. \tag{181}$$

The proportionality constant ε is called the *molar extinction coefficient*; its value depends on the wavelength and on the nature of the material absorbing the radiation. Figure 9.2 is a typical absorption *spectrum* having peaks at the wavelengths where the material absorbs radiation.

It should now be apparent why the extinction coefficient has a significantly large value only at particular wavelengths. We know now that resonance absorption can only occur if $h\nu = hc/\lambda = |E_2 - E_1|$, where E_1 and E_2 are allowed energies of the system. The maxima in the spectrum thus occur at those values of λ or ν that correspond to the differences between the stationary-state energy levels of the material.

Example 103 Assuming that the energy transitions may actually occur via the absorption of electromagnetic radiation, calculate the frequency and the wavelength of the radiation needed in each of the following cases:

a) the change in the translational energy of the argon atom in Example 76.
b) the change in the electronic energy of the hydrogen atom in Example 79 ($n = 1 \to n = 2$).
c) the rotational transition in Example 96 ($J = 1 \to J = 4$).
d) the vibrational transition in Example 97 ($v = 0 \to v = 4$).

Answer: In all these cases we use $\nu = \Delta E/h$.

a) $\Delta E = 2.5 \times 10^{-31}$ erg

$$\nu = (2.5 \times 10^{-31}\ \text{erg})/(6.6 \times 10^{-27}\ \text{erg-sec}) = 3.8 \times 10^{-5}\ \text{sec}^{-1}$$

$$\lambda = c/\nu = (3.0 \times 10^{10}\ \text{cm sec}^{-1})/(3.8 \times 10^{-5}\ \text{sec}^{-1}) = 7.9 \times 10^{14}\ \text{cm}$$

b) $\Delta E = 1.65 \times 10^{-11}$ erg

$$\nu = (1.65 \times 10^{-11})/(6.63 \times 10^{-27}) = 2.5 \times 10^{15}\ \text{sec}^{-1}$$

$$\lambda = (3.0 \times 10^{10})/(2.5 \times 10^{15}) = 1.2 \times 10^{-5}\ \text{cm}$$

c). $\Delta E = 2.2 \times 10^{-13}$ erg

$$\nu = (2.2 \times 10^{-13})/(6.6 \times 10^{-27}) = 3.3 \times 10^{13}\ \text{sec}^{-1}$$

$$\lambda = (3.0 \times 10^{10})/(3.3 \times 10^{13}) = 9.1 \times 10^{-4}\ \text{cm}$$

d) $\Delta E = 1.72 \times 10^{-12}$ erg

$$\nu = (1.72 \times 10^{-12})/(6.63 \times 10^{-27}) = 2.6 \times 10^{14}\ \text{sec}^{-1}$$

$$\lambda = (3.0 \times 10^{10})/(2.6 \times 10^{14}) = 1.2 \times 10^{-4}\ \text{cm}$$

The recording and interpretation of spectra such as that shown in Fig. 9.2 constitute an important part of the spectroscopist's work.

Our calculations have shown that ΔE values for electronic energy levels are quite different from those for vibrational levels, which in turn are of a different order of magnitude compared to those for rotational levels. Since ΔE is different for different phenomena, the frequency or wavelength of the radiation used must of course be suited to the energy requirement if a chosen phenomenon is to be observed. Table 18.1 classifies the wavelengths of electromagnetic radiation into subdivisions, in terms of both the type of radiation and the type of molecular energy levels whose changes correlate with radiation of that particular range of wavelengths. The divisions are somewhat arbitrary and there is considerable overlap of regions, but the divisions are appropriate for most cases.

The techniques for producing the radiation, resolving it so that the absorption of each monochromatic wavelength may be individually studied, and measuring the amount of radiation absorbed may be quite different for the various wavelengths. Spectrometers with very ingenious components and sophisticated capabilities have been designed and are constantly being improved.[2]

Figure 18.2 is a schematic representation of a typical absorption spectrometer. The basic components are the *source*, a *monochromator*, the

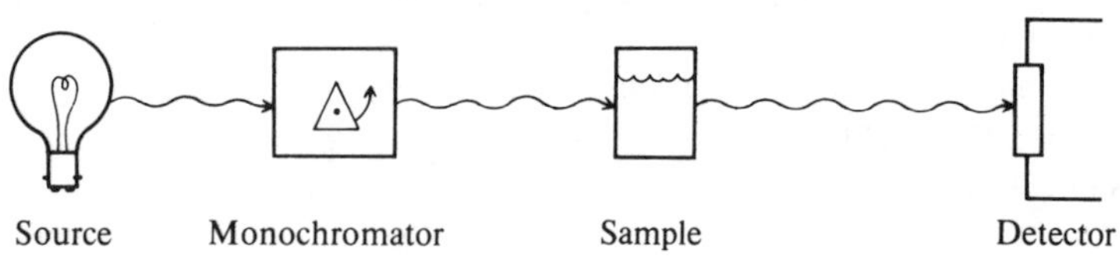

Fig. 18.2. Schematic diagram of the principal components of a typical spectrometer.

[2] For further information on the design and operation of spectrometers see C. N. Banwell, *Fundamentals of Molecular Spectroscopy*, New York: McGraw-Hill (1973).

Table 18.1. The Electromagnetic Radiation Spectrum

λ, cm	10^4	10^3	10^2	10	1	10^{-1}	10^{-2}	10^{-3}	10^{-4}	10^{-5}	10^{-6}	10^{-7}	10^{-8}	10^{-9}
Type of electromagnetic radiation	←		radio			→	infrared		visible	ultra-violet		x-rays		γ-rays
					microwave									
Molecular phenomenon	nmr		esr		rotation					electronic transitions		x-ray diffraction		
						vibration								

sample chamber, and the *detector*. There are usually other optical components for focusing and directing the radiation beam, as well as a device for recording the spectrum.

An incandescent bulb with tungsten filament may be used as a source of radiation in the visible range, while a hydrogen or argon lamp could be used for wavelengths in the ultraviolet region. A heated silicon carbide rod (called a Globar) serves as a source of infrared radiation.

Radiation sources normally produce a wide range of wavelengths. The monochromator separates out the particular wavelength to be used. In some cases a filter which absorbs all wavelengths except the desired one is used. Diffraction gratings and prisms resolve the radiation incident upon them into its component wavelengths (see Fig. 18.3a). The absorption of a succession of wavelengths by the sample may be measured by rotating the prism or grating (Fig. 18.3b).

The sample holder must be made from a material which does not significantly absorb the wavelengths being studied. Quartz or glass is used for ultraviolet radiation while cells made of NaCl are used for visible radiation.

The detector may be a photomultiplier tube, photovoltaic cell, thermocouple, or photographic plate. The detected signal is often amplified and fed to a strip recorder for a permanent record or in some cases to an oscilloscope.

Let us now consider what information concerning molecules may be obtained by a study of the data of spectroscopy.

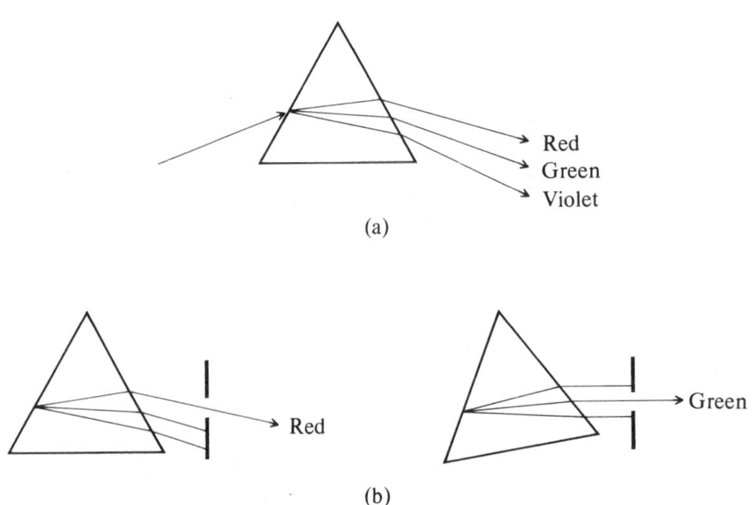

Fig. 18.3. (a) Resolution of the spectral components of visible light by a prism. (b) The prism may be oriented so that a single wavelength passes through a fixed slit.

18.5. Molecular Rotation; Microwave Spectroscopy

When a molecule changes its rotational energy level, an energy effect of the order of magnitude of 10^2 joules mole^{-1} (or about 10^{-15} ergs molecule^{-1}) is involved. This corresponds (via Eq. 337) to a frequency in the microwave region of the electromagnetic spectrum.

In order for the molecule to undergo a change in rotational energy by absorbing energy from the radiation field, the molecule must possess a permanent dipole moment. Consequently homonuclear diatomic molecules such as H_2 and N_2 cannot change their rotational energy in this manner and do not absorb microwave radiation. Neither do linear polyatomic molecules like CO_2 and acetylene nor such nonlinear polyatomic molecules as benzene, CH_4 and SF_6. None of these molecules have permanent dipole moments. Nonsymmetrical diatomics such as HCl and linear polyatomics such as HCN, as well as symmetric molecules such as CH_3Cl and all asymmetric molecules, will exhibit absorption spectra in the microwave region. All of these molecules have permanent dipole moments.

The selection rules for rotation allow the quantum number J to change by only one unit, that is $\Delta J = \pm 1$. For absorption by a diatomic rigid rotor, the energy change (see Eq. 319) will be:

$$\begin{aligned}\Delta E = E_{\text{rot}_2} - E_{\text{rot}_1} &= [J_2(J_2+1)(h^2/8\pi^2 I)] - [J_1(J_1+1)(h^2/8\pi^2 I)]\\ &= [J_2(J_2+1) - J_1(J_1+1)](h^2/8\pi^2 I).\end{aligned}$$

Now since $J_2 = J_1 + 1$ is required for absorption,

$$\begin{aligned}\Delta E &= [(J_1+1)(J_1+2) - J_1(J_1+1)](h^2/8\pi^2 I)\\ \Delta E &= 2(J_1+1)(h^2/8\pi^2 I).\end{aligned} \tag{339}$$

The frequency of radiation needed is $\nu = |\Delta E|/h$,

$$\nu = 2(J_1+1)(h/8\pi^2 I) = 2(J_1+1)B. \tag{340}$$

What does such a spectrum look like? It would be composed of absorptions for $J = 0$ to $J = 1$, $J = 1$ to $J = 2$, $J = 2$ to $J = 3$, etc.

$$\begin{aligned}\nu_{0\to 1} &= 2(0+1)B = 2B\\ \nu_{1\to 2} &= 2(1+1)B = 4B\\ \nu_{2\to 3} &= 2(2+1)B = 6B\\ \nu_{3\to 4} &= 2(3+1)B = 8B, \text{ etc.}\end{aligned}$$

Thus we would expect a series of evenly spaced absorption peaks, each separated from its nearest neighbors by $\Delta\nu = 2B = 2h/8\pi^2 I$.

From the measurement of the spacing between the lines of a microwave spectrum of a diatomic molecule, we can easily determine the moment of inertia and from this the equilibrium bond length r_e. For diatomics which are not rigid rotors the spacing between the lines is not constant but decreases as J increases. At low values of J, however, the spacing is still essentially equal to $2B$ and is easily measured. Thus for diatomic molecules the microwave spectrum is an excellent source of bond length data.

Example 104 The rotation spectrum of HCl vapor consists of a series of relatively evenly spaced lines when the chlorine is exclusively the ^{35}Cl isotope. The spacing between the first two lines of the spectrum, in terms of wavelength, is 4.7×10^{-2} cm. Calculate the bond length in HCl.

Answer: The difference in the frequency of any two adjacent rotational absorptions is equal to $2B$. Converting our wavelength difference to a frequency difference,

$$2B = \Delta\nu = c/\Delta\lambda = (3.0 \times 10^{10} \text{ cm sec}^{-1})/(4.7 \times 10^{-2} \text{ cm})$$
$$B = 3.2 \times 10^{11} \text{ sec}^{-1}.$$

Recalling that $B = h/8\pi^2 I$, we can now calculate I:

$$I = 6.6 \times 10^{-27} \text{ erg-sec}/8\pi^2(3.2 \times 10^{11} \text{ sec}^{-1})$$
$$= 2.6 \times 10^{-40} \text{ erg-sec}^2.$$

For a diatomic molecule the moment of inertia is $I = \mu r_e^2$, where r_e is the equilibrium internuclear separation. Thus

$$r_e^2 = (2.6 \times 10^{-40} \text{ erg-sec}^2)\bigg/\left[\frac{35 \times 1}{35 + 1}\bigg/6.02 \times 10^{23}\right] \text{gm}$$
$$= 1.61 \times 10^{-16} \text{ cm}^2$$
$$r_e = 1.3 \text{ Å}.$$

For linear polyatomic molecules, B and I can also be determined from the microwave spectrum. In this case the calculation of the r_e values is not so straightforward. The molecule now has more than one bond length to be determined, and $I = \sum_i m_i r_i^2$. This problem is overcome by using the method of *isotopic substitution*. For instance, if it is desired to determine the bond lengths in the molecule HCN, microwave spectra of HCN and the deuterium-substituted molecule DCN may be obtained. From these, the moments of inertia I_{HCN} and I_{DCN} can be determined. Recall that

$$I_{HCN} = m_H r_H^2 + m_C r_C^2 + m_N r_N^2 \tag{341a}$$

$$I_{DCN} = m_D r_D^2 + m_C r_C^2 + m_N r_N^2, \tag{341b}$$

where r_H is the distance of the hydrogen atom from the center of mass, etc.

Assuming that the change in mass has little effect on the bond lengths ($r_{CH} = r_{DH}$), and realizing that

$$r_{CH} = r_H - r_C$$
$$r_{CN} = r_C + r_N,$$

Eqs. (341a) and (341b) can be rewritten in terms of r_{CH} and r_{CN}. With two equations and two unknowns, the values of the bond lengths may be obtained (usually with some difficulty).

The spectra for nonlinear polyatomic molecules are much more complicated. Their interpretation is difficult—in some cases impossible.

18.6. Molecular Vibration; Infrared Spectroscopy

Differences in vibrational energy levels are of the magnitude 10^4 joules mole^{-1} (about 10^{-13} erg molecule^{-1}). This corresponds to radiation found in the infrared region of the electromagnetic spectrum.

The dipole moment of a molecule must be different in its initial and final states in order for the molecule to change its vibrational states by absorbing infrared radiation. For a harmonic oscillator the selection rules require that the quantum number v may change only by ± 1 in such a transition. For absorption by a diatomic harmonic oscillator the energy change (see Eq. 324) is

$$\Delta E = E_{\text{vib}_2} - E_{\text{vib}_1} = [h\nu_0(v_2 + \tfrac{1}{2})] - [h\nu_0(v_1 + \tfrac{1}{2})]$$
$$= h\nu_0(v_2 - v_1),$$

but since $v_2 = v_1 + 1$ for absorption,

$$\Delta E = h\nu_0$$
$$h\nu = h\nu_0$$
$$\nu = \nu_0,$$

and the frequency of the radiation absorbed is equal to the natural frequency of the harmonic oscillator. The radiation absorbed for all transitions, that is $v = 0 \rightarrow v = 1$, $v = 1 \rightarrow v = 2$, $v = 2 \rightarrow v = 3$, etc., is the same:

$$\nu_{0\rightarrow 1} = \nu_0 = \frac{1}{2\pi}\sqrt{\kappa/\mu}$$

$$\nu_{1\rightarrow 2} = \nu_0 = \frac{1}{2\pi}\sqrt{\kappa/\mu}$$

$$\nu_{2\rightarrow 3} = \nu_0 = \frac{1}{2\pi}\sqrt{\kappa/\mu}, \text{ etc.}$$

The vibrational absorption spectrum of a harmonic oscillator thus consists of a single line. From the frequency of that line the force constant κ of the bond can be calculated. This constant is directly related to the ease with which the bond can be stretched and is therefore a measure of the bond strength.

Example 105 The natural frequency (determined from the vibrational spectrum) of the N—H bond in $(CH_3)_2NH$ is 1.02×10^{14} sec^{-1}. What should the stretching frequency of the N—D bond in $(CH_3)_2ND$ be?

Answer: From above,

$$\nu_{0_{NH}} = \frac{1}{2\pi}\sqrt{\kappa_{NH}/\mu_{NH}}$$

$$\nu_{0_{ND}} = \frac{1}{2\pi}\sqrt{\kappa_{ND}/\mu_{ND}}.$$

The difference in mass will not make the force constants of ND and NH significantly different. Therefore

$$\nu_{0_{NH}}/\nu_{0_{ND}} = \sqrt{\mu_{ND}/\mu_{NH}}.$$

The reduced masses (N is Avogadro's number) are:

$$\mu_{NH} = [44 \times 1/(44+1)]/N = 1.62 \times 10^{-24}\text{ g}$$
$$\mu_{ND} = [44 \times 2/(44+2)]/N = 3.17 \times 10^{-24}\text{ g}$$

and

$$\nu_{0_{ND}} = \nu_{0_{NH}}\sqrt{\mu_{NH}/\mu_{ND}} = 1.02 \times 10^{14}(1.62 \times 10^{-24}/3.17 \times 10^{-24})^{1/2}$$
$$= 7.3 \times 10^{13}\text{ sec}^{-1}.$$

Real diatomic molecules do not vibrate harmonically. One effect of the anharmonic motion is to relieve the restriction $\Delta v = \pm 1$. An absorption corresponding to $\Delta v = +1$ is called a *fundamental*, while those for $\Delta v > 1$ are called *overtones*. The intensities of overtone absorptions are usually quite weak in comparison with the fundamental.

Anharmonic motion also causes vibrational energy levels to become more closely spaced (ΔE becomes smaller) as v increases. This is usually not a severe problem, however. The difference in energy between the lowest vibrational state ($v = 0$) and the next state ($v = 1$) is large compared to the thermal energy (kT) available at usual temperatures ($\sim$25°C). This means that in colliding with each other, few molecules gain enough energy to allow them to change from the $v = 0$ ground state. Virtually all the molecules are therefore in the $v = 0$ state unless the temperature is elevated. Consequently the only transition which will occur by absorbing radiation is $v = 0 \rightarrow v = 1$, and consequently only one intense absorption peak is observed.

Since ΔE_{rot} is significantly less than ΔE_{vib}, a spectrum in the microwave region which involves only changes in rotational states may be obtained. A spectrum in the infrared, however, displays the vibration-rotation spectrum of the molecule rather than the pure vibration absorption. This involves the simultaneous change of v and J, the selection rules allowing $\Delta v = +1$ and $\Delta J = \pm 1$ simultaneously. Figure 18.4 illustrates schematically the vibration-rotation spectrum of HCl. Note that there is no peak corresponding to the pure vibration transition with $\Delta v = +1$, $\Delta J = 0$ and $\nu = \nu_0$. This transition is in general forbidden. Note that the peaks are located symmetrically about ν_0, each separated from its next neighbor by $\Delta \nu = 2B$. The infrared spectrum may therefore be a source of both ν_0 and B, from which κ and r_e can be obtained.

Since ΔE_{rot} is of the same order of magnitude as the thermal energy kT, a significant number of molecules will be in rotational states with $J > 0$. Thus for rotational transitions, unlike those for vibrations, transitions other than $J = 0 \rightarrow J = 1$ occur. The relative intensity of the peak corresponding to a particular transition reflects the relative population of the initial rotational state.

Peak	$J_A \rightarrow J_B$	E_{rot}	E_{vib} ($v = 0 \rightarrow v = 1$)	ν
1	$0 \rightarrow 1$	$2hB$	$h\nu_0$	$\nu_0 + 2B$
2	$1 \rightarrow 2$	$4hB$	$h\nu_0$	$\nu_0 + 4B$
3	$2 \rightarrow 3$	$6hB$	$h\nu_0$	$\nu_0 + 6B$
4	$3 \rightarrow 4$	$8hB$	$h\nu_0$	$\nu_0 + 8B$
5	$1 \rightarrow 0$	$-2hB$	$h\nu_0$	$\nu_0 - 2B$
6	$2 \rightarrow 1$	$-4hB$	$h\nu_0$	$\nu_0 - 4B$
7	$3 \rightarrow 2$	$-6hB$	$h\nu_0$	$\nu_0 - 6B$
8	$4 \rightarrow 3$	$-8hB$	$h\nu_0$	$\nu_0 - 8B$

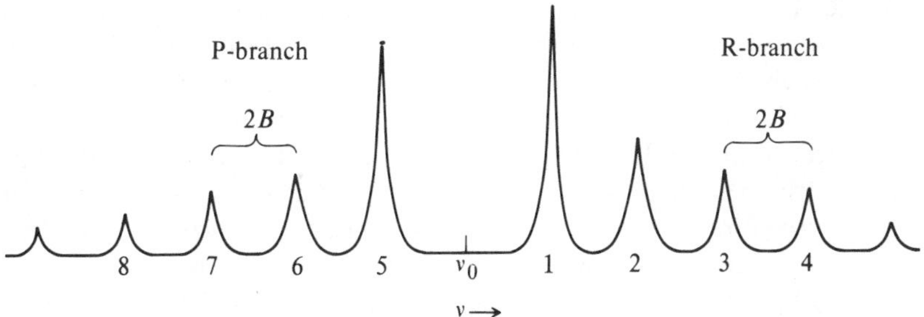

Fig. 18.4. Vibration-rotation spectrum in the infrared.

For a nonrigid rotor the spacings of the lines in spectra like Fig. 18.4 will not be constant. The lines in the *R-branch* ($\nu > \nu_0$) become more closely spaced as J increases, while the lines in the *P-branch* ($\nu < \nu_0$) are spaced farther apart as the magnitude of J increases.

The vibration spectrum of polyatomic molecules is of course much more complex than for diatomics. There will be $3n - 5$ vibrational absorptions for linear molecules, $3n - 6$ for nonlinear molecules. Each of these will have the general appearance of Fig. 18.4 since a $\Delta J = \pm 1$ change may occur simultaneously with each of the $\Delta v_j = +1$ changes. The rotational peaks may also be much more complex in the nonlinear cases, whether symmetric or asymmetric. The difficulty is compounded by the presence of overtones of each of the $\Delta v_j = +1$ fundamentals. The possibility of a detailed analysis of such complicated spectra to yield a set of bond distances and bond strengths is rather remote. The complexity of the infrared spectra of polyatomic molecules is the basis of one of their most useful applications, however.

Since the allowed absorptions seen in infrared spectroscopy depend on the details of the molecular structure, spectra are characteristic. Molecules whose structures differ only slightly may have spectra which differ by a significant degree. The presence of a substance in a mixture can be detected by the presence of its characteristic pattern of peaks in the spectrum. A molecule's spectrum is literally its fingerprint or "mug shot," which can be used as a powerful identifying device.

Infrared spectra have been used extensively in clinical practice. Solvent extracts of bacteria and viruses possess characteristic spectra which can be used for their detection. Toxicologists use infrared spectra to detect the presence of poisons in the organs of the deceased. Pathologists routinely use spectra in urine and blood studies. And one of the most common methods of monitoring the atmosphere for pollutants is the on-site use of infrared spectroscopy.

It was pointed out in Chapter 16 that some normal modes are made up primarily of the vibration of a single bond or of a small group of atoms in a molecule. The wavelength absorbed by the vibration of an O—H bond will always be approximately the same, regardless of the radical to which the O—H group is attached. The presence or absence of an absorption near the wavelength where the O—H bond vibration usually appears confirms or refutes the presence of an O—H group in the molecule. The same may be said of the characteristic frequencies of other groups such as C=O and C=C. These are called *group frequencies* and can be of considerable help in establishing molecular structure. Table 18.2 lists a number of group frequencies. In this table the value of each absorption is given in terms of the *wave number* $\bar{\nu}$. This latter quantity has units of cm^{-1}; its conversion to wavelength is simply

$$\lambda = 1/\bar{\nu}. \tag{342}$$

Table 18.2. Selected Group Absorption Frequencies

Group	*Approximate wave number,* cm^{-1}
OH	3600
NH_2, NH	3450
(benzene ring C–H)	3045
C≡C	2200
C=O	1700
C=C	1650
CF	1200
CCl	700
CBr	550
CI	500

Example 106 In studies of the cholesterol present in the blood, it is sometimes necessary to distinguish between the free cholesterol

H_3C CH_3 CH_3 HO

and that which is esterified with fatty acids

H_3C CH_3 CH_3 O R—C—O

How could the vibrational spectra of these two molecules be used to distinguish between them?

Answer: The spectrum of free cholesterol should display an hydroxide absorption but no carbonyl absorption. For the ester, the reverse should be the case. Table 18.2 indicates that we should expect the hydroxide absorption in the neighborhood of 3600 cm^{-1} and that due to carbonyl near 1700 cm^{-1}. Figure 18.5 gives the infrared spectrum of cholesterol (a) and of cholesterol acetate (b) in the region 1800–1600 cm^{-1}. The strong carbonyl group frequency is not present in the former but is in the latter.

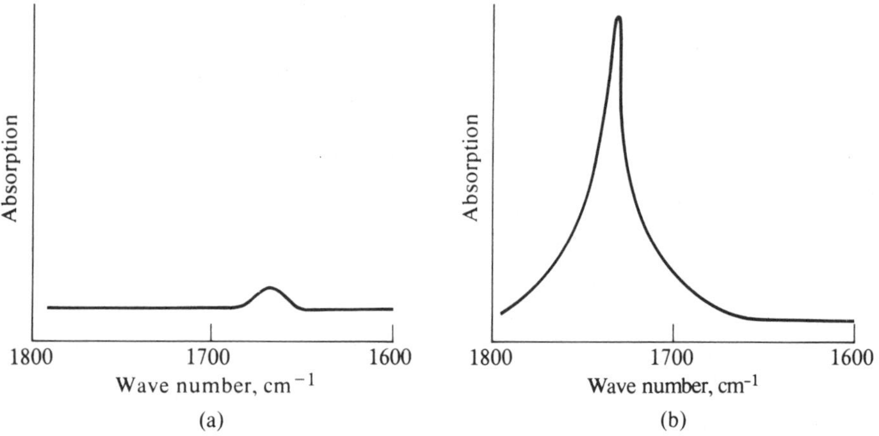

Fig. 18.5. Absorption spectra of cholesterol (a) and cholesterol acetate (b).

18.7. Electronic Spectra; Visible and Ultraviolet Spectroscopy

For an electron to absorb energy it must move into an atomic or molecular orbital which does not already contain two electrons. An inner shell electron must therefore absorb enough energy to excite it into a partially filled valence orbital or into an empty orbital above the valence orbitals. This requires rather energetic radiation which may even be in the x-ray range. For the excitation of valence electrons less energy is required (of the order of magnitude of 10^{-10} erg molecule^{-1}) and the radiation is generally in the ultraviolet. In some cases the energy levels may be spaced so closely that the absorbed radiation is in the visible range. Examples of this are electronic absorptions in the extended conjugated π systems of organic dyes (see Problem 18.6).

The energy differences between electronic energy levels are obviously greater than those for vibration and rotation. When the electronic quantum number changes in the absorption of ultraviolet radiation, the vibrational and rotational quantum numbers may also change. Similar to the vibration-rotation spectra discussed in the previous section, we obtain lines in the electronic spectrum, called *fine structure*, which are due to changes in the vibration and the rotation energy levels. The selection rules for infrared absorption no longer hold, but there is some limitation to the vibrational transitions which may occur simultaneously with an electronic transition. Figure 18.6 illustrates the *Franck-Condon principle*, also called the *vertical rule*. This principle says that even though the molecule is vibrating, an electronic transition occurs so quickly that the instantaneous internuclear separation in the excited electronic state is the same as it was in the initial state. The transition therefore occurs between vibrational states with large probabilities at the same internuclear distance. The transition illustrated in the figure involves an electronic energy change from the $n = 1$ to $n = 2$ state and a vibrational change from $v = 0$ to $v = 2$.

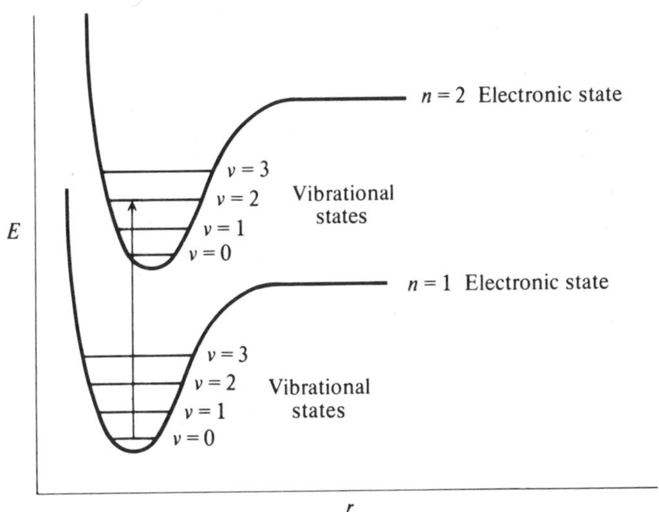

Fig. 18.6. The Franck-Condon principle.

From the spacing of the absorption maxima of an electronic spectrum we may be able to determine ν_0 for the vibrational modes (including those whose absorptions in the infrared are forbidden) and the moment of inertia I for the molecular rotations.

Emission electronic spectra are also important. Heat energy is used to excite atoms or molecules, which, in returning to more stable energy states, emit the excess energy in the form of radiation. The emission spectrum obtained may be used in the same manner as an absorption spectrum. Emission spectra are used to determine the identity of materials in samples of unknown composition. Radiation spectra from the stars have been used to determine both their chemical constituents and the states of those constituents. Emission of radiation following heating is also the basis of incandescent lighting.

18.8. Raman Scattering

When a molecular sample is placed in an intense beam of monochromatic radiation of a frequency which does not correspond to the energy difference between any two of the allowed stationary states of the molecule, the nonresonance phenomenon of *scattering* may be observed. A relatively small amount of the radiation is absorbed, but since the state to which the molecules are excited is not a stationary state they immediately give up energy to return to a stable stationary state. This energy, given off as radiation, is emitted in essentially all directions and can be detected by a device placed perpendicular to the exciting radiation beam (Fig. 18.7).

The excited molecule may emit radiation of exactly the same frequency as that which excited it. It would then return to its initial state. This type of scattering is called *Rayleigh scattering*. Figure 18.8(a) illustrates that the excited molecule may also emit radiation of a longer or shorter wavelength and return to a stable state which is different from its initial state. This is called *Raman scattering*.

In a scattering spectrum (Fig. 18.8b) we find an emission peak of frequency equal to the excitation frequency ν_1. This corresponds to Rayleigh scattering and will account for most of the scattered radiation (which in turn is only a small fraction of the incident radiation). Displaced from this line will be

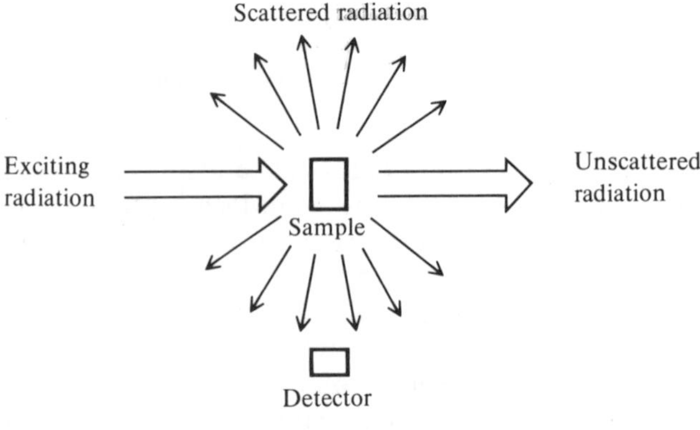

Figure 18.7

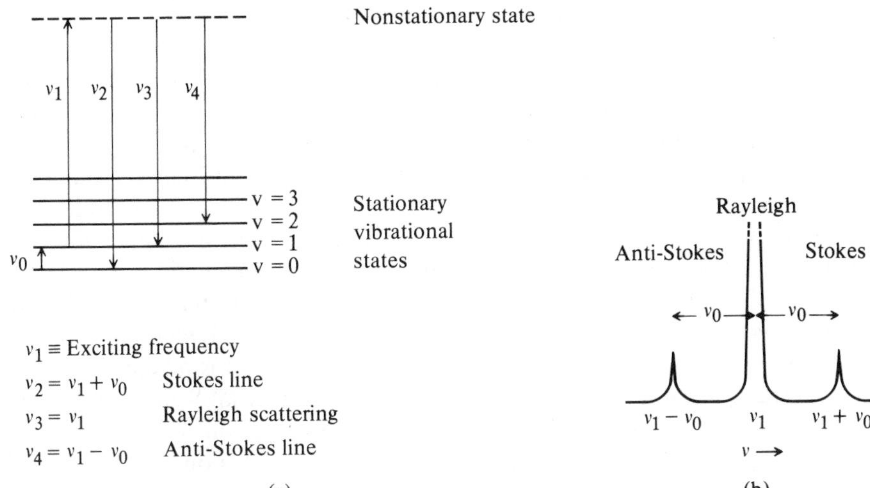

Figure 18.8

smaller peaks corresponding to emissions which return the molecule to states other than its initial state. The illustration of Fig. 18.8 is in terms of vibrational states. A similar situation exists for rotation, when the system returns to a state with a different J from the initial state. When ΔJ or $\Delta v > 0$, the line is called a *Stokes line*. When ΔJ or $\Delta v < 0$, it is an *anti-Stokes line*.

The selection rules are different for Raman scattering than for infrared and microwave absorption. The rotational spectra of molecules with no dipole moments (such as H_2 and CO_2) may be obtained from the Raman spectra. For linear molecules the allowed rotational transitions are $\Delta J = \pm 2$ instead of $\Delta J = \pm 1$. In many cases vibrational transitions forbidden in the infrared (see Section 18.6) are allowed in the Raman, and vice versa. In many cases Raman spectra may therefore be used to obtain the important parameters of rotation and vibration, I and ν_0, for rotational or vibrational transitions which do not occur in the microwave and infrared.

One of the major difficulties of Raman spectroscopy has been the fact that only a small fraction of the incident radiation is scattered, and of this only a small portion is of a different frequency from the incident radiation and therefore contains useful information. The recent development of lasers has provided sources of high intensity monochromatic radiation in the visible. Raman spectra may be enhanced considerably by the use of such a source.

Before considering another consequence of nonresonance scattering, let us examine two more resonance phenomena.

18.9. Electron Spin Resonance

We have previously discussed the fact that an electron has a spin function in addition to its spatial wave function. It may be considered to be spinning about its axis with the two spin states α and β resulting from spinning in opposite directions (clockwise and counterclockwise). Normally these two states are of identical energy. A spinning charge, however, produces a magnetic field: the electron acts as a tiny magnet. The magnetic polarity (that is, the relative position of the north and south magnetic poles of the magnet) of a charge spinning clockwise is opposite to that of a charge spinning counterclockwise. An external magnetic field interacts with these electronic magnets; in one spin state the electronic magnet is aligned with the field, in the other it is aligned opposite to the external field. When there is an external magnetic field, the two spin states of an electron are therefore of different energies. The magnitude of this difference depends on the strength of the external field and is given by

$$\Delta E = 2\mu_B H_0, \tag{343}$$

where μ_B is the *Bohr magneton* (9.27×10^{-21} erg gauss^{-1}) and H_0 is the magnitude of the magnetic field strength in gauss.

Energy may be absorbed from electromagnetic radiation via the interaction of the magnetic field component of the radiation with the electron's magnetic dipole. The frequency of the radiation which produces the change of the electron spin from the low energy state to the high energy state is

$$\nu = 2\mu_B H_0/h \tag{344}$$

Radiation of about 10-cm wavelength is generally used, which means that a field strength of about 10^3 gauss is necessary.

In order for an *electron spin resonance* spectrum to be observed there must be one or more unpaired electrons. The esr method is therefore used to detect the presence of unpaired electrons such as in transition metal complexes (see Section 15.8) and in organic free radicals. Fine structure which results from the interaction of electron and nuclear spins may also yield useful molecular information.

Example 107 The presence of unpaired electrons in systems undergoing photosynthesis has been detected by esr spectroscopy. One mechanism of photosynthesis which has been proposed requires that unpaired electrons be produced as the initiating step of the reaction sequence— the step in which the system absorbs photons. Another mechanism proposes that the unpaired electrons are produced by a reaction step which occurs later in the photochemical reaction sequence. How may we decide by experiment which of these two proposed mechanisms is more likely?

Answer: Perhaps the most straightforward procedure would be to determine whether there is a time lag between the exposure of the system to light and the appearance of an esr signal. Experimentally it has been found that there is indeed a time lag in at least one photosynthetic system (B. Commoner, D. H. Kohl, and J. Townsend, *Proceedings of the National Academy of Sciences of the United States* **50**, 638, 1963). For this particular system the second mechanism is more likely.

18.10. Nuclear Magnetic Resonance

Nuclei also have spin wave functions and, being charged, they also act as magnets. When there is an external magnetic field with which the nuclear magnetic moment can interact, the energies of the different spin states are different. Like the electron, a nucleus can absorb energy from an electromagnetic field through the interaction of the nuclear magnetic moment and the magnetic component of the radiation.

The magnitudes of the spin of all nuclei are not the same. One result is that the energy separations between the spin states of various nuclei are different. In general this energy separation is

$$\Delta E = g_N \mu_N H_0, \tag{345}$$

where g_N is the *nuclear g factor* for the particular nucleus (for the ^{14}N isotope $g_N = 0.40$, whereas for the proton, ^1H, it is 5.58), μ_N is the *nuclear magneton* (5.05×10^{-24} erg gauss^{-1}), and H_0 is the magnitude of the magnetic field. The hydrogen nucleus consisting of a single proton has been the most thoroughly studied. For this case Eq. (345) reduces to

$$\Delta E = 2.82 \times 10^{-23} H_0 \text{ erg}, \tag{346}$$

and the frequency of radiation absorbed is

$$\nu = 4.26 \times 10^3 H_0 \text{ sec}^{-1}. \tag{347}$$

With a magnetic field of 10,000 gauss the frequency absorbed by a hydrogen nucleus in its transition between its two spin states would be 4.26×10^7 sec^{-1} or a wavelength of 700 cm. This frequency is in the radio-wave region of electromagnetic radiation.

The existence of two effects makes *nuclear magnetic resonance* (nmr) spectroscopy (the study of the absorption spectra of nuclear spin transitions) very useful. These are the *chemical shift* and *spin-spin splitting*.

Equation (347) applies exactly only to a *bare* proton. In a molecule the hydrogen nucleus is influenced by the electrons in the orbital around it. The presence of an external magnetic field causes these electrons to move in such a manner that an induced magnetic field opposed to the external field is set up.

As a result the effective magnetic field H which reaches the nucleus is less than the external field H_0:

$$H = H_0 - \Delta, \tag{348}$$

where Δ is the shielding effect. The frequency of the radiation absorbed by a proton in a molecule is therefore

$$\nu = 4.26 \times 10^3 H = 4.26 \times 10^3 (H_0 - \Delta). \tag{349}$$

The size of Δ depends on the density of electrons around the proton. If we compare an O—H bond and a C—H bond we know that the electron density is greater around the proton in the C—H case since oxygen is more electronegative than carbon. In both cases the proton will feel a lower magnetic field than the external field because of the presence of electrons in its vicinity. However a proton bound to oxygen experiences a larger field (lower Δ) than one bound to a carbon, because the electron density is greater in the latter case.

Figure 18.9 is the nmr spectrum of ethanol. Whereas other spectra we have discussed have been plots of absorption (or emission) intensity versus the frequency (or wavelength) of the radiation, this spectrum is a plot of the intensity versus the external magnetic field H_0. The difference in energy ΔE of the two spin states is directly proportional to the external magnetic field H_0 (Eq. 346). We could set the field strength (and therefore ΔE), then vary the radiation frequency, observing an absorption when $\nu = |\Delta E|/h$. Alternatively we could choose a radiation frequency and vary the field (and therefore ΔE) until the resonance condition $\Delta E = h\nu$ is reached. In either case absorption maxima are obtained when the radiation frequency and the magnitude of the field are properly matched—that is, an absorption spectrum

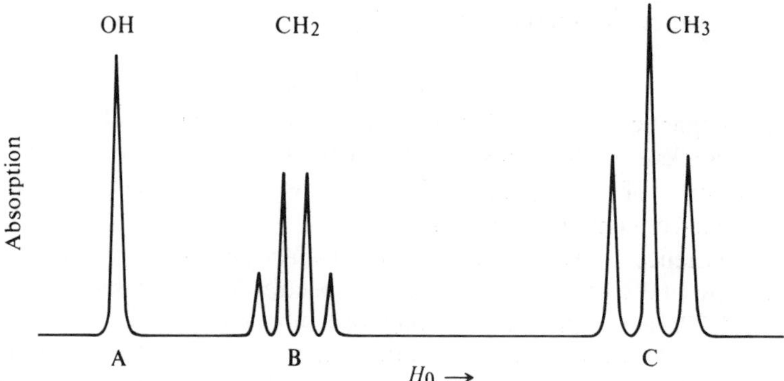

Fig. 18.9. The nuclear magnetic resonance spectrum of ethanol, CH_3CH_2OH.

is obtained in both cases. Fixing the frequency and varying the field is easier experimentally, so it is the method of choice.

The peaks of the ethanol spectrum occur in three groups, one for each of the three types of hydrogen atoms of the molecule: OH, CH_2, and CH_3. The total or integrated intensity of an absorption peak (the area under the peak) is proportional to the number of protons which absorb at that field strength. Most nuclear magnetic resonance spectrometers are capable of giving this integrated intensity directly. The ratio of this quantity for the three groups of peaks of the ethanol spectrum is A: B: C = 1: 2: 3. This means that if the molecule has one hydrogen of type A it must have two of type B and three of type C. We may therefore immediately assign peak A as the OH proton, the B peaks as the two CH_2 protons, and the C peaks as the three CH_3 protons.

With a fixed radiation frequency ν, all three types of protons will absorb when *the magnetic field which reaches them* ($H = H_0 - \Delta$) is exactly the same. But for a nucleus with a large amount of shielding Δ, a larger external field H_0 is required for the nucleus to feel a given magnetic field H than for a nucleus with a small amount of shielding. The lower the external field at which a peak appears in the nmr spectrum, the smaller the amount of shielding.

In the spectrum of ethanol, peak A (corresponding to the OH proton) appears at lowest field, next the CH_2 peaks, and finally the CH_3 peaks at highest field. Oxygen is more electronegative than carbon; thus the electron density is smaller about the hydrogen nucleus attached to oxygen than around those attached to the carbons. With fewer electrons, the shielding is less and the proton absorbs at a lower value of the external magnetic field. In our discussion of relative acidities in Section 15.10, we learned that a highly electronegative atom makes its neighboring atoms more electronegative. The carbon atom attached directly to the oxygen is more electronegative than the carbon which is one atom removed from the oxygen. Consequently the hydrogens attached to the more electronegative carbon are shielded less and therefore absorb the fixed radiation at a lower external field. This checks with the assignment of these peaks based on their integrated intensities.

The absorption of radiation at different values of external field strength due to different chemical environment is called *chemical shift*.

Now we must explain why there are three peaks for the CH_3 protons, four peaks for the CH_2 protons and only one for the OH proton. The presence of nonequivalent protons nearby affects the magnetic field felt by a proton. This is because protons act as magnets. When a proton's magnetic north-south axis is aligned in the same way as that of the external magnetic field, the external field is enhanced. When the proton's axis is opposite to that of the external field, its field diminishes the effect of the external field.

Figure 18.10(a) shows schematically the four ways in which the two protons on CH_2 may be arranged relative to the external field and the net effect on the field. If each of the four possible arrangements is assumed to be equally probable, each of the four configurations will exist in one-fourth of the molecules in a sample. In a fourth of the molecules the field felt by the CH_3 protons will be increased by the neighboring CH_2 protons; in another fourth the field will be decreased; and in half the molecules, the field will be unaffected. As a result the CH_3 protons absorb at three slightly different values of H_0: half of them at the field which would have been observed if there were no neighboring protons, a fourth at a slightly lower field, and another fourth at a slightly higher field. This is clearly illustrated in the spectrum (Fig. 18.9).

Our treatment here is somewhat oversimplified. This effect is actually transferred through the molecular bonds, not through space. It rapidly diminishes with distance so that usually only protons on neighboring atoms are affected. The phenomenon is known as *spin-spin coupling*.

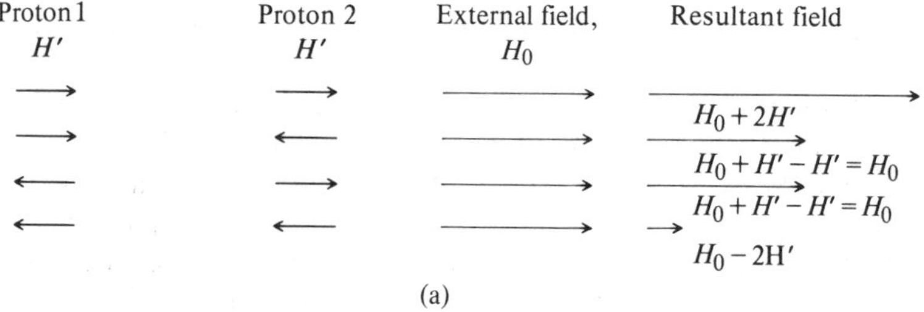

(a)

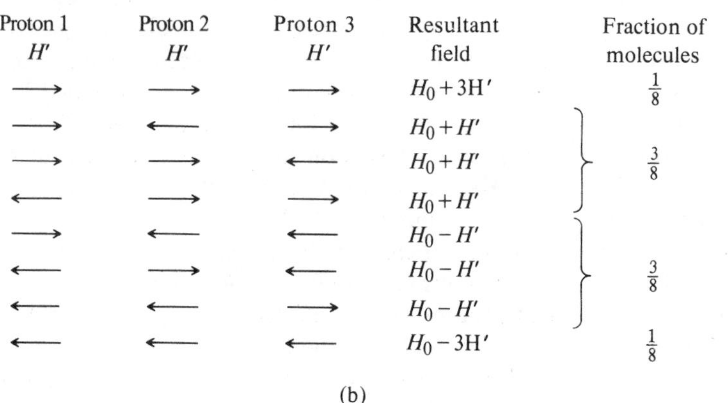

(b)

Figure 18.10

An analysis similar to the above may be carried out for the effect of the CH_3 protons on the CH_2 absorption. Figure 18.10(b) outlines the results. The CH_2 proton absorption is split into four peaks with an intensity ratio of 1 : 3 : 3 : 1. Again this is observed in the ethanol spectrum.

In general the absorption peak of a proton is split into $n + 1$ peaks by n adjacent protons, provided these adjacent protons are all in the same chemical environment and are different chemically from the absorbing proton.

Example 108 Predict the nmr spectrum of propane ($CH_3—CH_2—CH_3$).

Answer: The hydrogens of the two methyl groups are all equivalent and are chemically different from the $—CH_2—$ protons. We should therefore expect two groups of peaks, one for the six CH_3 protons and one for the two CH_2 protons. The CH_2 protons are shielded less and thus appear at a lower external field strength. The six equivalent methyl protons split the CH_2 peak into seven ($n + 1 = 7$) peaks, while the two methylene protons split the CH_3 peak into a triplet ($n + 1 = 3$). The net result is shown schematically in Fig. 18.11.

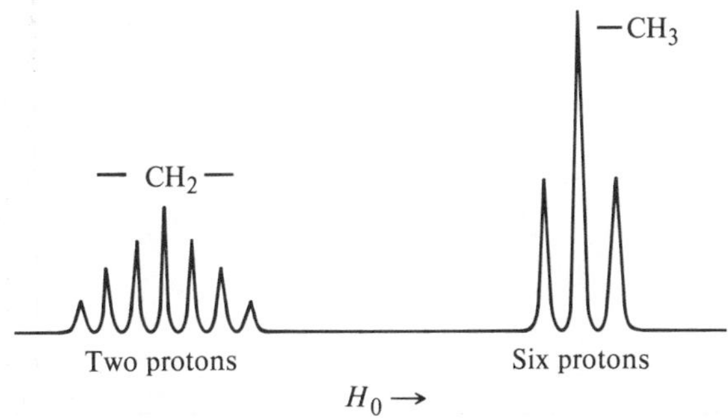

Fig. 18.11. Predicted nmr spectrum of $CH_3—CH_2—CH_3$.

Several questions are immediately apparent. Why is the OH proton absorption in ethanol only one peak instead of three? And why did we only consider the splitting caused by the CH_3 protons and not the OH proton when we discussed the CH_2 absorption? Protons attached to highly electronegative atoms are very mobile; they move readily between molecules. This exchange occurs so rapidly compared with the time required for absorption of radiation in nmr spectroscopy that the spin-spin coupling effects are not observable. If this exchange could be slowed down sufficiently we might observe the splitting. Indeed this can be accomplished by cooling, and at low enough temperatures the OH proton absorption becomes three peaks and the CH_2 becomes eight. (Each of the four peaks is split into two by the coupling of the OH proton.)

The chemical shift and spin-spin coupling can be valuable tools in the identification of unknown substances.

Example 109 Suppose the chemical analysis of a substance yields the formula C_2H_4O. On the basis of its nmr spectrum (Fig. 18.12), identify this material. The integrated intensity ratio is $A : B = 1 : 3$.

Answer: Since we know there are four hydrogen atoms per molecule, the integrated intensity ratio tells us that there is one of type A and three of type B. The two types of protons are on adjacent atoms since proton A splits the three protons of group B into a doublet ($n + 1 = 2$) and the three protons of B split the A proton into a quartet ($n + 1 = 4$).

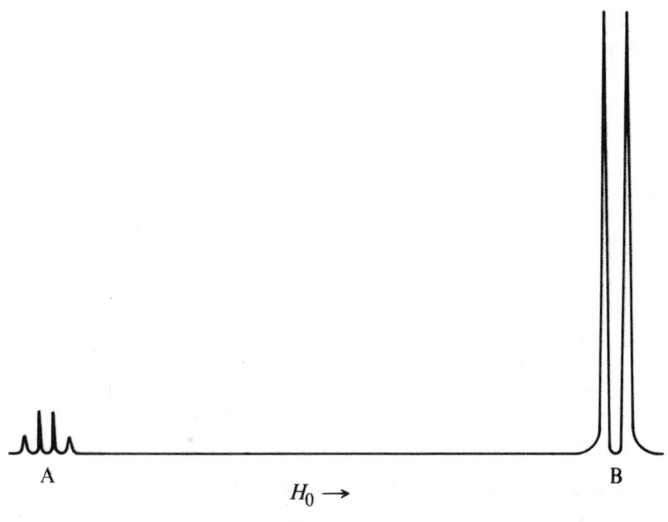

Figure 18.12

It does not take a great deal of effort at this point to propose the structure

$$\underset{(B)}{CH_3}-C\overset{\displaystyle O}{\underset{\displaystyle \underset{(A)}{H}}{\diagup\!\!\diagdown}}$$

(acetaldehyde) for this molecule. The aldehyde oxygen is very electronegative and causes the shielding of proton B to be smaller than that for the A protons. This proton's absorption thus occurs at lower field.

The nuclear magnetic resonance spectrum of a molecule is characteristic; like the molecule's infrared spectrum, it is a fingerprint (see Section 18.6) and can be used to detect the presence of a molecular species in a mixture. Nmr spectroscopy is finding increasing use in clinical and other analytical studies.

18.11. X-ray Crystallography

In Sections 18.3 and 18.8 we discussed the nonresonance scattering of electromagnetic radiation by atoms and molecules. When radiation of a wavelength which does not correspond to a transition between allowed stationary states is incident upon a system, part of the radiation is scattered by the atoms or molecules of the system. Most of this scattered radiation is of the same wavelength as the incident radiation.

When the molecules of the system are arranged randomly relative to each other as in a gas, the scattered radiation from one scatterer bears no relation to that from the other scatterers. If they are arranged in an orderly, three-dimensional array, however, as in a crystalline solid (see Chapter 17), separated by constant repeat distances in every direction, the radiation scattered from the various molecules will interact constructively in some directions, destructively in others. The net effect is the *diffraction* of some of the incident radiation by the array of molecules. The directions in which the radiation is diffracted are well defined so that if a photographic film is placed around the diffracting crystal a large number of exposure spots are obtained when the film is developed. Each of the spots is caused by one of the rays of diffracted radiation. The direction of each diffracted ray, relative to the incident radiation and the crystal, can be related to a set of three integer indices[3] designated $hk\ell$. Figure 18.13 is a typical x-ray diffraction photograph.

The intensities $I_{hk\ell}$ of the various diffraction maxima are different, the intensity of the radiation diffracted in a particular direction being related to the three-dimensional structure of the molecules which make up the crystal. The electron density ρ in a crystal is periodic. Its value along a line in any direction in the crystal follows a pattern which repeats over and over again. Figure 18.14 illustrates this pattern for one such direction in an idealized crystalline form of CO_2. Any such translationally repeating, or periodic, function may be written mathematically as a Fourier sum, a sum of

[3] For a discussion of the method of determining the indices for particular diffraction maxima, as well as discussions of other aspects of crystallography, see D. E. Sands, *Introduction to Crystallography*, Menlo Park, Calif.: W. A. Benjamin (1969).

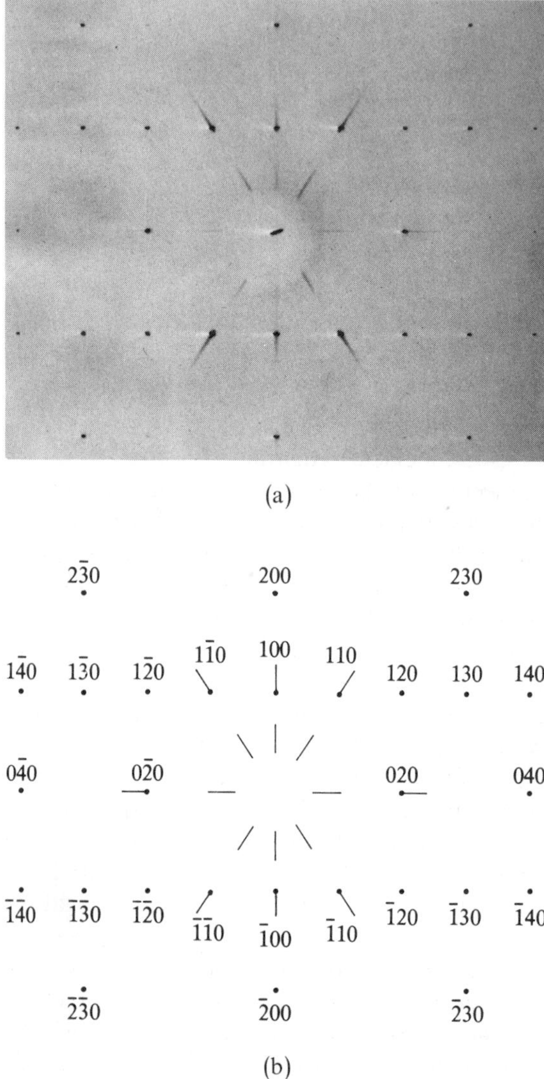

Fig. 18.13. An x-ray diffraction photograph of a crystal of a derivative of the anticancer agent emorene without indices (a) and with indices associated with the diffraction spots (b). The notation $\bar{2}$ means -2.

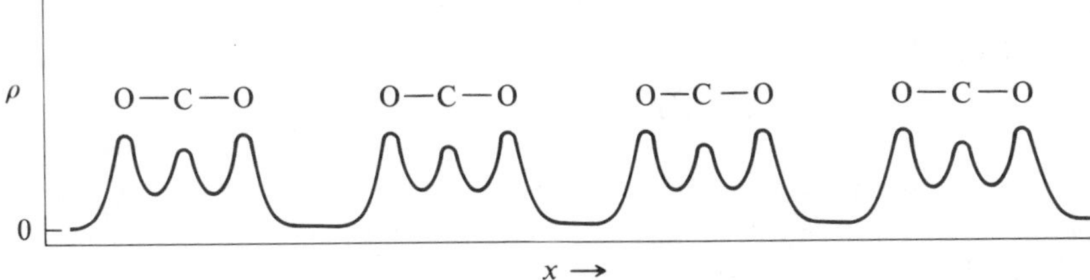

Fig. 18.14. The electron density as a periodic function of position.

sine and cosine terms, or a sum of complex exponential terms. The electron density ρ at any point in the crystal (x, y, z) can be shown[3] to be equal to

$$\rho(x, y, z) = \sum_{hk\ell} F_{hk\ell}\, e^{-2\pi i(hx+ky+\ell z)}, \tag{350}$$

where $i = \sqrt{-1}$ and $hk\ell$ are the indices mentioned previously. The sum is over all possible combinations of the values of these three integers. A most important quantity in this equation is the *structure factor* $F_{hk\ell}$. It is equal to

$$F_{hk\ell} = \sum_n f_n e^{2\pi i(hx_n+ky_n+\ell z_n)}, \tag{351}$$

where the sum is over all the atoms in a single unit cell (see Section 17.7). The position (x_n, y_n, z_n) is the fractional position (see Example 110) in the unit cell of the nth atom, and f_n is the *atomic scattering factor* of the nth atom. For our purposes, the scattering factor can be assigned a value equal to the number of electrons the nth atom possesses.

At this point we should note the circularity of the above discussion. If we knew the set of $F_{hk\ell}$ values, Eq. (350) could be used to calculate the electron density of the crystal. Knowing the electron density, we would know the molecular structure (Fig. 18.14). Alternatively, in order to calculate $F_{hk\ell}$ using Eq. (351) we would need to know the position of all the atoms of the molecule, that is the molecular structure.

This is where the intensities of the diffraction maxima come into the picture. It can be shown that the intensity of a particular diffraction maximum $I_{hk\ell}$ is related to the corresponding structure factor $F_{hk\ell}$:

$$I_{hk\ell} = (F_{hk\ell})(F^*_{hk\ell}), \tag{352}$$

where $F^*_{hk\ell}$ is the complex conjugate of $F_{hk\ell}$ (see Section 13.4). It should be evident from Eq. (351) that $F_{hk\ell}$ may be a complex number. Equation (351) can be written as

$$F_{hk\ell} = A_{hk\ell}\, e^{i\beta_{hk\ell}}, \tag{353}$$

where A and β are non-negative real numbers (that is, they do not contain $\sqrt{-1}$). The intensity is then equal to

$$I_{hk\ell} = (Ae^{i\beta})(Ae^{-i\beta}) = A_{hk\ell}^2 . \tag{354}$$

By taking the square root of the intensity we obtain the *amplitude* $A_{hk\ell}$ of the structure factor but not its *phase* $\beta_{hk\ell}$. In order to determine the electron density and therefore the structure of the molecules making up the scattering crystal, we must determine $\beta_{hk\ell}$.

There are a number of ways of getting around this so-called *phase problem* and determining the structure from scattering data. We shall discuss only the one which is historically most important, the *heavy atom method.*

The magnitude of the contribution a particular atom of the molecule will make to the structure factor is related to the number of electrons it possesses (see Eq. 351). When a molecule contains an atom with a significantly larger number of electrons than the other atoms (an iodine atom attached to an organic molecule, for example), this atom will in general contribute most to the structure factor sum.

A method exists for determining the position of such a heavy atom in a crystal.[3] Knowing its position (x_H, y_H, z_H), we can calculate its contribution to the structure factors:

$$(F_{hk\ell})_H = f_H e^{2\pi i(hx_H + ky_H + \ell z_H)} = A'_{hk\ell} e^{i\beta'_{hk\ell}}.$$

The assumption is then made that the heavy atom phase β' does not differ radically from β, the true total phase of the structure factor, and the approximate structure factors

$$F'_{hk\ell} = \sqrt{I_{hk\ell}}\, e^{i\beta'_{hk\ell}}$$

are used in Eq. (350) to calculate an approximate electron density. If this assumption is roughly true for a large number of the structure factors, the resulting electron density calculation will yield a structure which more or less resembles the true molecular structure. From this calculation the approximate positions of some of the lighter atoms are determined. These positions are then combined with the position of the heavy atom to calculate new phases which should be closer to the real values. A new approximate electron density calculation is then made. This should resemble the real structure more closely than the first so that further light-atom positions can be determined. The procedure is repeated until all the atoms are located. Computers are of great assistance with the volume of calculations involved.

In order for diffraction to yield useful data, the wavelength of the radiation used must be of the same order of magnitude as the spacings between the molecules. Since these spacings are in the Ångstrom range (10^{-8} cm), the radiation must be in the x-ray region.

The majority of the detailed information about molecular structure which is now known has come from x-ray crystallographic studies. These have ranged from determining the structures of diatomic molecules to determining those of proteins.

Because the diffraction pattern of a crystalline material is characteristic, unknown solids may be identified by comparing the diffraction photographs of the unknown with those of known materials. For this purpose diffraction data for thousands of organic and inorganic substances have been measured and tabulated. This technique has been employed to identify the constituents of gallstones, kidney stones, and other body calculi. Using x-ray diffraction data, identifying the thyroid stones of Problem 2.3 as calcite ($CaCO_3$) was quite straightforward.

Example 110 Assume that the crystal structure of BaO is such that Fig. 18.15 is the unit cell and that the positions of the atoms (relative to the unit cell origin at 0) are, in Å:

Atom	x'	y'	z'
Ba (1)	2	2	1.5
Ba (2)	4	6	2.5
O (1)	2	2	2.5
O (2)	4	6	1.5

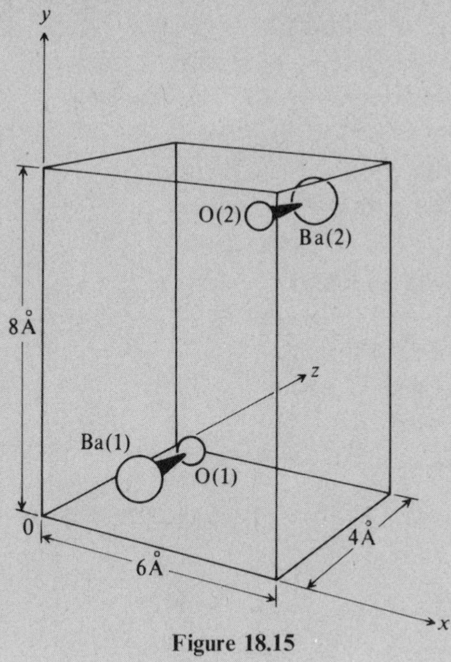

Figure 18.15

a) Calculate β for the diffraction maxima with indices (i) 1, 0, 0 and (ii) 3, 2, 9.
b) Calculate β_H, using the barium positions, for the same two maxima.
c) Comment on the applicability of the heavy-atom method to the solution of this crystal structure.

Answer: The fractional coordinates to be used in Eq. (351) are $(x'/6)$, $(y'/8)$, and $(z'/4)$:

Atom	x	y	z
Ba (1)	0.333	0.250	0.375
Ba (2)	0.667	0.750	0.625
O (1)	0.333	0.250	0.625
O (2)	0.667	0.750	0.375

Equation (351) may also be written

$$F_{hk\ell} = \sum_n f_n[\cos 2\pi(hx_n + ky_n + \ell z_n) + i \sin 2\pi(hx_n + ky_n + \ell z_n)].$$

a, i)
$$\begin{aligned}
F_{100} &= 56[\cos 2\pi(1 \cdot 0.333 + 0 \cdot 0.250 + 0 \cdot 0.375) \\
&\quad + i \sin 2\pi(1 \cdot 0.333 + 0 \cdot 0.250 + 0 \cdot 0.375)] \\
&\quad + 56[\cos 2\pi(1 \cdot 0.667 + 0 \cdot 0.750 + 0 \cdot 0.625) \\
&\quad + i \sin 2\pi(1 \cdot 0.667 + 0 \cdot 0.750 + 0 \cdot 0.625)] \\
&\quad + 8[\cos 2\pi(1 \cdot 0.333 + 0 \cdot 0.250 + 0 \cdot 0.625) \\
&\quad + i \sin 2\pi(1 \cdot 0.333 + 0 \cdot 0.250 + 0 \cdot 0.625)] \\
&\quad + 8[\cos 2\pi(1 \cdot 0.667 + 0 \cdot 0.750 + 0 \cdot 0.375) \\
&\quad + i \sin 2\pi(1 \cdot 0.667 + 0 \cdot 0.750 + 0 \cdot 0.375)]
\end{aligned}$$

$$\begin{aligned}
F_{100} &= 56[\cos (0.666\pi) + i \sin (0.666\pi)] \\
&\quad + 56[\cos (1.334\pi) + i \sin (1.334\pi)] \\
&\quad + 8[\cos (0.666\pi) + i \sin (0.666\pi)] \\
&\quad + 8[\cos (1.334\pi) + i \sin (1.334\pi)] \\
&= 56(-0.500 + 0.866i) \\
&\quad + 56(-0.500 - 0.866i) \\
&\quad + 8(-0.500 + 0.866i) \\
&\quad + 8(-0.500 - 0.866i) \\
&= 56(-1.000) + 8(-1.000) \\
&= -64.000
\end{aligned}$$

Now since $F_{100} = A_{100} e^{i\beta_{100}}$ and A is positive:

$$A_{100} = 64.00,$$
$$e^{i\beta_{100}} = -1,$$
$$\beta_{100} = \pi.$$

a, ii) $$\begin{aligned} F_{329} &= 56[\cos 2\pi(3\cdot 0.333 + 2\cdot 0.250 + 9\cdot 0.375) \\ &\quad + i \sin 2\pi(3\cdot 0.333 + 2\cdot 0.250 + 9\cdot 0.375)] \\ &\quad + 56[\cos 2\pi(3\cdot 0.667 + 2\cdot 0.750 + 9\cdot 0.625) \\ &\quad + i \sin 2\pi(3\cdot 0.667 + 2\cdot 0.750 + 9\cdot 0.625)] \\ &\quad + 8[\cos 2\pi(3\cdot 0.333 + 2\cdot 0.250 + 9\cdot 0.625) \\ &\quad + i \sin 2\pi(3\cdot 0.333 + 2\cdot 0.250 + 9\cdot 0.625)] \\ &\quad + 8[\cos 2\pi(3\cdot 0.667 + 2\cdot 0.750 + 9\cdot 0.375) \\ &\quad + i \sin 2\pi(3\cdot 0.667 + 2\cdot 0.750 + 9\cdot 0.375)] \\ &= 56[\cos(9.75\pi) + i\sin(9.75\pi)] \\ &\quad + 56[\cos(18.25\pi) + i\sin(18.25\pi)] \\ &\quad + 8[\cos(14.25\pi) + i\sin(14.25\pi)] \\ &\quad + 8[\cos(13.75\pi) + i\sin(13.75\pi)] \\ &= 56(0.707 - 0.707i) + 56(0.707 + 0.707i) \\ &\quad + 8(0.707 + 0.707i) + 8(0.707 - 0.707i) \\ &= 79.18 + 11.31 \\ F_{329} &= 90.49 \\ A_{329} &= 90.49 \\ e^{i\beta_{329}} &= +1 \\ \beta_{329} &= 0. \end{aligned}$$

b, i) From (a, i):

$$\begin{aligned} F_{H_{100}} &= 56(-0.500 + 0.866i) \\ &\quad + 56(-0.500 - 0.866i) \\ &= 56(-1.000) \\ &= -56.000 \\ A_{H_{100}} &= 56.00 \\ e^{i\beta_H} &= -1 \\ \beta_{H_{100}} &= \pi. \end{aligned}$$

b, ii) From (b, i):

$$\begin{aligned} F_{H_{329}} &= 79.18 \\ e^{i\beta_H} &= +1 \\ \beta_{H_{329}} &= 0. \end{aligned}$$

c) In both cases β_H agrees with β. If the agreement is as good with the other diffraction maxima, the heavy-atom approach would be quite successful.

Problems

18.1 Calculate the molar extinction coefficient of the amino acid tryptophan at a wavelength of 2750 Å if a 10^{-3} molar solution in a 1-cm cell reduces the power of an incident beam of this wavelength by 28%.

18.2 It has been suggested throughout this chapter that absorption spectroscopy of various types is a powerful tool for identifying molecules. For instance, the following two groups appear often in naturally occurring steroids:

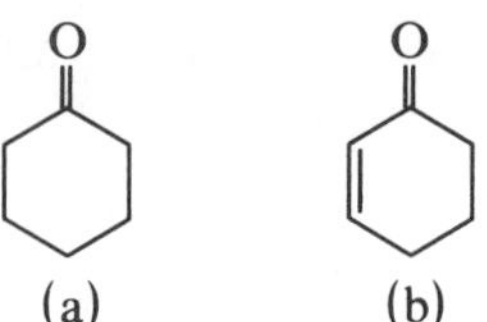

Steroids in which one of the two is present absorb radiation at about 1720 cm^{-1}, while those with the other group absorb at about 1680 cm^{-1}. Which group absorbs at 1720 cm^{-1} and which at 1680 cm^{-1}? Explain your choice.

18.3 The following stretching frequencies are observed for bonds of biochemical interest:

OH	3700 cm^{-1}
NH	3500
CH	3000
SH	2600
PH	2400

Explain the relative frequencies in terms of bond characteristics.

18.4 Why should the following sequence of colors be expected?

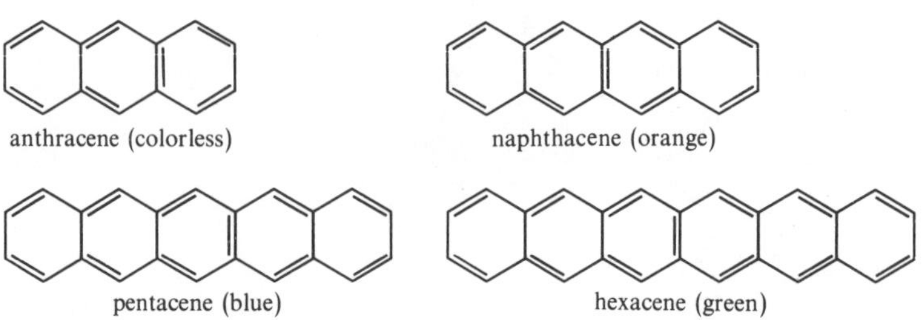

anthracene (colorless) naphthacene (orange)

pentacene (blue) hexacene (green)

18.5 The absorption spectrum of a dilute solution in a nonpolar solvent of a steroid known to possess a single hydroxide group exhibits a single absorption peak in the region between 2.60 microns and 3.30 microns, located at $\lambda = 2.80$ microns. The spectrum of a somewhat more concentrated solution possesses two peaks in this region, one at 2.80 microns and one at 3.15 microns. When the spectrum of an even more concentrated solution is obtained, both these peaks are still present but the peak at $\lambda = 2.80$ microns has diminished in intensity (size), while the one at $\lambda = 3.15$ microns has increased. Explain.

18.6 In Example 73 the energies of the electrons belonging to the delocalized pi system of vitamin A_1 were calculated by approximating the conjugated system as a one-dimensional box. The same was done for vitamin A_2 in Problem 13.7. A better treatment was elicited in Problem 15.13 in terms of molecular orbitals. The agreement of the particle-in-a-box approximation with reality may be checked by comparing the electronic absorption spectrum of each molecule with the wavelength which this treatment predicts for a transition from the highest occupied wave function ($n = 6$ for vitamin A_1) to the lowest unoccupied wave function ($n = 7$ for vitamin A_1; see Example 73). Vitamin A_1 absorbs radiation with wavelength 3280 Å, while vitamin A_2 absorbs a wavelength of 3690 Å. Compare the observed and calculated values and comment on the appropriateness of the particle-in-a-box approximation. (The electronic absorption of vitamins A_1 and A_2 is of great importance. The visual pigments rhodopsin and porphyropsin, which are derived from these vitamin molecules, are largely responsible for human sight.)

18.7 Flavoproteins sometimes contain the organic free radical semiquinone as well as Fe^{++} ions and/or Mo^{+5} ions. Explain how these three may be detected and identified by one of the spectroscopic methods discussed in this chapter.

18.8 How would the spectrum given in Fig. 18.11 differ for 1,3-dibromo propane ($CH_2Br—CH_2—CH_2Br$)?

18.9 The molecular formula of caffeine is $C_8H_{10}N_4O_2$. Its nuclear magnetic resonance spectrum appears in Fig. 18.16. The integrated intensities are A: B: C: D = 1: 3: 3: 3. Propose a structure for caffeine.

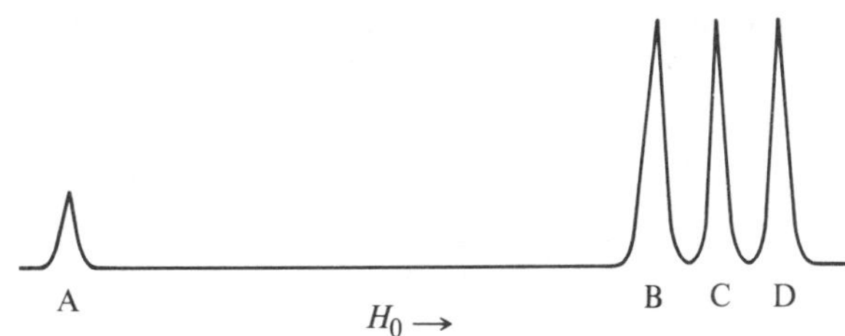

Fig. 18.16. The nmr spectrum of caffeine.

18.10 Identify the molecule with formula $C_8H_{14}O_4$ which gives the nmr spectrum shown in Fig. 18.17. The integrated intensities are A: B: C = 2: 2: 3. The compound also has an infrared absorption peak at about 1700 cm^{-1}.

18.11 In Section 18.6 we mentioned that only the $v = 0$ vibrational state is normally occupied, whereas molecules generally exhibit a variety of rotational states. The ratio of the number of molecules in an energy state with energy E_j and degeneracy g_j to the number in a state with E_i and g_i is given by

$$\frac{n_j}{n_i} = \frac{g_j}{g_i} e^{-(E_j - E_i)/kT}. \tag{355}$$

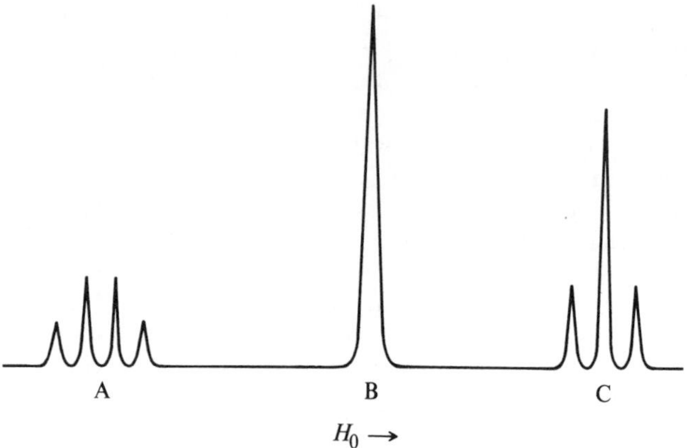

Fig. 18.17. The nmr spectrum of $C_8H_{14}O_4$.

(a) Calculate the ratio of the number of hydrogen molecules with rotational state $J = 1$ to the number with $J = 0$ (see Example 96) at 25°C. (b) Determine the ratio of the number of carbon monoxide molecules in the vibrational state corresponding to $v = 1$ to the number with $v = 0$ (see Example 97) at 25°C.

18.12 At what temperature will the number of carbon monoxide molecules in the $v = 1$ vibrational state be equal to 0.1 times the number in the $v = 0$ state? (See Problem 18.11.) Comment on the effect temperature changes may have on spectra.

Appendixes

APPENDIX I: SELECTED NUMERICAL CONSTANTS

Name	*Symbol*	*Numerical value*
Avogadro's number	N	6.022×10^{23} mole^{-1}
Boltzmann's constant	k	1.381×10^{-16} erg deg^{-1}
Electronic charge	e	1.602×10^{-19} coulomb
		4.803×10^{-10} esu
Electronic mass	m	9.110×10^{-28} g
Faraday's constant	F	9.649×10^{4} coulomb equivalent^{-1}
		23.06 kcal volt^{-1} equivalent^{-1}
Gas constant	R	0.08206 liter-atm deg^{-1} mole^{-1}
		8.314×10^{7} erg deg^{-1} mole^{-1}
		1.987 cal deg^{-1} mole^{-1}
		6.237×10^{4} cm^{3} torr deg^{-1} mole^{-1}
Gravitational constant	g	980.6 cm sec^{-2}
Planck's constant	h	6.626×10^{-27} erg-sec
Speed of light	c	3.00×10^{10} cm sec^{-1}

APPENDIX II: UNITS AND CONVERSION FACTORS

For the conversion stated, multiply by the factor shown. For the reverse conversion, multiply by the reciprocal factor.

To convert	*Factor*	*Reciprocal*
Inches to centimeters	2.540	0.3937
Miles to kilometers	1.609	0.6214
Square inches to square centimeters	6.452	0.1550
U.S. gallons to liters	3.785	0.2642
Fluid ounces to cubic centimeters	29.57	0.03381
Cubic feet to liters	28.32	0.03532
Avoirdupois ounces to grams	28.35	0.03527
Pounds to kilograms	0.4536	2.205
Pounds per square inch to atmospheres	0.06805	14.70
Atmospheres to dynes per square centimeter	1.013×10^6	9.870×10^{-7}
Torr to dynes per square centimeter	1333	7.501×10^{-4}
Atmospheres to torr	760	1.316×10^{-3}
Calories to joules	4.184	0.2390
Joules to ergs	1×10^7	1×10^{-7}
Debye to coulomb-centimeters	3.34×10^{-28}	2.99×10^{27}
Liter-atmospheres to ergs	1.013×10^9	9.87×10^{-10}
Liter-atmospheres to calories	24.206	0.0413

Also

$1 \text{ erg} = 1 \text{ gm cm}^2 \text{ sec}^{-2}$

$1 \text{ dyne} = 1 \text{ gm-cm sec}^{-2}$

$1 \text{ watt} = 10^7 \text{ erg sec}^{-1}$

APPENDIX III: THE PERIODIC TABLE

1*s*	1 H	
2*s*	3 Li	4 Be
3*s*	11 Na	12 Mg
4*s*	19 K	20 Ca
5*s*	37 Rb	38 Sr
6*s*	55 Cs	56 Ba
7*s*	87 Fr	88 Ra

3*d*	21 Sc	22 Ti	23 V	24 Cr	25 Mn	26 Fe	27 Co	28 Ni	29 Cu	30 Zn
4*d*	39 Y	40 Zr	41 Nb	42 Mo	43 Tc	44 Ru	45 Rh	46 Pd	47 Ag	48 Cd
5*d*	57– La	72 Hf	73 Ta	74 W	75 Re	76 Os	77 Ir	78 Pt	79 Au	80 Hg
6*d*	89– Ac	103								

						2 He
2*p*	5 B	6 C	7 N	8 O	9 F	10 Ne
3*p*	13 Al	14 Si	15 P	16 S	17 Cl	18 Ar
4*p*	31 Ga	32 Ge	33 As	34 Se	35 Br	36 Kr
5*p*	49 In	50 Sn	51 Sb	52 Te	53 I	54 Xe
6*p*	81 Tl	82 Pb	83 Bi	84 Po	85 At	86 Rn
7*p*						

4*f*	58 Ce	59 Pr	60 Nd	61 Pm	62 Sm	63 Eu	64 Gd	65 Tb	66 Dy	67 Ho	68 Er	69 Tm	70 Yb	71 Lu
5*f*	90 Th	91 Pa	92 U	93 Np	94 Pu	95 Am	96 Cm	97 Bk	98 Cf	99 Es	100 Fm	101 Md	102	103 Lr

APPENDIX IV: ATOMIC WEIGHTS OF THE ELEMENTS

Element	Symbol	Atomic Number	Atomic Weight
Actinium	Ac	89	(227)
Aluminum	Al	13	26.98
Americium	Am	95	(243)
Antimony	Sb	51	121.75
Argon	Ar	18	39.948
Arsenic	As	33	74.92
Astatine	At	85	(210)
Barium	Ba	56	137.34
Berkelium	Bk	97	(249)
Beryllium	Be	4	9.012
Bismuth	Bi	83	208.98
Boron	B	5	10.81
Bromine	Br	35	79.909
Cadmium	Cd	48	112.40
Calcium	Ca	20	40.08
Californium	Cf	98	(251)
Carbon	C	6	12.011
Cerium	Ce	58	140.12
Cesium	Cs	55	132.91
Chlorine	Cl	17	35.453
Chromium	Cr	24	52.00
Cobalt	Co	27	58.93
Copper	Cu	29	63.54
Curium	Cm	96	(247)
Dysprosium	Dy	66	162.50
Einsteinium	Es	99	(254)
Mercury	Hg	80	200.59
Molybdenum	Mo	42	95.94
Neodymium	Nd	60	144.24
Neon	Ne	10	20.183
Neptunium	Np	93	(237)
Nickel	Ni	28	58.71
Niobium	Nb	41	92.91
Nitrogen	N	7	14.007
Nobelium	No	102	(253)
Osmium	Os	76	190.2
Oxygen	O	8	15.9994
Palladium	Pd	46	106.4
Phosphorus	P	15	30.974
Platinum	Pt	78	195.09
Plutonium	Pu	94	(242)
Polonium	Po	84	(210)
Potassium	K	19	39.102
Praseodymium	Pr	59	140.91
Promethium	Pm	61	(147)
Protactinium	Pa	91	(231)
Radium	Ra	88	(226)
Radon	Rn	86	(222)
Rhenium	Re	75	186.23
Rhodium	Rh	45	102.91
Rubidium	Rb	37	85.47
Ruthenium	Ru	44	101.1

Element	Symbol	Atomic Number	Atomic Weight	Element	Symbol	Atomic Number	Atomic Weight
Erbium	Er	68	167.26	Samarium	Sm	62	150.35
Europium	Eu	63	151.96	Scandium	Sc	21	44.96
Fermium	Fm	100	(253)	Selenium	Se	34	78.96
Fluorine	F	9	19.00	Silicon	Si	14	28.09
Francium	Fr	87	(223)	Silver	Ag	47	107.870
Gadolinium	Gd	64	157.25	Sodium	Na	11	22.9898
Gallium	Ga	31	69.72	Strontium	Sr	38	87.62
Germanium	Ge	32	72.59	Sulfur	S	16	32.064
Gold	Au	79	196.97	Tantalum	Ta	73	180.95
Hafnium	Hf	72	178.49	Technetium	Tc	43	(99)
Helium	He	2	4.003	Tellurium	Te	52	127.60
Holmium	Ho	67	164.93	Terbium	Tb	65	158.92
Hydrogen	H	1	1.0080	Thallium	Tl	81	204.37
Indium	In	49	114.82	Thorium	Th	90	232.04
Iodine	I	53	126.90	Thulium	Tm	69	168.93
Iridium	Ir	77	192.2	Tin	Sn	50	118.69
Iron	Fe	26	55.85	Titanium	Ti	22	47.90
Krypton	Kr	36	83.80	Tungsten	W	74	183.85
Lanthanum	La	57	138.91	Uranium	U	92	238.03
Lawrencium	Lr	103	(257)	Vanadium	V	23	50.94
Lead	Pb	82	207.19	Xenon	Xe	54	131.30
Lithium	Li	3	6.939	Ytterbium	Yb	70	173.04
Lutetium	Lu	71	174.97	Yttrium	Y	39	88.91
Magnesium	Mg	12	24.312	Zinc	Zn	30	65.37
Manganese	Mn	25	54.94	Zirconium	Zr	40	91.22
Mendelevium	Md	101	(256)				

APPENDIX V: FOUR-PLACE COMMON LOGARITHMS

N	0	1	2	3	4	5	6	7	8	9
10	.0000	.0043	.0086	.0128	.0170	.0212	.0253	.0294	.0334	.0374
11	.0414	.0453	.0492	.0531	.0569	.0607	.0645	.0682	.0719	.0755
12	.0792	.0828	.0864	.0899	.0934	.0969	.1004	.1038	.1072	.1106
13	.1139	.1173	.1206	.1239	.1271	.1303	.1335	.1367	.1399	.1430
14	.1461	.1492	.1523	.1553	.1584	.1614	.1644	.1673	.1703	.1732
15	.1761	.1790	.1818	.1847	.1875	.1903	.1931	.1959	.1987	.2014
16	.2041	.2068	.2095	.2122	.2148	.2175	.2201	.2227	.2253	.2279
17	.2304	.2330	.2355	.2380	.2405	.2430	.2455	.2480	.2504	.2529
18	.2553	.2577	.2601	.2625	.2648	.2672	.2695	.2718	.2742	.2765
19	.2788	.2810	.2833	.2856	.2878	.2900	.2923	.2945	.2967	.2989
20	.3010	.3032	.3054	.3075	.3096	.3118	.3139	.3160	.3181	.3201
21	.3222	.3243	.3263	.3284	.3304	.3324	.3345	.3365	.3385	.3404
22	.3424	.3444	.3464	.3483	.3502	.3522	.3541	.3560	.3579	.3598
23	.3617	.3636	.3655	.3674	.3692	.3711	.3729	.3747	.3766	.3784
24	.3802	.3820	.3838	.3856	.3874	.3892	.3909	.3927	.3945	.3962
25	.3979	.3997	.4014	.4031	.4048	.4065	.4082	.4099	.4116	.4133
26	.4150	.4166	.4183	.4200	.4216	.4232	.4249	.4265	.4281	.4298
27	.4314	.4330	.4346	.4362	.4378	.4393	.4409	.4425	.4440	.4456
28	.4472	.4487	.4502	.4518	.4533	.4548	.4564	.4579	.4594	.4609
29	.4624	.4639	.4654	.4669	.4683	.4698	.4713	.4728	.4742	.4757
30	.4771	.4786	.4800	.4814	.4829	.4843	.4857	.4871	.4886	.4900
31	.4914	.4928	.4942	.4955	.4969	.4983	.4997	.5011	.5024	.5038
32	.5051	.5065	.5079	.5092	.5105	.5119	.5132	.5145	.5159	.5172
33	.5185	.5198	.5211	.5224	.5237	.5250	.5263	.5276	.5289	.5302
34	.5315	.5328	.5340	.5353	.5366	.5378	.5391	.5403	.5416	.5428
35	.5441	.5453	.5465	.5478	.5490	.5502	.5514	.5527	.5539	.5551
36	.5563	.5575	.5587	.5599	.5611	.5623	.5635	.5647	.5658	.5670
37	.5682	.5694	.5705	.5717	.5729	.5740	.5752	.5763	.5775	.5786
38	.5798	.5809	.5821	.5832	.5843	.5855	.5866	.5877	.5888	.5899
39	.5911	.5922	.5933	.5944	.5955	.5966	.5977	.5988	.5999	.6010
40	.6021	.6031	.6042	.6053	.6064	.6075	.6085	.6096	.6107	.6117
41	.6128	.6138	.6149	.6160	.6170	.6180	.6191	.6201	.6212	.6222
42	.6232	.6243	.6253	.6263	.6274	.6284	.6294	.6304	.6314	.6325
43	.6335	.6345	.6355	.6365	.6375	.6385	.6395	.6405	.6415	.6425
44	.6435	.6444	.6454	.6464	.6474	.6484	.6493	.6503	.6513	.6522
45	.6532	.6542	.6551	.6561	.6571	.6580	.6590	.6599	.6609	.6618
46	.6628	.6637	.6646	.6656	.6665	.6675	.6684	.6693	.6702	.6712
47	.6721	.6730	.6739	.6749	.6758	.6767	.6776	.6785	.6794	.6803
48	.6812	.6821	.6830	.6839	.6848	.6857	.6866	.6875	.6884	.6893
49	.6902	.6911	.6920	.6928	.6937	.6946	.6955	.6964	.6972	.6981
50	.6990	.6998	.7007	.7016	.7024	.7033	.7042	.7050	.7059	.7067
51	.7076	.7084	.7093	.7101	.7110	.7118	.7126	.7135	.7143	.7152
52	.7160	.7168	.7177	.7185	.7193	.7202	.7210	.7218	.7226	.7235
53	.7243	.7251	.7259	.7267	.7275	.7284	.7292	.7300	.7308	.7316
54	.7324	.7332	.7340	.7348	.7356	.7364	.7372	.7380	.7388	.7396

N	0	1	2	3	4	5	6	7	8	9
55	.7404	.7412	.7419	.7427	.7435	.7443	.7451	.7459	.7466	.7474
56	.7482	.7490	.7497	.7505	.7513	.7520	.7528	.7536	.7543	.7551
57	.7559	.7566	.7574	.7582	.7589	.7597	.7604	.7612	.7619	.7627
58	.7634	.7642	.7649	.7657	.7664	.7672	.7679	.7686	.7694	.7701
59	.7709	.7716	.7723	.7731	.7738	.7745	.7752	.7760	.7767	.7774
60	.7782	.7789	.7796	.7803	.7810	.7818	.7825	.7832	.7839	.7846
61	.7853	.7860	.7868	.7875	.7882	.7889	.7896	.7903	.7910	.7917
62	.7924	.7931	.7938	.7945	.7952	.7959	.7966	.7973	.7980	.7987
63	.7993	.8000	.8007	.8014	.8021	.8028	.8035	.8041	.8048	.8055
64	.8062	.8069	.8075	.8082	.8089	.8096	.8102	.8109	.8116	.8122
65	.8129	.8136	.8142	.8149	.8156	.8162	.8169	.8176	.8182	.8189
66	.8195	.8202	.8209	.8215	.8222	.8228	.8235	.8241	.8248	.8254
67	.8261	.8267	.8274	.8280	.8287	.8293	.8299	.8306	.8312	.8319
68	.8325	.8331	.8338	.8344	.8351	.8357	.8363	.8370	.8376	.8382
69	.8388	.8395	.8401	.8407	.8414	.8420	.8426	.8432	.8439	.8445
70	.8451	.8457	.8463	.8470	.8476	.8482	.8488	.8494	.8500	.8506
71	.8513	.8519	.8525	.8531	.8537	.8543	.8549	.8555	.8561	.8567
72	.8573	.8579	.8585	.8591	.8597	.8603	.8609	.8615	.8621	.8627
73	.8633	.8639	.8645	.8651	.8657	.8663	.8669	.8675	.8681	.8686
74	.8692	.8698	.8704	.8710	.8716	.8722	.8727	.8733	.8739	.8745
75	.8751	.8756	.8762	.8768	.8774	.8779	.8785	.8791	.8797	.8802
76	.8808	.8814	.8820	.8825	.8831	.8837	.8842	.8848	.8854	.8859
77	.8865	.8871	.8876	.8882	.8887	.8893	.8899	.8904	.8910	.8915
78	.8921	.8927	.8932	.8938	.8943	.8949	.8954	.8960	.8965	.8971
79	.8976	.8982	.8987	.8993	.8998	.9004	.9009	.9015	.9020	.9025
80	.9031	.9036	.9042	.9047	.9053	.9058	.9063	.9069	.9074	.9079
81	.9085	.9090	.9096	.9101	.9106	.9112	.9117	.9122	.9128	.9133
82	.9138	.9143	.9149	.9154	.9159	.9165	.9170	.9175	.9180	.9186
83	.9191	.9196	.9201	.9206	.9212	.9217	.9222	.9227	.9232	.9238
84	.9243	.9248	.9253	.9258	.9263	.9269	.9274	.9279	.9284	.9289
85	.9294	.9299	.9304	.9309	.9315	.9320	.9325	.9330	.9335	.9340
86	.9345	.9350	.9355	.9360	.9365	.9370	.9375	.9380	.9385	.9390
87	.9395	.9400	.9405	.9410	.9415	.9420	.9425	.9430	.9435	.9440
88	.9445	.9450	.9455	.9460	.9465	.9469	.9474	.9479	.9484	.9489
89	.9494	.9499	.9504	.9509	.9513	.9518	.9523	.9528	.9533	.9538
90	.9542	.9547	.9552	.9557	.9562	.9566	.9571	.9576	.9581	.9586
91	.9590	.9595	.9600	.9605	.9609	.9614	.9619	.9624	.9628	.9633
92	.9638	.9643	.9647	.9652	.9657	.9661	.9666	.9671	.9675	.9680
93	.9685	.9689	.9694	.9699	.9703	.9708	.9713	.9717	.9722	.9727
94	.9731	.9736	.9741	.9745	.9750	.9754	.9759	.9763	.9768	.9773
95	.9777	.9782	.9786	.9791	.9795	.9800	.9805	.9809	.9814	.9818
96	.9823	.9827	.9832	.9836	.9841	.9845	.9850	.9854	.9859	.9863
97	.9868	.9872	.9877	.9881	.9886	.9890	.9894	.9899	.9903	.9908
98	.9912	.9917	.9921	.9926	.9930	.9934	.9939	.9943	.9948	.9952
99	.9956	.9961	.9965	.9969	.9974	.9978	.9983	.9987	.9991	.9996

APPENDIX VI: ANSWERS TO SELECTED PROBLEMS

Chapter 1

1.3 (c) 23.6, (f) 49.5, (g) 19.2 cal mole^{-1} deg^{-1}
1.4 0.231 cal g^{-1} deg^{-1}
1.5 2 atoms
1.7 890 cal at 25°; 1113 cal at 100° for all three
(a) 1.86×10^{10} and 2.32×10^{10} cm^2 sec^{-2} at 25° and 100° respectively
(b) 3.68×10^9 and 4.60×10^9
(c) 8.87×10^8 and 1.11×10^9
1.8 $W + Q = W = \Delta E = nC_V\,\Delta T = nC_V(T_f - T_i)$
1.11 750 watts
1.14 4.79×10^{20} molecules, 3.86×10^{-3} watts
1.15 (a) $W = 0$, $Q = \Delta E = 750$ cal
(b) $W = -500$ cal, $Q = 1250$ cal, $\Delta E = 750$ cal

Chapter 2

2.3 $\Delta H = 33.08$ kcal, $\Delta E = 33.67$ kcal
2.4 (a) -209.2 kcal, (b) ~30 moles, (c) 40%
2.7 (a) 111 Cal (kcal), (b) 7.5 oz
2.9 1.15 volumes of CO_2 to each volume of O_2
2.10 7525 cal mole^{-1}
2.13 -12.70 kcal mole^{-1}

Chapter 3

3.3 (a) $\Delta S^\circ = 128.1$ cal deg^{-1}, $\Delta G^\circ = -54.5$ kcal
3.4 (a) 7.36 cal deg^{-1}, (b) -1925 cal
3.6 (a) $\Delta G^\circ = 688.1$ kcal, $\Delta S^\circ = -61.88$ cal deg^{-1}, $\Delta H^\circ = 669.6$ kcal,
$\Delta E^\circ = 669.6$ kcal
3.8 (a) -17.90 kcal, (c) -16.51 kcal
3.11 2^{-n}
3.13 No, he will win 5 of every 9 throws.

Chapter 4

4.2 23.2 cal deg^{-1}
4.4 (a) 1571 cal, (b) -12.53 kcal, (c) -35.24 cal deg^{-1}, (d) 614 cal
4.6 72.8 ml
4.9 5292 cal mole^{-1}
4.11 (a) 2.32 atm, (b) -4.4°C
4.12 (b) -0.0072 deg atm^{-1}, (d) $+0.096$ deg atm^{-1}
4.15 (a) yes, (b) In a room of very low humidity, most of the pentahydrate will disappear; in a room of high humidity, the trihydrate will become pentahydrate.

Chapter 5

5.1 (b) (1) 2.25, (2) 1710
(c) (1) -480 cal mole^{-1}, (2) -4410 cal mole^{-1}
(d) $\Delta G = -70$ cal mole^{-1} in both cases
5.5 $\Delta G_2^\circ = 2(\Delta G_1^\circ)$ and $K_2 = (K_1)^2$
5.6 $p_{CO} = 2.52 \times 10^{-6}$ atm
5.7 (a) dissociation, (b) CH_4 and CCl_4 = 43.3 mole percent each, CH_2Cl_2 = 13.4 mole percent
5.9 (a) $K_p = 0.002031$, (b) 1.18%
5.11 (a) no change, (b) decrease, (c) decrease, (d) increase
5.12 (b) 4.6×10^{17} molecules
5.14 1.64×10^{-10} atm
5.15 (a) 1124 cal, (b) 0.282 atm

Chapter 6

6.1 (a) 62 meters, (b) 0.80 atm, (d) 122 cm^3 at body temp and 2 atm
6.2 (c) $\Delta G^\circ = -13.33$ kcal, $K = 5.93 \times 10^9$
6.6 (a) $\Delta G^\circ = 6.49$ kcal, $K = 1.87 \times 10^{-5}$
6.7 (a) 4.10 kcal, (b) $-\infty$, (c) essentially 100%
6.8 (1) (a) $\Delta G^\circ = 5.62$ kcal, $K = 8.0 \times 10^{-5}$, (b) 8.35 kcal
6.9 (b) $\gamma = 0.93$, $a = 0.87$

Chapter 7

7.2 (a) 100.2°C, (b) -0.6°C
7.3 (b) 7.58 and 8.55 atm, (c) 0.97 atm
7.4 (b) $i = 1.22$, $\alpha = 0.22$
(c) $i = 1.95$, $\alpha = 0.95$

7.5 KCl: $i = 2$, $\gamma = 0.97$; $Pb(NO_3)_2$: $i = 3$, $\gamma = 0.88$
7.9 (b) 48900 g $mole^{-1}$
7.10 $i = 0.5$, propionic acid is virtually totally associated into dimers
7.11 (a) 209.9 g $mole^{-1}$, (b) 0.63

Chapter 8

8.2 0.40 volt, −9.22 kcal
8.3 5.86 kcal
8.5 (a) $\varepsilon^\circ = -0.02$ volt, $\Delta G^\circ = 0.92$ kcal, $\Delta G = -8.08$ kcal
(b) (1) 0.34 volt, −15.68 kcal, (2) −6.0 kcal, (3) 96 molal
8.7 (a) 1.23×10^{-10}, (c) 8.13 molal, (d) 1.23×10^{-5} molal
8.9 −2.75 kcal
8.10 (a) 1.95×10^{-10}
8.13 5.125

Chapter 9

9.1 $[A]_0 - [A]_t = [A]_0 - \frac{1}{2}[A]_0 = \frac{1}{2}[A]_0 = ak_0\tau$; $\tau = [A]_0/2ak_0$
9.3 first order, $k = 0.65$ hr^{-1}
9.5 0.845, 44.0 days
9.7 second order, 10^7 liters $mole^{-1}$ sec^{-1}
9.9 81.1×10^{-7} sec
9.11 1355 B.C.
9.13 $R = k[COCl_2][Cl_2]^{1/2}$, $k = 0.3$ $liter^{1/2}$ $mole^{-1/2}$ sec^{-1}
9.15 0 min: 1.21×10^4; 20 min: 6.95×10^3; 1 hr: 2.36×10^3; 2 hr: 5.11×10^2; 6 hr: 78.9; 1 day: 71.8; 3 days: 56.9; 7 days: 35.8; 10 days: 25.3

Chapter 10

10.1 (a) $K_\alpha = 1.45 \times 10^4$, $K_\beta = 2.44 \times 10^4$
(b) aldehyde: 1.43×10^{-5}; α-glucose: 0.207; β-glucose: 0.348 mole $liter^{-1}$
10.3 Use steady state approximation for $[C_2H_5]$, $[CH_3]$, and [H].
10.5 sucrose + $H^+ \overset{K}{\rightleftharpoons}$ sucrose · H^+
sucrose · H^+ + $H_2O \overset{k}{\rightarrow}$ glucose + fructose
$R = k[\text{sucrose} \cdot H^+][H_2O] = kK[\text{sucrose}][H^+][H_2O]$
10.7 Assume the two equilibria are rapidly established and are maintained throughout the reaction; alternatively, the steady state approximation and the assumption $k_4 \gg k_5[Cl_2]$ lead to the same results.

10.9 $O_3 \overset{K}{\rightleftharpoons} O_2 + O$

$O + O_3 \xrightarrow{k} 2O_2$

$R = k[O][O_3] = kK[O_3]^2/[O_2]$

Chapter 11

11.1 $k = 2.5 \times 10^{-5}$ liter mole^{-1} sec^{-1}
11.3 $E_a = 34.3$ kcal mole^{-1}, $A = 1.31 \times 10^{17}$ sec^{-1}
11.5 27.57 kcal
11.6 98.9 kcal

Chapter 12

12.1 $K_m = 1.33 \times 10^{-2}$, $v_{max} = 8.16 \times 10^{-5}$
12.3 $K_m = 1.91 \times 10^{-3}$, $v_{max} = 9.96 \times 10^{-6}$
12.6 β-phenylpropionic acid

12.8 $\frac{1}{v} = \frac{1}{v_{max}}(1 + K_I[I]_0)\left(\frac{K_m}{[S]_0} + 1\right)$

12.9 $\frac{1}{v} = \frac{1}{v_{max}}\left(1 + K_I'[I]_0 + \frac{K_m}{[S]_0}\right)$

Chapter 13

13.1 (a) 7.64×10^{-9}, (b) 7.32×10^{-9}, (c) 6.54×10^{-9} cm
13.3 7.4×10^{-32} cm
13.5 (a) 2.84×10^{-12} to 4.68×10^{-12} erg quantum^{-1}
(b) 8 molecules quantum^{-1}
13.7 (a) 1.68×10^{-7} cm, (b) 12 electrons, (c) less, because E is an inverse function of L^2, (d) 3.9×10^{-11} erg
13.9 2.17×10^{-11} erg sec^{-1}, 2.17×10^{-18} watt

Chapter 14

14.4 (a) 312 kcal mole^{-1}, (b) 78 kcal mole^{-1}
14.7 (b) using $E_{ip} = 2\pi^2 Z'^2 \mu e^4/h^2 n^2$: $Z'_{3s} = 1.85$, $Z'_{2p_1} = 3.74$, $Z'_{2p_2} = 4.60$, $Z'_{2p_3} = 5.40$, (c) $\sigma_{2p} = 0.83$
14.8 (a) 0.53×10^{-8} cm, (b) $2s$: 2.78×10^{-8} cm; $2p$: 2.12×10^{-8} cm, (c) 0.32, (d) 0.68

Chapter 15

15.1 $\psi_{MO} = 1s_H + \lambda 1s_{He} = \sigma_{1s}$, $\psi_{HeH^+} = \sigma_{1s}^2$

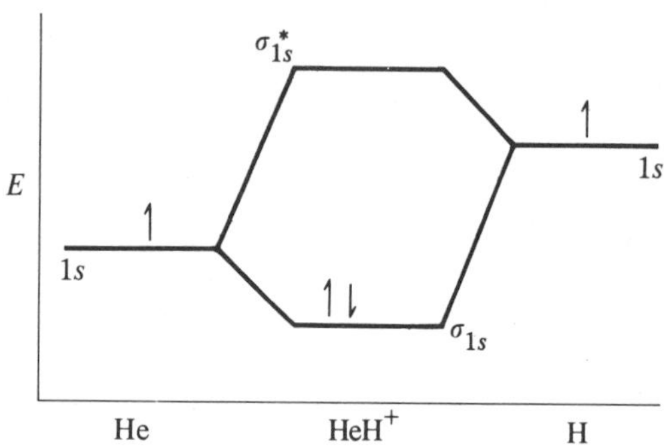

15.2 σ bond formed by overlap of filled sp^3 atomic orbital on N with an empty sp^3 orbital on B (dative or coordinate covalent bond)

15.5 NH_3: nonplanar, BF_3: planar with bond angles of 120°

15.7 (a) 2—3: sp^3—sp^3 σ bond
6—7: sp^2—sp^2 σ bond and p—p π bond
7—8: sp^2—sp^3 σ bond
(b) 2—3 > 7—8 > 6—7

15.10 Yes. If N is sp^2 hybridized, two of its valence electrons participate in σ bond formation with adjacent carbon atoms, two are paired in the third nonbonding sp^2 hybrid orbital, and the fifth valence electron is in the p_z orbital where it can participate in conjugation with the p_z electrons of the sp^2 hybridized carbon atoms.

15.12 Boron is sp^3 hybridized with one electron in each of three of the four hybrid orbitals. Each boron forms two normal σ bonds with hydrogen atoms. This leaves each boron with two sp^3 hybrid orbitals and one electron. A three-centered MO is formed by the simultaneous overlap of an sp^3 orbital on each of the boron atoms with the 1s orbital of one of the central hydrogens. A second three-centered MO is formed similarly. There are four electrons available to reside in these two MO's: one from each boron and one from each hydrogen.

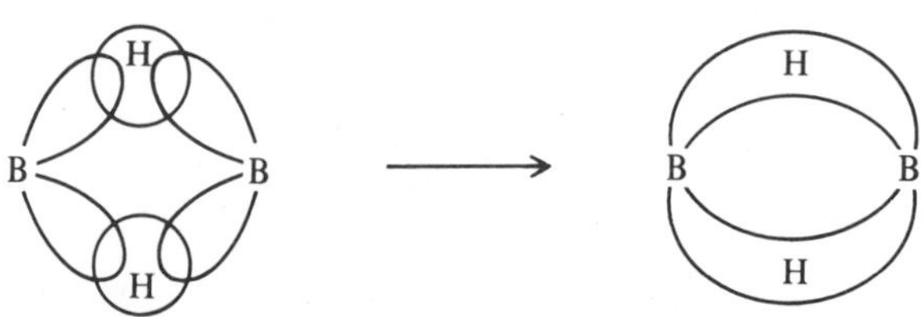

Chapter 16

16.1 (a) 6.79×10^{-39} g cm^2, (c) $I_x = I_y = I_z = 1.25 \times 10^{-37}$ g cm^2

16.2

J	E_{rot}, erg $\times 10^{16}$	*Degeneracy*
1	7.68	3
2	23.05	5
3	46.09	7
4	76.82	9
5	115.20	11

16.3 0, 1.64×10^{-16}, 4.92×10^{-16} erg
16.6 H_2: 5.20×10^5 dyne cm^{-1}
HD: 5.09×10^5
D_2: 5.19×10^5
16.8 Its length would increase by 5.13×10^{-2} cm.
16.10 (c) 3 translational, 3 rotational, and 66 vibrational
16.11 6-7

Chapter 17

17.1 chemical adsorption
17.3 For the substance which can form intermolecular hydrogen bonds: (a) higher, (b) higher, (c) may be either higher or lower, (d) higher, (e) higher, (f) higher, (g) lower, (h) lower, (i) lower.
17.5 The energy of interaction between a nonpolar solute and a polar solvent is not great enough to offset the energy required to separate the polar solvent molecules. The energy required to separate nonpolar solvent molecules is much less, however.
17.7 Atoms and molecules with a larger number of less tightly bound electrons are more polarizable. Thus (a) polarizability increases as we go down a column of the periodic table and (b) triple bonds (N_2) are more polarizable than double bonds (O_2), which are more polarizable than single bonds (H_2).
17.10 (a) Hydrocarbons are nonpolar, whereas water is polar.
(b) Their ionic ends are soluble in a polar solvent, while their hydrocarbon ends are insoluble.
(c) There is a hydrophobic region in the central region of the micelle (oil and hydrocarbon portion of soap) and a hydrophilic surface (ions and polar water molecules).
(d) The surface of each micelle is charged, and the micelles are insulated from each other by a layer of strongly associated water molecules.

(e) Equilibrium is shifted to the right by increased H^+ concentration

$$RCOO^- + H^+ \rightleftharpoons RCOOH,$$

decreasing the interaction with polar solvent.

(f) Ca and Mg salts of soaps are insoluble.

17.12 (a) longer than normal, (b) shorter than normal, (c) longer than normal

17.13 (a) -9.4×10^{-18} erg (13.6×10^{-2} cal mole^{-1})
(b) -6.0×10^{-16} erg (8.6 cal mole^{-1})

Chapter 18

18.1 1.43×10^2 cm^{-1} molar^{-1}

18.3 bond strengths: OH > NH > CH > SH > PH

18.5 dilute, no hydrogen bonding, OH absorbs at 2.80 μ; less dilute, some OH groups are hydrogen bonded with weaker bonds and absorption at higher wavelength, 3.15 μ; more concentrated, number of hydrogen-bonded OH groups increases, increasing the intensity of the 3.15 absorption and decreasing the 2.80 peak.

18.8 There would be five peaks at low field instead of seven, and the three high-field peaks would be less intense.

18.9

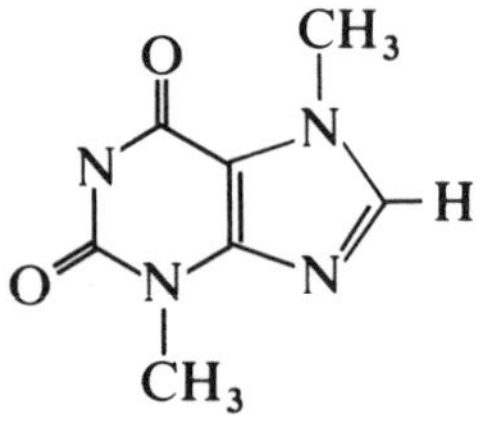

18.10 $CH_3CH_2O\overset{\overset{O}{\|}}{C}CH_2CH_2\overset{\overset{O}{\|}}{C}OCH_2CH_3$

Index

Index